METHODS IN MOLECULAR BIOLOGY

For further volumes:
http://www.springer.com/series/7651

For over 35 years, biological scientists have come to rely on the research protocols and methodologies in the critically acclaimed *Methods in Molecular Biology* series. The series was the first to introduce the step-by-step protocols approach that has become the standard in all biomedical protocol publishing. Each protocol is provided in readily-reproducible step-by-step fashion, opening with an introductory overview, a list of the materials and reagents needed to complete the experiment, and followed by a detailed procedure that is supported with a helpful notes section offering tips and tricks of the trade as well as troubleshooting advice. These hallmark features were introduced by series editor Dr. John Walker and constitute the key ingredient in each and every volume of the *Methods in Molecular Biology* series. Tested and trusted, comprehensive and reliable, all protocols from the series are indexed in PubMed.

Potassium Channels

Methods and Protocols

Edited by

Simone Furini

Department of Electrical, Electronic and Information Engineering "Guglielmo Marconi", University of Bologna, Cesena, Italy

Humana Press

Editor
Simone Furini
Department of Electrical, Electronic
and Information Engineering
"Guglielmo Marconi"
University of Bologna
Cesena, Italy

ISSN 1064-3745 ISSN 1940-6029 (electronic)
Methods in Molecular Biology
ISBN 978-1-0716-3817-0 ISBN 978-1-0716-3818-7 (eBook)
https://doi.org/10.1007/978-1-0716-3818-7

This Humana imprint is published by the registered company Springer Science+Business Media, LLC, part of Springer Nature.
The registered company address is: 1 New York Plaza, New York, NY 10004, U.S.A.

Paper in this product is recyclable.

Preface

Potassium channels play a ubiquitous role in all life forms, contributing to essential functions such as cell homeostasis and excitability. The diverse biological functions regulated by K+ channels underscore the high variability in their structural and functional characteristics, a diversity unparalleled by other ion channel families. Notably, in the human genome, the number of genes coding for K+ channels is approximately double that of Ca2+ channels and four times that of Na+ channels. Consequently, it is unsurprising that mutations in K+ channels are implicated in various hereditary diseases, making these proteins of significant pharmaceutical interest as both potential drug targets and contributors to side effects.

The biological roles of K+ channels and their unique biophysical properties have captivated generations of researchers who have adopted numerous experimental and computational tools to investigate the structural and functional properties of these membrane proteins.

Electrophysiological measurements with the patch-clamp technique are a common method for collecting data on channel functioning. The need to rapidly test numerous experimental conditions or protein-drug interactions has fostered the development of automated patch-clamp technologies and assays based on the analysis of Rb+ and Th+ fluxes. Extracting relevant information on channel behaviors from electrophysiological data benefits from ad-hoc experimental protocols, as in dynamic clamp, or from automatic algorithms based on kinetic models of K+ channels or machine learning approaches.

In the last 20 years, atomic structures of several K+ channels have been solved with atomic resolution, thanks to X-ray crystallography or Cryo-electron microscopy experiments. As of 2023, the Protein Data Bank includes more than 50 unique structures of K+ channels from different subfamilies, including, among others, delayed-rectified, inward-rectified, mechanosensitive, calcium-activated, and two-pore channels. Atomic structures are invaluable sources of information regarding how ion channels work at microscopic scales. However, to better understand the structure-function relationship, other techniques providing dynamic and/or low-resolution data in physiological conditions are needed. These include experimental techniques such as nuclear magnetic resonance, fluorescence resonance energy transfer, surface plasmon resonance, voltage clamp fluorimetry, and isothermal titration calorimetry. Moreover, computational models, particularly molecular dynamics simulations, are routinely used to complement experimental data at the atomic scale with dynamic information that could better inform on the structure-function relation.

This book presents a range of experimental and computational methods for the structural and functional analysis of K+ channels, in the Methods in Molecular Biology series format. We hope that this volume might be a valuable resource for researchers interested in the study of potassium channels.

Cesena, Italy *Simone Furini*

Contents

Contributors

CRYSTAL ARCHER • *University of Texas Health Science Center at San Antonio, San Antonio, TX, USA*

CHIARA BARTOLUCCI • *Department of Electrical, Electronic and Information Engineering 'Guglielmo Marconi', University of Bologna, Cesena, Italy*

ALEKSANDER BIES • *Faculty of Science and Technology, Institute of Computer Science, University of Silesia, Sosnowiec, Poland*

TINATIN I. BRELIDZE • *Department of Pharmacology and Physiology, Georgetown University Medical Center, Washington, DC, USA*

VLADISLAV I. CHUBINSKIY-NADEZHDIN • *Institute of Cytology, Russian Academy of Sciences, St. Petersburg, Russian Federation*

ROGER D. CONE • *Life Sciences Institute, University of Michigan, Ann Arbor, MI, USA; Department of Molecular and Integrative Physiology, School of Medicine, University of Michigan, Ann Arbor, MI, USA*

ANA COUTINHO • *iBB, Institute for Bioengineering and Biosciences, Instituto Superior Té cnico, Universidade de Lisboa, Lisbon, Portugal; Associate Laboratory i4HB, Institute for Health and Bioeconomy at Instituto Superior Técnico, Universidade de Lisboa, Lisbon, Portugal; Departamento de Química e Bioquímica, Faculdade de Ciências, Universidade de Lisboa, Lisbon, Portugal*

CAMDEN M. DRIGGERS • *Department of Chemical Physiology and Biochemistry, School of Medicine, Oregon Health and Science University, Portland, OR, USA*

MARÍA LUISA DURÁN-PASTÉN • *Laboratorio Nacional de Canalopatías, Instituto de Fisiología Celular, Universidad Nacional Autónoma de México, México City, Mexico*

ASSMAA ELSHEIKH • *Department of Chemical Physiology and Biochemistry, School of Medicine, Oregon Health and Science University, Portland, OR, USA; Department of Medical Biochemistry, Tanta University, Tanta, Egypt*

DAVID FEDIDA • *Faculty of Medicine, Department of Anesthesiology, Pharmacology and Therapeutics, University of British Columbia, Vancouver, BC, Canada*

CARLOS A. H. FERNANDES • *UMR 7590, CNRS, Muséum National d'Histoire Naturelle, IRD, Institut de Minéralogie, Physique des Matériaux et de Cosmochimie, IMPMC, Sorbonne Université, Paris, France*

LUIS E. GIMENEZ • *Life Sciences Institute, University of Michigan, Ann Arbor, MI, USA*

GRIGORY GLUKHOV • *Faculty of Biology, Shenzhen MSU-BIT University, Shenzhen, China*

ANNA GLUKHOVA • *Faculty of Biology, Shenzhen MSU-BIT University, Shenzhen, China*

KITTY HENDRIKS • *Research Unit Molecular Biophysics, Leibniz-Forschungsinstitut für Molekulare Pharmakologie (FMP), Berlin, Germany*

CIRIA C. HERNANDEZ • *Life Sciences Institute, University of Michigan, Ann Arbor, MI, USA; Center for Chemical Genomics, Life Sciences Institute, University of Michigan, Ann Arbor, MI, USA*

MASAYUKI IWAMOTO • *Department of Molecular Neuroscience, Faculty of Medical Sciences, University of Fukui, Fukui, Japan*

PAREESA KAMGAR-DAYHOFF • *Department of Pharmacology and Physiology, Georgetown University Medical Center, Washington, DC, USA*

MARIA KARLOVA • *Faculty of Biology, Moscow Lomonosov University, Moscow, Russia*

EKATERINA KRAVCHUK • *Faculty of Biology, Moscow Lomonosov University, Moscow, Russia*
ADAM LANGE • *Research Unit Molecular Biophysics, Leibniz-Forschungsinstitut für Molekulare Pharmakologie (FMP), Berlin, Germany; Institut für Biologie, Humboldt-Universität zu Berlin, Berlin, Germany*
SASCHA LANGE • *Research Unit Molecular Biophysics, Leibniz-Forschungsinstitut für Molekulare Pharmakologie (FMP), Berlin, Germany*
EMMANUELLE LOGETTE • *Blue Brain Project, Ecole Polytechnique Fédérale de Lausanne, Geneva, Switzerland*
ENOCH LUIS • *Laboratorio Nacional de Canalopatías, Instituto de Fisiología Celular, Universidad Nacional Autónoma de México, México City, Mexico*
YUKA MATSUKI • *Department of Anesthesiology and Reanimatology, Faculty of Medical Sciences, University of Fukui, Fukui, Japan*
MARIA I. MCKILLOP • *Department of Pharmacology and Physiology, Georgetown University Medical Center, Washington, DC, USA*
SHIGETOSHI OIKI • *Biomedical Imaging Research Center, University of Fukui, Fukui, Japan*
CARL ÖSTER • *Research Unit Molecular Biophysics, Leibniz-Forschungsinstitut für Molekulare Pharmakologie (FMP), Berlin, Germany*
JOSÉ ANTONIO POVEDA • *Instituto de Investigación, Desarrollo e Innovación en Biotecnología Sanitaria de Elche, Universidad Miguel Hernández, Elche, Spain*
RAJNISH RANJAN • *Blue Brain Project, Ecole Polytechnique Fédérale de Lausanne, Geneva, Switzerland*
MARÍA LOURDES RENART • *Instituto de Investigación, Desarrollo e Innovación en Biotecnología Sanitaria de Elche, Universidad Miguel Hernández, Elche, Spain*
MONIKA RICHTER-LASKOWSKA • *Lukasiewicz Research Network – Krakow Institute of Technology, The Centre for Biomedical Engineering, Kraków, Poland*
LUCA SALA • *Department of Biotechnology and Biosciences, University of Milano-Bicocca, Milan, Italy; Istituto Auxologico Italiano IRCCS, Center for Cardiac Arrhythmias of Genetic Origin and Laboratory of Cardiovascular Genetics, Milan, Italy*
DANIEL SASTRE • *Faculty of Medicine, Department of Anesthesiology, Pharmacology and Therapeutics, University of British Columbia, Vancouver, BC, Canada*
MAX E. SCHOENING • *Department of Biomedical Engineering, Washington University in St. Louis, St. Louis, MO, USA*
SHOW-LING SHYNG • *Department of Chemical Physiology and Biochemistry, School of Medicine, Oregon Health and Science University, Portland, OR, USA*
JONATHAN R. SILVA • *Department of Biomedical Engineering, Washington University in St. Louis, St. Louis, MO, USA*
OLGA S. SOKOLOVA • *Faculty of Biology, Shenzhen MSU-BIT University, Shenzhen, China; Faculty of Biology, Moscow Lomonosov University, Moscow, Russia*
PRAKRITI TIWARI • *Department of Oncology, Georgetown University Medical Center, Washington, DC, USA; Department of Biology, College of Arts & Sciences, Georgetown University, Washington, DC, USA*
PURUSHOTTAM B. TIWARI • *Department of Oncology, Georgetown University Medical Center, Washington, DC, USA*
ELIZAVETA TRIFONOVA • *Faculty of Biology, Moscow Lomonosov University, Moscow, Russia*
PAULINA TRYBEK • *Faculty of Science and Technology, Institute of Physics, University of Silesia in Katowice, Chorzow, Poland*
VALERIA Y. VASILEVA • *Institute of Cytology, Russian Academy of Sciences, St. Petersburg, Russian Federation*

Catherine Vénien-Bryan • *UMR 7590, CNRS, Muséum National d'Histoire Naturelle, IRD, Institut de Minéralogie, Physique des Matériaux et de Cosmochimie, IMPMC, Sorbonne Université, Paris, France*
Agata Wawrzkiewicz-Jalowiecka • *Department of Physical Chemistry and Technology of Polymers, Silesian University of Technology, Gliwice, Poland*

Chapter 1

Asymmetric Lipid Bilayers and Potassium Channels Embedded Therein in the Contact Bubble Bilayer

Yuka Matsuki, Masayuki Iwamoto, and Shigetoshi Oiki

Abstract

Cell membranes are highly intricate systems comprising numerous lipid species and membrane proteins, where channel proteins, lipid molecules, and lipid bilayers, as continuous elastic fabric, collectively engage in multi-modal interplays. Owing to the complexity of the native cell membrane, studying the elementary processes of channel–membrane interactions necessitates a bottom-up approach starting from forming simplified synthetic membranes. This is the rationale for establishing an in vitro membrane reconstitution system consisting of a lipid bilayer with a defined lipid composition and a channel molecule. Recent technological advancements have facilitated the development of asymmetric membranes, and the contact bubble bilayer (CBB) method allows single-channel current recordings under arbitrary lipid compositions in asymmetric bilayers. Here, we present an experimental protocol for the formation of asymmetric membranes using the CBB method. The KcsA potassium channel is a prototypical model channel with huge structural and functional information and thus serves as a reporter of membrane actions on the embedded channels. We demonstrate specific interactions of anionic lipids in the inner leaflet. Considering that the local lipid composition varies steadily in cell membranes, we 'present a novel lipid perfusion technique that allows rapidly changing the lipid composition while monitoring the single-channel behavior. Finally, we demonstrate a leaflet perfusion method for modifying the composition of individual leaflets. These techniques with custom synthetic membranes allow for variable experiments, providing crucial insights into channel–membrane interplay in cell membranes.

Key words Contact bubble bilayer, Membrane leaflet, Membrane perfusion, Leaflet perfusion, Asymmetric composition, Intact membrane, Single-channel recording

1 Introduction

The cell membrane is a highly complex and chaotic system [1] containing hundreds of lipid species (lipidomics) [2, 3] and membrane proteins (proteomics) [4]. While crowded [5], lipids and membrane proteins diffuse in the fluidic membrane, permitting interactions with various species and yielding a large number of paired interactions. Sometimes, they assemble inhomogeneously as mosaic membranes [6]. Some lipids assemble into rafts [7, 8], and

Simone Furini (ed.), *Potassium Channels: Methods and Protocols*, Methods in Molecular Biology, vol. 2796,
https://doi.org/10.1007/978-1-0716-3818-7_1,

membrane proteins often form clusters [9]. Unlike soluble proteins, for which water solvates in a homogeneous 3D space, membrane proteins are solvated in a lipid slab, where the embedded protein and surrounding lipid bilayer reconcile their shape by conformational changes and membrane deformation [10]. The bilayer matrix elastically deforms its curvature and thickness under the influence of membrane tension and external forces [11, 12], such as by organized cytoskeletons [13, 14]. In addition, an inhomogeneous electric field is generated in the bilayer matrix, originating from external and internal membrane sources (transmembrane and membrane dipole potentials [15, 16]).

The cell membrane is further complicated by the trans-bilayer asymmetry [17, 18]. Preferred accumulations of sphingomyelin in the outer leaflet of the eukaryotic plasma membrane and phosphatidylserine (PS) and phosphatidylethanolamine (PE) in the inner leaflet are generated and maintained by the energy-consuming active transport of lipids (flippases and floppases). By contrast, the scramblase and spontaneous flip-flop dissipate toward the equilibrium chemical composition [19–22]. A recent study has revealed a striking asymmetry of phospholipid acyl chains in the plasma membrane [23]. The inner leaflet comprises highly unsaturated lipids, whereas the outer leaflet lipids are mostly saturated. Accordingly, the asymmetry of the membrane covers the entire membrane thickness, including the lipid headgroups, backbone linkages (ester or ether), and acyl chains (saturation and length), forming a bilayer 3D anisotropic structure. This compositional asymmetry further complicates the inhomogeneous physical features of the bilayer, such as spontaneous membrane curvature [24], asymmetric tension, and lipid packing density [24–27], thus exerting an inhomogeneous force on the embedded proteins (the force-from-lipid principle) [28, 29]. For example, rich unsaturated acyl chains yield loosely packed inner leaflets that facilitate lipid diffusion [23], permitting more freedom of conformational changes in the inner half of the channel's transmembrane domain. In addition, the oil droplets bulge and bud from the ER membrane under asymmetric tension (*see* Fig. 1) [25, 30, 31]. The lipid bilayer affords the highly heterogeneous and anisotropic environment, in which channel proteins have adapted and survived.

In the present work, we focus on channel proteins as representatives of integral membrane proteins. Similar to other transporters, the channel permits ion transfer across membranes by forming a membrane-spanning pore. Channel proteins allow single-channel current recordings that characterize single-molecule functional behavior. The channel reminds us of the importance of the membrane potential while frequently overlooking the fact that all membrane proteins are influenced under the high electric field (10^7 V/m).

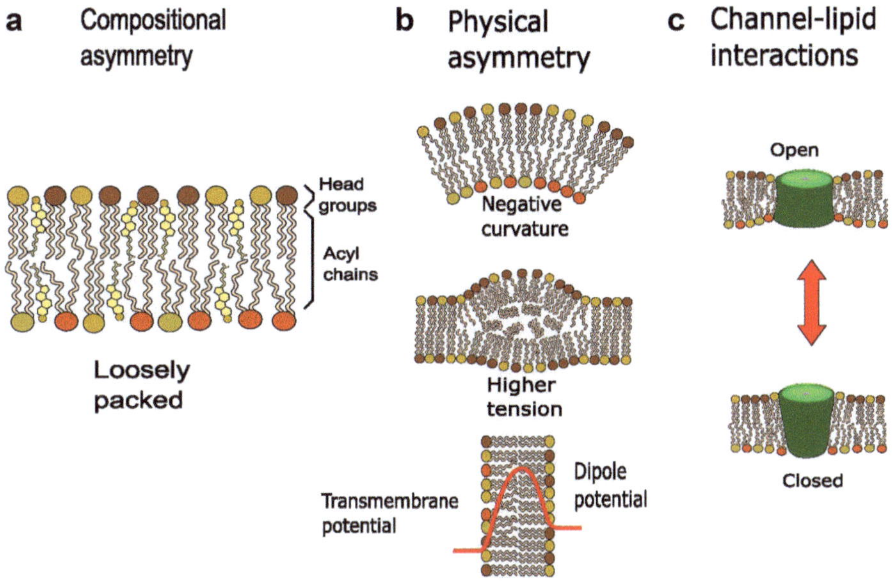

Fig. 1 Asymmetric membranes and outcomes thereof
(**a**) Compositional asymmetry. The two leaflets have distinct lipid compositions, such as abundant sphingomyelin in the outer leaflet and PS prevalent in the inner leaflet with negative charges. The inner leaflet has a high content of unsaturated acyl chains. These compositional asymmetries yield physical asymmetry. Having highly unsaturated acyl chains, the inner leaflet is loosely packed and more diffusive. (**b**) Physical asymmetry. (Upper) Based on the shape rule [82], having either columnar, cone, or reverse cone shape of a lipid molecule, accumulating specific lipids in a leaflet shapes either positive or negative curvature. (Middle) In the ER membrane, triglycerides accumulate between two leaflets and bud a lipid droplet on one side because of uneven leaflet tension. (Lower) Apart from the transmembrane potential generated from charge separation across the membrane, the dipole potential is generated by the oriented dipole of lipid molecules and water. (**c**) Channel–lipid interaction. Each channel species has a different width of the hydrophobic band, toward which the lipid bilayer accommodates its thickness near the channel interface, generating membrane deformation energy (hydrophobic mismatch). With different physical features, the deformation should be asymmetric in the leaflets

The channel–membrane interplay reaches a broad range of chemical and physical interactions. Channel proteins have a hydrophobic surface in the membrane interface, where exposed hydrophobic residues are solvated by membrane lipids for secure embedding. Channel proteins have adapted their structure and function to the asymmetric membrane environment along both evolutional time and lifetime scales. Channel proteins have accommodated to the membrane even though lipid compositions and the physical nature of prokaryotes and eukaryotes differ substantially [32]. The crystal structures of channel proteins of different origins share structural similarities [33, 34]. A typical eukaryotic channel protein is synthesized in the ER membrane having nearly symmetric compositions, where a nascent protein folds and accommodates and is then sorted to the plasma membrane, where it functions in an asymmetric environment. Channel proteins undergo global conformational changes upon gating [35], and changes in the hydrophobic bands of the channels induce local deformation of the surrounding membranes. At the same time, conformational changes to the open-gated structure yield an ionic current, thus changing the membrane potential. In return, a high electric field across the thin membrane affects the conformational changes of the channel proteins. Recent developments in structural biology have led to the accumulation of membrane-embedded structures of various membrane proteins [36, 37], and those in asymmetric membranes are recently accumulating [36]. However, the structural information of the channel under the membrane potential is poor [38–40]. We are still at the starting point of understanding complex channel–membrane systems.

To understand the channel–membrane interaction, one prefers to examine in controlled and simplified environments in vitro [41, 42], and we have learned about specific lipid–channel interactions from these models and synthetic membranes [43, 44]. Synthetic membranes, such as liposomes, supported bilayers, nanodisks, and planar lipid bilayers (PLBs), offer controlled environments for investigating channel–membrane interactions through custom design [36, 45, 46]. Recent advances have facilitated the formation of asymmetric membranes in various systems [43, 47–51]. Herein, we present the contact bubble bilayer (CBB) method [52–54] for forming asymmetric membranes. This method is distinguished from others by the dynamic manipulability of the bilayer, such as immediate electrolyte compositional changes [52] and tension applications [55]. In biological membranes, the vicinity of a channel protein varies persistently, and the channel responds steadily to local changes in lipid composition under varying physiological conditions. However, rapid changes in lipid composition and recording of immediate responses of channel proteins have not been achieved owing to the present experimental limitations. We developed a perfusion method for lipid compositional changes using an unprecedented access route in the CBB [56]. Furthermore, we extended the membrane perfusion method to leaflet perfusion, which has never before been achieved. Synthetic membranes make a step forward in understanding complex biological membranes by focusing on the specific channel–membrane interplay.

In this chapter, we first note the frequently overlooked membrane intactness and then summarize the background of asymmetric membrane formation. Next, we present a protocol for the asymmetric formation using the CBB method. Subsequently, we use the membrane perfusion method to examine the dynamic response of the channel to changes in the lipid composition. Finally, we demonstrate a leaflet perfusion method and present detailed methods.

1.1 Membrane Intactness

Before investigating the asymmetric membranes, we raise the issue of the membrane macroscopic features expressed here as intactness, which is frequently overlooked in synthetic membranes. The membrane is topologically intact as long as it maintains the integrity of a continuous membrane without defects. Membrane intactness is crucial for membrane function, ensuring the maintenance of asymmetric aqueous compositions necessary for driving ion conduction across the channel and generating the membrane potential. Even a single defect (an aqueous route formed in the lipid bilayer) across the entire membrane area deteriorates the membrane potential.

Biological membranes maintain their intactness, whereas forming a synthetic membrane without defects is challenging. For example, liposomes often face difficulties maintaining ion gradients across the entire population. Nanodisks are isolated membranes, and the membrane intactness and potential are not relevant, even though various types of asymmetric membranes are formed in the nanodisks [45]. In this respect, the CBB and other lipid bilayer techniques, such as the PLB [57, 58] and droplet interface bilayer (DIB), offer macroscopic membrane intactness. For instance, bacterial, mitochondrial, and plant membranes bear −200 mV of a membrane potential, which is challenging for other synthetic membranes but is readily applied to the CBB.

1.2 Generation of the Asymmetric Membrane in Biology and Technology

We begin by considering the origin of membrane asymmetry. In the cell, membranes successive from a parent cell are modified in their asymmetric compositions. Active flippases and floppases generate asymmetry by consuming energy (*see* Fig. 2, left), while scramblases and spontaneous flip-flops dissipate asymmetry. The fine-tuning between active and passive transport offers membrane asymmetry, but asymmetric lipid composition varies under various physiological conditions and is utilized for signal transduction [59].

In the de novo synthesis of lipid bilayers in liposomes, membrane lipid molecules (phospholipid, etc.), having very small critical micelle concentration, automatically assemble into a lipid bilayer in an aqueous solution via the hydrophobic effect [60]. To introduce asymmetry in liposomes, β-cyclodextrin (β-CD) adds and withdraws specific lipids to and from the outer leaflet of the liposomes, modifying the outer leaflet composition [61, 62]. Here we introduce alternative synthetic methods for generating asymmetric membranes, applying distinct principles and elaborated manipulations (*see* Fig. 2, right).

1.3 Principles and Methods for Asymmetric Membrane Formation in Synthetic Membranes

In 1972, an asymmetric membrane was formed in a PLB by using the folding method [63]. Monolayer formation at the air–water interface was developed using Langmuir–Blodgett technology [64]. In the folding method, monolayers of different lipid compositions at the air–water interface were separately formed into two compartments, which are brought together in the hole by raising the water levels (*see* Fig. 3a). Since then, the same surface chemistry

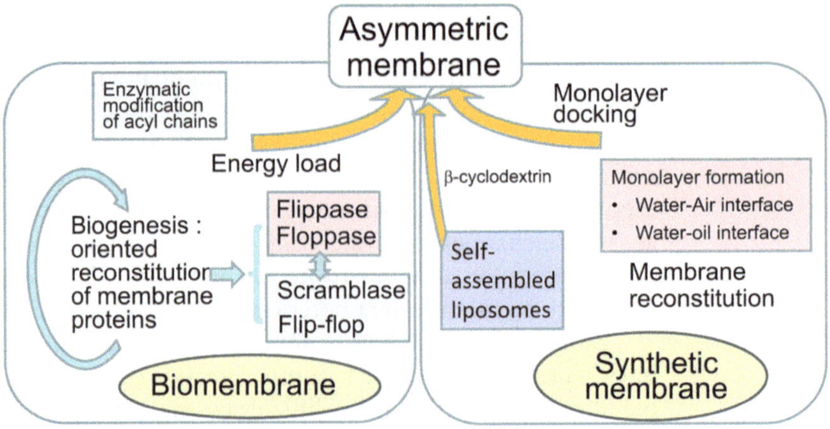

Fig. 2 Generation of asymmetric membranes in biology and technology
In the biomembranes (left), the membrane asymmetry grows progressively from the nearly symmetric ER membrane to the highly asymmetric plasma membrane. The oriented reconstitution of the active transporters (flippases and floppases) in the ER membrane is the first step in asymmetry generation. The compositional asymmetry is developed by active flippases and floppases. In the synthetic membranes (right), asymmetric membranes can be generated in two ways. Liposomes spontaneously form through the self-assembly of phospholipids, while they are symmetric in composition. β-CD extracts specific lipids from the outer leaflet of liposomes, thus generating asymmetry. Alternatively, forming two different monolayers followed by docking is another method. Various methods are available for forming monolayers either in the water–air or water–oil interface

principles have been applied to the formation of asymmetric membranes [54, 65]. Two monolayers of different compositions are formed at either air–water or oil–water interface, which are then mechanically assembled into bilayers (*see* Fig. 3).

As shown in Fig. 3b, two types of monolayers are formed at the water-in-oil droplet and bulk oil–water interfaces [48, 49]. When water droplets pass through the oil–water interface, an asymmetric bilayer is formed, with the outer leaflet having different composition. Figure 3c shows water-in-oil droplets in which lipids are dispersed in either the oil or aqueous phase. These lipids spontaneously transfer to the water–oil interface, lining the water droplets. The lipid monolayer is oriented with the hydrophobic side facing the exterior. Thus, when docked together, a bilayer is formed in the contact area [66]. When two water-in-oil droplets contain different lipid species, an asymmetric bilayer is formed upon docking [47]. In the CBB, a water bubble is blown from a fine pipette, and the bubble size is maintained by holding the bubble pressure. Different lipids spontaneously line the bubbles, and the two bubbles are docked using manipulators. Unlike other methods, the bubble pressure is arbitrarily controlled, thus changing the monolayer and bilayer tensions [55]. The bilayer in CBB is an open system. Each membrane leaflet is a part of the system thermodynamically equilibrated with the bulk monolayer phase of each bubble. Thus, the bilayer is governed by the bulk monolayers.

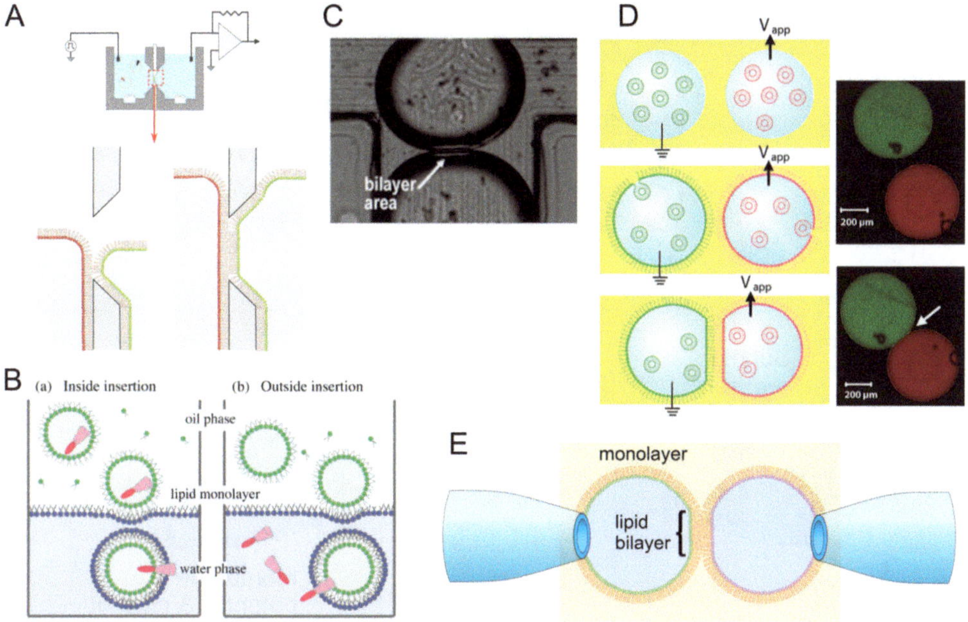

Fig. 3 Methods for forming asymmetric lipid bilayers
The docking of two monolayers of distinct lipid compositions underlies the principle for asymmetric membrane formation. Thus, the methods differ where the monolayers are formed and how they are docked. (**a**) The folding method utilizes the air–water interface for the monolayer formation, similar to the Langmuir–Blodgett method. Two monolayers are folded in a small aperture between two compartments. (**b**) In the droplet transfer method, water-in-oil droplets lined with a monolayer are transferred across a monolayer at the oil–water interface. Adapted with permission from J. Am. Chem. Soc. 2011, 133, 30, 11,774–11,779. Copyright 2011 American Chemical Society. (**c**) The lipid bilayer was formed as water-in-oil droplets apposed together. Adapted with permission from Anal. Chem. 2006, 78, 24, 8169–8174. Copyright 2006 American Chemical Society. (**d**) Asymmetric membrane using DIB. Adapted with permission from J. Am. Chem. Soc. 2008, 130, 18, 5878–5879. Copyright 2008 American Chemical Society. (**e**) In the CBB method, a water bubble is blown from a pipette into the oil phase, and lipids dispersed in the bubble are spontaneously transferred to the oil–water interface, stabilizing the bubble. Two bubbles are docked mechanically

1.4 CBB as a Platform for the Chemical and Physical Interplay of the Lipid Bilayer

The CBB is a versatile technique for creating synthetic membranes of asymmetric lipid compositions [43]. The CBB method was inspired by the water-in-oil droplet method proposed by Funakoshi et al. (*see* Fig. 3c) [66], which was distributed by Bayley et al. as a DIB (*see* Fig. 3d) [67]. We considered that the generation of water-in-oil droplets by blowing the electrolyte solution into the oil phase from a pipette (*see* Fig. 3e) is feasible because we had fully experienced the patch-clamp [68, 69] and tip-dip [70, 71] methods, as well as the conventional PLB method [58, 72]. The bubbles are stably maintained by applying the holding pressure, allowing for stable measurement of the electrical properties of the membrane and channels. We can make asymmetric membranes in the CBB method much more easily than the folding method [43]. Pipette-

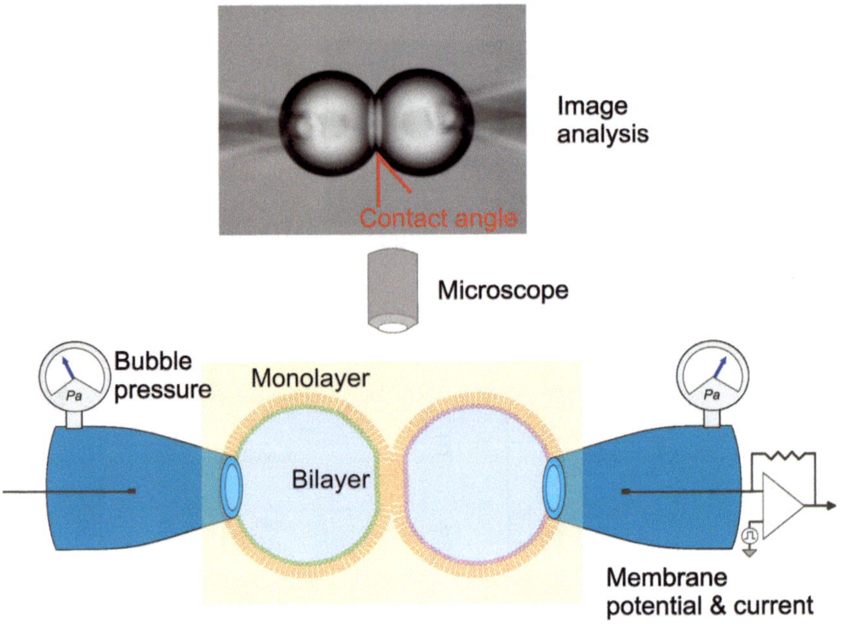

Fig. 4 Asymmetric membrane formation in the contact bubble bilayer method
Liposomes of different lipid compositions are dispersed in two bubbles separately. Meanwhile, the lipids are transferred to the oil–water interface, by which the monolayer tension is decreased, and the bubbles are blown easily and maintained steadily. Then, two bubbles are mechanically docked via motor drive manipulators. In this system, the bubble pressure and geometry are measured, and the monolayer and bilayer tension are readily evaluated

based methods allow mechanical manipulability of the bubbles and the membrane, being promising for functional channel studies. Meanwhile, the surface chemistry of the water-in-oil droplet has been developed [26, 46, 73], and we have established a bilayer tension control system in the CBB [44, 55]. This free manipulability distinguishes the CBB from other lipid bilayer methods.

The principles of tension manipulation and evaluation are briefly described [54]. The Young–Laplace principle is used to evaluate the monolayer tension, $P = 2\ \gamma_{mono}/R$, where γ_{mono} is the monolayer tension and R is the bubble radius [65, 74]. The bubble pressure below 1000 Pa is measured with a fine pressure gauge, and the radius is measured using an image analysis program by fitting an ellipse to the bubble images. The bubble-lining monolayer is continuous with one of the bilayer leaflets and thus chemically and mechanically equilibrated. Accordingly, the leaflet tension is evaluated from the monolayer tension by the Young principle, $\gamma_{leaf} = \gamma_{mono} \cos[\theta]$, where γ_{leaf} is the leaflet tension and θ is the contact angle between the monolayer and the leaflet (*see* Fig. 4). Consequently, the bilayer tension $\gamma_{bileyer} = \gamma_{leaf}^{L} + \gamma_{leaf}^{R}$. Manipulation of the pneumatic microinjector for bubble pressure and monitoring of the pressure and bubble geometry permit dynamic mechanical manipulation and evaluation of the bilayer tension.

2 Materials

2.1 Equipment

1. Inverted microscope.
2. Digital camera.
3. Patch-clamp amplifier.
4. AD converter.
5. Motor-driven micromanipulator × 2.
6. Micromanipulator for perfusion pipette × 2.
7. Pneumatic microinjector × 2 (IM-11-2, Narishige) (*see* **Note 1**).
8. Fine pressure gauge (DP103, Validyne Engineering, Los Angeles, CA, USA) (*see* **Note 2**).
9. Pressure measurement amplifier.
10. Microvolume perfusion system.
11. Micropipette puller.
12. Microforge.
13. A thin-wall borosilicate capillary glass.
14. Pipette holder equipped with a pressure port × 2.
15. AgCl wire electrode × 2.
16. Hole slide glass.
17. Tuberculin syringe.
18. Injection needle (25 G × 60 mM).
19. Personal computer for data acquisition and image analysis.

2.2 Chemicals

1. 1-Palmitoyl-2-oleoyl-glycero-3-phosphocholine.
2. 1-Palmitoyl-2-oleoyl-sn-glycero-3-phospho-(1′-rac-glycerol).
3. Hexadecane.
4. Potassium chloride.
5. Succinic acid.
6. 2-[4-(2-Hydroxyethyl)-1-piperazinyl]ethanesulfonic acid (HEPES).
7. Siliconizing reagent.
8. KcsA potassium channel (see reference [72] for preparation).

3 Methods

3.1 Formation of Asymmetric CBB

Here, we present a brief protocol for the formation of asymmetric membranes in the CBB. A detailed protocol is published using a video guide [53].

1. Fill 100 μL of hexadecane (*see* **Note 3**) into the shallow well of the siliconized hole slide glass (*see* **Note 4**).
2. Fill half the length of a micropipette with the electrolyte solution using a tuberculin syringe (*see* **Note 5**).
3. Set the micropipette onto a micropipette holder having a pressure port, allowing the Ag/AgCl wire electrode to soak in the pipette electrolyte solution.
4. Connect one of the micropipette holders to the head stage of the patch-clamp amplifier and the other to the electrical ground.
5. Connect a microinjector to the pressure port of the micropipette holder.
6. Place 1 μL of the electrolyte solution on the deck of a hole slide glass to create an electrolyte dome. Insert two electrodes to adjust the electrode offset potential.
7. Create a dome for the liposome-containing solution on the deck, and aspirate it into the pipette using a microinjector (*see* **Notes 6** and 7).
8. Operate the microinjector such that the flow of the liposome solution in and out of the pipette tip stops. Subsequently, set the pressure offset of the fine pressure gauge to zero (*see* **Note 8**).
9. Repeat the procedure for the other pipette in the liposome solution with different lipid compositions.
10. Manipulate the micromanipulator, and dip the tip of the micropipette into hexadecane in the shallow well.
11. Blow a water bubble slowly by increasing the pressure until the bubble reaches the desired size (e.g., 50 μm in diameter), and then maintain the same pressure (holding pressure) (*see* **Note 9**).
12. Contact the bubbles using manipulators (*see* Fig. 4) (*see* **Note 10**).
13. Fine-tune the pressure to maintain the bubble size because the size may gradually change even at a constant intra-bubble pressure.
14. If the bilayer breaks and the bubbles coalesce, restart the procedure from **Step** 7. Otherwise, membrane asymmetry cannot be guaranteed.
15. Set the membrane potential to the appropriate value using a patch-clamp amplifier while waiting for the channel current to emerge (*see* Fig. 6).

16. Confirm the offset potential whether it stays null or not at the end of the experiments by breaking the CBB. Applying a high membrane potential (electrical breakdown, using Zap on the amplifier) renders the two bubbles to be merged.
17. Correct the liquid junction potential [75] when asymmetric electrolyte solutions are used, such that the calculated value is added to the applied membrane potential to obtain the true membrane potential (*see* **Note 11**).

3.2 Evaluation of Membrane Asymmetry

If the formed bilayer is intact, the chemical composition of the leaflet remains asymmetric because the spontaneous flip-flop is slow [22]. Membrane asymmetry is quantitatively evaluated if the density of the charged lipids is different in the two leaflets, which generate asymmetric surface potentials [76].

3.2.1 Asymmetric Surface Potentials

Surface potential is a vital membrane property of lipid bilayers; thus, evaluating the surface potential is essential. The membrane potential is measured between the two bulk electrolyte solutions, which is the electric counterpart of the driving force for ion flux across the channel. However, all membrane proteins respond to the intramembrane electric field rather than the membrane potential (*see* Fig. 5a). For example, when the surface potential is modified, the voltage dependency of the voltage-gated channel shifts [77]. Surface potential also affects ion permeation. When evaluating the ion selectivity of a channel having both cation and anion permeation, surface potential must be taken into consideration. The negative surface potential accumulates cations while depleting anions in the vicinity of the membrane relative to the bulk electrolyte solution, thus biasing the apparent ion selectivity [78].

The dependence on the membrane electric field is not limited to membrane proteins. The physical features of a membrane, such as its thickness and electric capacitance, are affected by the electric field. When the surface potentials of the two sides differ substantially, which is often the case for the cell membrane, the differences in the surface potentials can be evaluated (*see* Fig. 5a) [76, 79].

3.2.2 Evaluation of the Asymmetric Surface Potentials

1. Measure the membrane capacitance using the ramp potential at different membrane potentials.
2. Evaluate the membrane area using image analysis. The membrane area expands upon voltage application.
3. Calculate the specific capacitance (C_{spec}; μF/cm^2), and plot it as a function of the voltage.
4. Fit C_{spec} with a parabolic function, and the voltage for the minimal value represents the surface potential (*see* Fig. 5b).

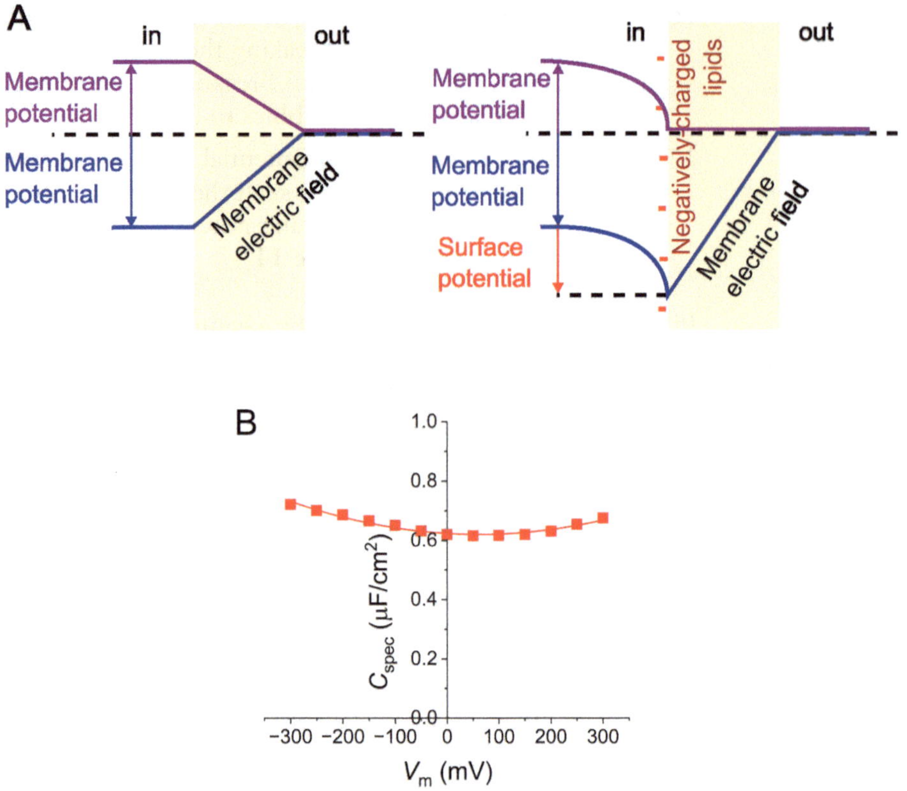

Fig. 5 Surface potential and the transmembrane potential
(**a**) The potential profile in the absence and presence of the surface potential. As far as the surface potential is fixed, the membrane electric field changes in parallel with the membrane potential. The surface potential overwhelms on the intracellular side (right panel). When the membrane potential is balanced with the surface potential ($V_m = -V_{surface}$), the membrane electric field becomes zero. (**b**) Surface potential evaluation through measurements of the membrane capacitance. The membrane capacitance shows the minimum when the membrane electric field is zero, having the thickest membrane

3.2.3 The KcsA Channel in Asymmetric Membranes

Negatively charged lipids have been known as a prerequisite to maintaining KcsA channel activity [43, 80]. To identify the leaflet of negatively charged lipids responsible for channel activity, we performed asymmetric membrane experiments. To examine the negatively charged lipids to the activation gating of the KcsA channel, we examined the non-inactivating mutant (E71A) reconstituted into the asymmetric membranes (*see* Fig. 6) [52]. The channel is fully activated in the symmetric membrane of PG. By contrast, the channel exhibits brief openings in the symmetric PC membrane, indicating the crucial role of PG for the activation gate. When PG is present in the inner leaflet while PC is present in the outer leaflet, the channel remains active, similar to a symmetric PG membrane. In reverse asymmetry, where PG is present in the outer leaflet, channel activity is similar to that in the symmetric PC

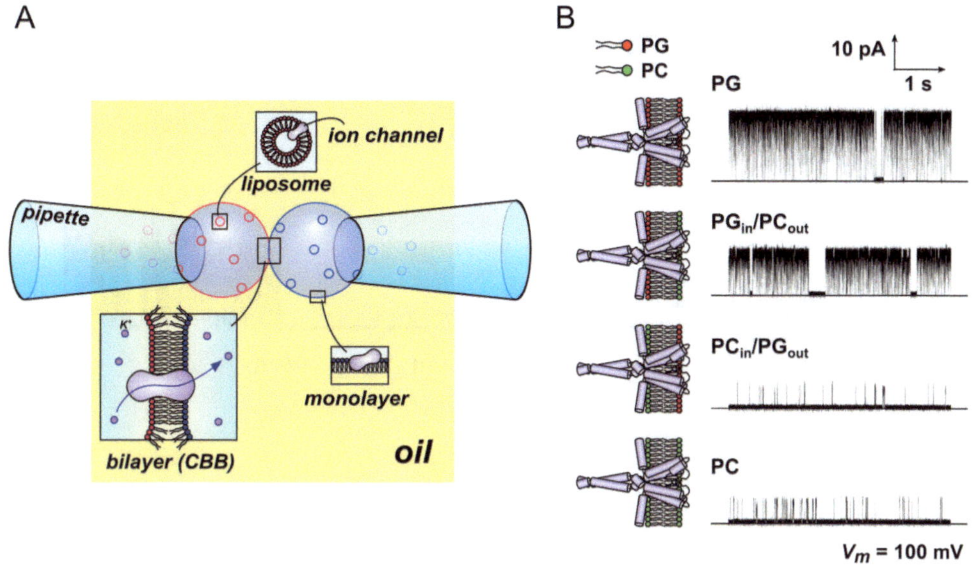

Fig. 6 Channel activities in asymmetric membranes
(**a**) Asymmetric membrane in CBB. Lipid bilayers are formed with different phospholipids. The PG liposomes are exclusively included in one of the bubbles, while another bubble contains PC. (**b**) Single-channel current of the non-inactivating mutant, E71A, in different lipid bilayers at the membrane potential of +100 mV. The channel is oriented with the cytoplasmic side facing the left bubble. The channel stays open most of the time when PG is present in the inner leaflet [52]. The single-channel conductance differs substantially in different membrane compositions

membrane. These results unequivocally demonstrate that the inner leaflet PG is responsible for maintaining high channel activity. Moreover, a charged lipid sensor located at the inner membrane interface was identified as the N-terminal M0 helix [43].

In these single-channel data, the single-channel conductance varied substantially in the charged lipid membranes (*see* Fig. 6). The surface potential accumulates counterions at the vicinity of the membrane, modulating the conductance, while other mechanism has also been proposed [81].

3.3 Membrane Leaflet Perfusion

3.3.1 Lipid Perfusion

The lipid composition steadily changes around the channel in biomembranes. However, changing the lipid composition and examining the channel response in real time are not feasible in cell and synthetic membrane experiments. β-CD serves or withdraws lipids from the outer leaflets of liposomes [61], but the time course of this change is slow. In the CBB, rapid compositional changes in the membrane are readily attainable via an unprecedented supply route [56]. The interior of the hydrophobic bilayer is open to the bulk oil phase (*see* Fig. 7). Thus, spraying hydrophobic substances, such as lipids, from outside the bubble (in the oil phase) close to the bilayer permits easy transfer toward the bilayer interior (membrane perfusion) [56]. This unique access route is valid for hydrophobic

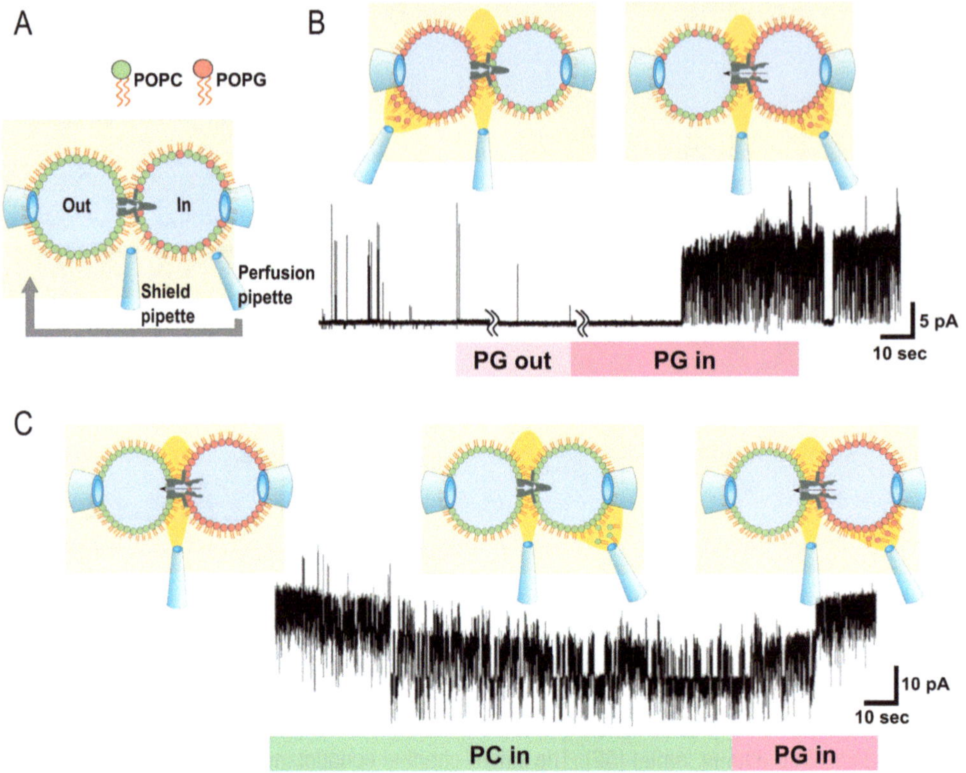

Fig. 7 Leaflet perfusion
(**a**) Leaflet perfusion setup. Two pipettes are set near the bubbles: one for the shielding, spraying hexadecane for preventing diffusion of the perfusing lipids toward another bubble across the bulk organic solvent. Another pipette containing specific lipids dispersed in hexadecane is used for the leaflet perfusion. The sprayed lipids are partitioned into the monolayer of the bubble, on which the lipids diffuse toward the relevant leaflet of the bilayer. The relevant lipid builds up in the leaflet and modifies the lipid composition exclusive to the relevant leaflet. The channel responds to this asymmetric membrane

substances by bypassing the aqueous phase. Previously, we performed cholesterol perfusion during single-channel recordings of KcsA channels and observed their activity changes immediately [55].

3.3.2 Leaflet Perfusion

Now, we perform challenging experiments: Perfusing phospholipids exclusively on one of the leaflets allows the examination of how changes in lipid species and their concentrations in one leaflet influence channel activity. For independent leaflet perfusion, the tip of the perfusion pipette was located far from the bilayer, and a hydrophobic substance was sprayed onto one bubble. The substance partitioned the monolayer, diffusing along the monolayer toward the leaflet. At the same time, this substance likely diffuses across the bulk organic solvent and reaches another bubble. Accordingly, an oil shield was used to prevent the diffusion of hydrophobic substances through the bulk oil phase. The oil shield pipette was located near the bilayer, spraying organic solvent continuously.

1. Set two perfusion pipettes close to the bubbles (*see* **Note 12**).
2. Place a pipette filled with hexadecane (shielding pipette) near the center of the CBB.
3. Set a perfusion pipette containing arbitrary lipids dispersed in hexadecane to the base of each bubble.
4. Spray hexadecane continuously from the shielding pipette to the bilayer throughout the experiment.
5. Record single-channel current.
6. Locate the perfusion pipette at the base of the bubbles on either side, and move it to the other side.

3.3.3 Leaflet Perfusion of PC Vs. PG

Figure 7 shows the single-channel data of the E71A mutant KcsA channel in a symmetric membrane with PC/PG = 3:1. In this experiment, the KcsA channel was oriented with its cytoplasmic side facing the bubbles on the right. First, we examined the sidedness of the effective leaflet perfusion. A pipette at the center was used to continuously spray hexadecane to prevent the leaflet perfusing solution from reaching the other bubbles. The perfusing pipette was first located at the left bubble (outer leaflet side), but PG perfusion failed to activate the channel. However, perfusion of the right bubble (inner leaflet side) successfully activated the channel. Upon the perfusion of PC to the inner leaflet (or right bubble), the channel current gradually decreased, whereas the perfusion of PG increased the current. Accordingly, the CBB allowed leaflet perfusion during single-channel current recordings, thereby extending the feasibility of detecting the dynamic behavior of channel activity upon leaflet compositional changes.

3.4 Conclusions and Outlook

Throughout their lifetime, channel proteins exhibit remarkable resilience in the tangled and ever-changing microenvironments of the cell membrane. They consistently respond to extramembranous and intramembranous stimuli, indicating that their interactions with the lipid bilayer are intricately encoded within their structures. However, the full extent of these interactions remains largely unexplored, with only a fraction of the encoded information deciphered thus far. In vitro experiments using synthetic bilayers have made significant progress in elucidating certain aspects of the channel–membrane interplay. Nonetheless, there exists a significant disparity between native and synthetic membranes. To bridge this gap, we are pursuing a bottom-up approach aimed at forming simplified synthetic membranes by extracting features of complex biological counterparts, to which channels are embedded.

The contact bubble bilayer (CBB) method has emerged as a versatile technique for creating synthetic membranes that possess macroscopic membrane intactness and allow for the manipulation of lipid composition. By leveraging the CBB method, we can engineer membranes to mimic the chemical and physical characteristics of natural membranes. A particularly promising avenue is the generation of asymmetric lipid bilayers, which offers valuable insights into the structure and functionality of biological membranes, interacting with channel proteins. Additionally, the CBB method, combined with bilayer and leaflet perfusion techniques, enables dynamic modifications of lipid compositions and facilitates the investigation of their impact on channel behavior and membrane properties.

In conclusion, the CBB method represents a significant advancement in the development of synthetic membranes. By utilizing this technique and exploring various lipid compositions, we can enhance our understanding of channel–membrane interactions and pave the way for the creation of more sophisticated synthetic membranes that closely resemble their natural counterparts. This will ultimately contribute to a deeper comprehension of biological processes and provide new avenues for the design and development of functional biomimetic systems.

4 Notes

1. This pneumatic microinjector is fine-controllable with manual actuation.
2. This fine pressure gauge measures the low-pressure range below 1000 Pa, which is a prerequisite for the size of the bubbles. For instance, when R is 50 μm and γ_{mono} is 2.5 mN/m, P is 100 Pa.
3. The melting point of hexadecane is 19.2 °C, and the experiments should be performed above 20 °C. Other organic solvents, such as decane, squalene, and squalane, can be used.
4. The surface of the shallow well of the hole slide glass is made siliconized for the filling hexadecane shaping convex surface through modified wettability [65].
5. The tip diameter of the bubble-forming pipette should be 30 to 50 μm. A glass pipette with this tip diameter is prepared as follows: First, prepare a glass pipette with a fine tip. Then, set the pipette on a microforge, and place an electric heating wire of the microforge in contact with the portion of the pipette that has a diameter of 30 to 50 μm. Heating the pipette for a short time will cause the pipette to crack after heating, and the pipette will break, forming the desired tip diameter.

6. The siliconization of the hole slide glass made the deck surface water repellant, allowing the dome formation.
7. The lipid concentration in the liposome solution should start at 2 mg/mL and increase or decrease the concentration as needed.
8. This operation compensates for two opposite actions: the capillary effect by the glass pipette and the hydrostatic pressure of the water column in the pipette.
9. If the bubble holding pressure is not stable, discard the bubbles into the hexadecane layer by quickly putting the pipette tip out into the air. Then dip the pipette tip into the hexadecane phase again to create a new bubble. Repeat this operation until the bubble holding pressure stabilizes.
10. We have formed asymmetric membranes with various lipid compositions. When the bilayer is unstable, increasing the lipid concentration in the electrolyte solution helps stabilize the membrane.
11. The liquid junction potential is calculated using JPCalc [75].
12. For stable and reliable perfusion, the tip diameter of the perfusion pipette should be 3–5 μm.

Acknowledgments

We appreciate Masako Takashima for her technical assistance in the experiment.

References

1. Sezgin E, Levental I, Mayor S, Eggeling C (2017) The mystery of membrane organization: composition, regulation and roles of lipid rafts. Nat Rev Mol Cell Bio 18:361–374. https://doi.org/10.1038/nrm.2017.16
2. Skotland T, Sandvig K (2022) Need for more focus on lipid species in studies of biological and model membranes. Prog Lipid Res 86: 101160. https://doi.org/10.1016/j.plipres.2022.101160
3. Dingjan T, Futerman AH (2021) The fine-tuning of cell membrane lipid bilayers accentuates their compositional complexity. BioEssays 43:2100021. https://doi.org/10.1002/bies.202100021
4. Whitelegge JP (2013) Integral membrane proteins and bilayer proteomics. Anal Chem 85: 2558–2568. https://doi.org/10.1021/ac303064a
5. Engelman DM (2005) Membranes are more mosaic than fluid. Nature 438:578–580. https://doi.org/10.1038/nature04394
6. Nicolson GL, de Mattos GF (2023) The fluid–mosaic model of cell membranes: a brief introduction, historical features, some general principles, and its adaptation to current information. Biochim Biophys Acta Biomembr 184135. https://doi.org/10.1016/j.bbamem.2023.184135
7. Simons K, Gerl MJ (2010) Revitalizing membrane rafts: new tools and insights. Nat Rev Mol Cell Biol 11:688–699. https://doi.org/10.1038/nrm2977
8. Jacobson K, Mouritsen OG, Anderson RGW (2007) Lipid rafts: at a crossroad between cell biology and physics. Nat Cell Biol 9:7–14. https://doi.org/10.1038/ncb0107-7

9. Dixon RE, Navedo MF, Binder MD, Santana LF (2022) Mechanisms and physiological implications of cooperative gating of clustered ion channels. Physiol Rev 102:1159–1210. https://doi.org/10.1152/physrev.00022.2021
10. Phillips R, Ursell T, Wiggins P, Sens P (2009) Emerging roles for lipids in shaping membrane-protein function. Nature 459:379–385. https://doi.org/10.1038/nature08147
11. Rangamani P (2022) The many faces of membrane tension: challenges across systems and scales. Biochim Biophys Acta Biomembr 1864:183897. https://doi.org/10.1016/j.bbamem.2022.183897
12. Iwamoto M, Oiki S (2021) Physical and chemical interplay between the membrane and a prototypical potassium channel reconstituted on a lipid bilayer platform. Front Mol Neurosci 14:634121. https://doi.org/10.3389/fnmol.2021.634121
13. Kessels MM, Qualmann B (2021) Interplay between membrane curvature and the actin cytoskeleton. Curr Opin Cell Biol 68:10–19. https://doi.org/10.1016/j.ceb.2020.08.008
14. Mogilner A, Manhart A (2016) Intracellular fluid mechanics: coupling cytoplasmic flow with active cytoskeletal gel. Annu Rev Fluid Mech 50:1–24. https://doi.org/10.1146/annurev-fluid-010816-060238
15. Sarkar P, Chattopadhyay A (2022) Membrane dipole potential: an emerging approach to explore membrane organization and function. J Phys Chem B 126:4415–4430. https://doi.org/10.1021/acs.jpcb.2c02476
16. Wang L (2012) Measurements and implications of the membrane dipole potential. Annu Rev Biochem 81:615–635. https://doi.org/10.1146/annurev-biochem-070110-123033
17. Marrink SJ, Corradi V, Souza PCT et al (2019) Computational modeling of realistic cell membranes. Chem Rev 119:6184–6226. https://doi.org/10.1021/acs.chemrev.8b00460
18. Van Meer G (2011) Dynamic transbilayer lipid asymmetry. Cold Spring Harb Perspect Biol 3:a004671. https://doi.org/10.1101/cshperspect.a004671
19. Clarke RJ, Hossain KR, Cao K (2020) Physiological roles of transverse lipid asymmetry of animal membranes. Biochim Biophys Acta Biomembr 1862:183382. https://doi.org/10.1016/j.bbamem.2020.183382
20. Contreras F-X, Sánchez-Magraner L, Alonso A, Goñi FM (2010) Transbilayer (flip-flop) lipid motion and lipid scrambling in membranes. FEBS Lett 584:1779–1786. https://doi.org/10.1016/j.febslet.2009.12.049
21. Pomorski TG, Menon AK (2016) Lipid somersaults: uncovering the mechanisms of protein-mediated lipid flipping. Prog Lipid Res 64:69–84. https://doi.org/10.1016/j.plipres.2016.08.003
22. Allhusen JS, Conboy JC (2017) The ins and outs of lipid flip-flop. Acc Chem Res 50:58–65. https://doi.org/10.1021/acs.accounts.6b00435
23. Lorent JH, Levental KR, Ganesan L et al (2020) Plasma membranes are asymmetric in lipid unsaturation, packing and protein shape. Nat Chem Biol 16:644–652. https://doi.org/10.1038/s41589-020-0529-6
24. Siggel M, Bhaskara RM, Hummer G (2019) Phospholipid scramblases remodel the shape of asymmetric membranes. J Phys Chem Lett 10:6351–6354. https://doi.org/10.1021/acs.jpclett.9b02531
25. Wang M, Yi X (2021) Bulging and budding of lipid droplets from symmetric and asymmetric membranes: competition between membrane elastic energy and interfacial energy. Soft Matter 17:5319–5328. https://doi.org/10.1039/d1sm00245g
26. Barlow NE, Kusumaatmaja H, Salehi-Reyhani A et al (2018) Measuring bilayer surface energy and curvature in asymmetric droplet interface bilayers. J R Soc Interface 15:20180610. https://doi.org/10.1098/rsif.2018.0610
27. Hossein A, Deserno M (2020) Spontaneous curvature, differential stress, and bending modulus of asymmetric lipid membranes. Biophys J 118:624–642. https://doi.org/10.1016/j.bpj.2019.11.3398
28. Anishkin A, Loukin SH, Teng J, Kung C (2014) Feeling the hidden mechanical forces in lipid bilayer is an original sense. Proc Natl Acad Sci 111:7898–7905. https://doi.org/10.1073/pnas.1313364111
29. Cox CD, Bavi N, Martinac B (2017) Origin of the force the force-from-lipids principle applied to piezo channels. Curr Top Membr 79:59–96. https://doi.org/10.1016/bs.ctm.2016.09.001
30. Gao M, Huang X, Song B-L, Yang H (2019) The biogenesis of lipid droplets: lipids take center stage. Prog Lipid Res 75:100989. https://doi.org/10.1016/j.plipres.2019.100989
31. M'barek KB, Ajjaji D, Chorlay A et al (2017) ER membrane phospholipids and surface tension control cellular lipid droplet formation. Dev Cell 41:591–604.e7. https://doi.org/10.1016/j.devcel.2017.05.012

32. van Meer G, Voelker DR, Feigenson GW (2008) Membrane lipids: where they are and how they behave. Nat Rev Mol Cell Biol 9: 112–124. https://doi.org/10.1038/nrm2330
33. Shealy RT, Murphy AD, Ramarathnam R et al (2003) Sequence-function analysis of the K^+-selective family of ion channels using a comprehensive alignment and the KcsA channel structure. Biophys J 84:2929–2942. https://doi.org/10.1016/s0006-3495(03)70020-4
34. Shaya D, Findeisen F, Abderemane-Ali F et al (2014) Structure of a prokaryotic sodium channel pore reveals essential gating elements and an outer ion binding site common to eukaryotic channels. J Mol Biol 426:467–483. https://doi.org/10.1016/j.jmb.2013.10.010
35. Shimizu H, Iwamoto M, Konno T et al (2008) Global twisting motion of single molecular KcsA potassium channel upon gating. Cell 132:67–78. https://doi.org/10.1016/j.cell.2007.11.040
36. Notti RQ, Walz T (2022) Native-like environments afford novel mechanistic insights into membrane proteins. Trends Biochem Sci 47: 561–569. https://doi.org/10.1016/j.tibs.2022.02.008
37. Natale AM, Deal PE, Minor DL (2021) Structural insights into the mechanisms and pharmacology of K2P potassium channels. J Mol Biol 433:166995–166995. https://doi.org/10.1016/j.jmb.2021.166995
38. Mandala VS, MacKinnon R (2022) Voltage-sensor movements in the Eag Kv channel under an applied electric field. Proc Natl Acad Sci 119:e2214151119. https://doi.org/10.1073/pnas.2214151119
39. Wisedchaisri G, Tonggu L, McCord E et al (2019) Resting-state structure and gating mechanism of a voltage-gated sodium channel. Cell 178:993–1003.e12. https://doi.org/10.1016/j.cell.2019.06.031
40. Yamakata A, Shimizu H, Oiki S (2015) Surface-enhanced IR absorption spectroscopy of the KcsA potassium channel upon application of an electric field. Phys Chem Chem Phys 17:21104–21111. https://doi.org/10.1039/c5cp02681d
41. Oiki S (2015) Channel function reconstitution and re-animation: a single-channel strategy in the postcrystal age. J Physiol 593:2553–2573. https://doi.org/10.1113/jp270025
42. Iwamoto M, Oiki S (2019) In bulla functional channel expression systems that mimic bacterial synthetic membranes. Methods Enzymol 621: 231–244. https://doi.org/10.1016/bs.mie.2019.02.011
43. Iwamoto M, Oiki S (2013) Amphipathic antenna of an inward rectifier K^+ channel responds to changes in the inner membrane leaflet. Proc Natl Acad Sci 110:749–754. https://doi.org/10.1073/pnas.1217323110
44. Iwamoto M, Oiki S (2021) Hysteresis of a tension-sensitive K^+ channel revealed by time-lapse tension measurements. JACS Au 1:467–474. https://doi.org/10.1021/jacsau.0c00098
45. Das KMP, Shih WM, Wagner G, Nasr ML (2020) Large nanodiscs: a potential game changer in structural biology of membrane protein complexes and virus entry. Front Bioeng Biotechnol 8:539. https://doi.org/10.3389/fbioe.2020.00539
46. Huang Y, Fuller GG, Suja VC (2022) Physicochemical characteristics of droplet interface bilayers. Adv Colloid Interf Sci 304:102666. https://doi.org/10.1016/j.cis.2022.102666
47. Hwang WL, Chen M, Cronin B et al (2008) Asymmetric droplet Interface bilayers. J Am Chem Soc 130:5878–5879. https://doi.org/10.1021/ja802089s
48. Yanagisawa M, Iwamoto M, Kato A et al (2011) Oriented reconstitution of a membrane protein in a giant unilamellar vesicle: experimental verification with the potassium channel KcsA. J Am Chem Soc 133:11774–11779. https://doi.org/10.1021/ja2040859
49. Pautot S, Frisken BJ, Weitz DA (2003) Engineering asymmetric vesicles. Proc Natl Acad Sci 100:10718–10721. https://doi.org/10.1073/pnas.1931005100
50. Liu P, Zabala-Ferrera O, Beltramo PJ (2021) Fabrication and electromechanical characterization of freestanding asymmetric membranes. Biophys J 120:1755–1764. https://doi.org/10.1016/j.bpj.2021.02.036
51. Marquardt D, Geier B, Pabst G (2015) Asymmetric lipid membranes: towards more realistic model systems. Membranes 5:180–196. https://doi.org/10.3390/membranes5020180
52. Iwamoto M, Oiki S (2015) Contact bubble bilayers with flush drainage. Sci Rep 5:9110. https://doi.org/10.1038/srep09110
53. Iwamoto M, Oiki S (2019) Lipid bilayer experiments with contact bubble bilayers for patch-clampers. J Vis Exp. https://doi.org/10.3791/58840
54. Oiki S, Iwamoto M (2018) Lipid bilayers manipulated through monolayer technologies for studies of channel-membrane interplay. Biol Pharm Bull 41:303–311. https://doi.org/10.1248/bpb.b17-00708

55. Iwamoto M, Oiki S (2018) Constitutive boost of a K^+ channel via inherent bilayer tension and a unique tension-dependent modality. Proc Natl Acad Sci 115:13117–13122. https://doi.org/10.1073/pnas.1812282115
56. Iwamoto M, Oiki S (2017) Membrane perfusion of hydrophobic substances around channels embedded in the contact bubble bilayer. Sci Rep 7:6857. https://doi.org/10.1038/s41598-017-07048-4
57. White SH (1986) The physical nature of planar bilayer membranes. In: Miller C (ed) Ion channel reconstitution. Plenum Press, New York, pp 3–35
58. Oiki S (2012) Planar lipid bilayer method for studying channel molecules. In: Okada Y (ed) Patch clamp techniques: from beginning to advanced protocols. Springer, Tokyo, pp 229–275
59. Leventis PA, Grinstein S (2010) The distribution and function of phosphatidylserine in cellular membranes. Annu Rev Biophys 39:407–427. https://doi.org/10.1146/annurev.biophys.093008.131234
60. White SH, von Heijne G, Engelman DM (2022) Cell boundaries: how membranes and their proteins work. Taylor & Francis, Boca Raton
61. Hammoud Z, Khreich N, Auezova L et al (2019) Cyclodextrin-membrane interaction in drug delivery and membrane structure maintenance. Int J Pharm 564:59–76. https://doi.org/10.1016/j.ijpharm.2019.03.063
62. Peyret A, Zhao H, Lecommandoux S (2018) Preparation and properties of asymmetric synthetic membranes based on lipid and polymer self-assembly. Langmuir 34:3376–3385. https://doi.org/10.1021/acs.langmuir.7b04233
63. Montal M, Mueller P (1972) Formation of bimolecular membranes from lipid monolayers and a study of their electrical properties. Proc Natl Acad Sci 69:3561–3566. https://doi.org/10.1073/pnas.69.12.3561
64. Oliveira ON, Caseli L, Ariga K (2022) The past and the future of langmuir and langmuir–blodgett films. Chem Rev 122:6459–6513. https://doi.org/10.1021/acs.chemrev.1c00754
65. de Gennes PG, Brochard-Wyart F, Quere D (2003) Capillarity and wetting phenomena: drops, bubbles, pearls, waves. Springer, New York
66. Funakoshi K, Suzuki H, Takeuchi S (2006) Lipid bilayer formation by contacting monolayers in a microfluidic device for membrane protein analysis. Anal Chem 78:8169–8174. https://doi.org/10.1021/ac0613479
67. Bayley H, Cronin B, Heron A et al (2008) Droplet interface bilayers. Mol BioSyst 4: 1191–1208. https://doi.org/10.1039/b808893d
68. Ando H, Kuno M, Shimizu H et al (2005) Coupled K^+–water flux through the HERG potassium channel measured by an osmotic pulse method. J Gen Physiol 126:529–538. https://doi.org/10.1085/jgp.200509377
69. Kuno M, Ando H, Morihata H et al (2009) Temperature dependence of proton permeation through a voltage-gated proton channel. J Gen Physiol 134:191–205. https://doi.org/10.1085/jgp.200910213
70. Oiki S, Koeppe RE, Andersen OS (1995) Voltage-dependent gating of an asymmetric gramicidin channel. Proc Natl Acad Sci 92: 2121–2125. https://doi.org/10.1073/pnas.92.6.2121
71. Oiki S, Danho W, Madison V, Montal M (1988) M2 delta, a candidate for the structure lining the ionic channel of the nicotinic cholinergic receptor. Proc Natl Acad Sci 85:8703–8707. https://doi.org/10.1073/pnas.85.22.8703
72. Iwamoto M, Shimizu H, Inoue F et al (2006) Surface structure and its dynamic rearrangements of the KcsA potassium channel upon gating and tetrabutylammonium blocking. J Biol Chem 281:28379–28386. https://doi.org/10.1074/jbc.m602018200
73. Taylor GJ, Venkatesan GA, Collier CP, Sarles SA (2015) Direct in situ measurement of specific capacitance, monolayer tension, and bilayer tension in a droplet interface bilayer. Soft Matter 11:7592–7605. https://doi.org/10.1039/c5sm01005e
74. Israelachvili JN (2011) Intermolecular and surface forces, 3rd edn. Academic Press, New York/London
75. Barry PH (1994) JPCalc, a software package for calculating liquid junction potential corrections in patch-clamp, intracellular, epithelial and bilayer measurements and for correcting junction potential measurements. J Neurosci Methods 51:107–116. https://doi.org/10.1016/0165-0270(94)90031-0
76. Alvarez O, Latorre R (1978) Voltage-dependent capacitance in lipid bilayers made from monolayers. Biophys J 21:1–17. https://doi.org/10.1016/s0006-3495(78)85505-2
77. Trapani JG, Korn SJ (2003) Effect of external pH on activation of the Kv1.5 potassium channel. Biophys J 84:195–204. https://doi.org/10.1016/s0006-3495(03)74842-5

78. Miedema H (2002) Surface potentials and the calculated selectivity of ion channels. Biophys J 82:156–159. https://doi.org/10.1016/s0006-3495(02)75382-4

79. Burtscher V, Hotka M, Freissmuth M, Sandtner W (2020) An electrophysiological approach to measure changes in the membrane surface potential in real time. Biophys J 118: 813–825. https://doi.org/10.1016/j.bpj.2019.06.033

80. Heginbotham L, Kolmakova-Partensky L, Miller C (1998) Functional reconstitution of a prokaryotic K^+ channel. J Gen Physiol 111: 741–749. https://doi.org/10.1085/jgp.111.6.741

81. Oakes V, Furini S, Domene C (1862) Effect of anionic lipids on ion permeation through the KcsA K^+-channel. Biochim Biophys Acta Biomembr 2020:183406. https://doi.org/10.1016/j.bbamem.2020.183406

82. Frolov VA, Shnyrova AV, Zimmerberg J (2011) Lipid polymorphisms and membrane shape. Cold Spring Harb Perspect Biol 3:a004747. https://doi.org/10.1101/cshperspect.a004747

Chapter 2

Detecting Bound Ions in Ion Channels by Solid-State NMR Experiments on ^{15}N-Labelled Ammonium Ions

Carl Öster, Sascha Lange, Kitty Hendriks, and Adam Lange

Abstract

Solid-state NMR allows for the study of membrane proteins under physiological conditions. Here we describe a method for detection of bound ions in the selectivity filter of ion channels using solid-state NMR. This method employs standard ^{1}H-detected solid-state NMR setup and experiment types, which is enabled by using ^{15}N-labelled ammonium ions to mimic potassium ions.

Key words Solid-state NMR, Ammonium, Ion channels, Membrane proteins, Ion binding

1 Introduction

Detailed information on ion binding and selectivity filter conformations in ion channels is often challenging to obtain using conventional structural biology techniques. Solid-state NMR is uniquely suitable to study membrane proteins under "near-native" conditions. The proteins are embedded into a lipid bilayer, and the experiments can be conducted under physiological buffer conditions and at room temperature. Here we describe a protocol for studying ion binding in potassium-selective and non-selective (sodium/potassium) cation channels. To enable the acquisition of high-resolution spectra and the assignment of ions bound in the specific ion binding sites, we use ^{15}N-labelled ammonium ions ($^{15}NH_4^+$) instead of potassium ions (K^+). Although ^{39}K is NMR active, it has a relatively low gyromagnetic ratio (ca 5% compared to ^{1}H and ca 50% compared to ^{15}N). Additionally, ^{39}K is a spin 3/2 nucleus and has a quadrupole moment, which results in line broadening and makes it very challenging for site-specific studies of ion binding in ion channels. $^{15}NH_4^+$, on the other hand, allows for the use of ^{1}H-detected solid-state NMR experiments. Generally, in order to obtain spectra with high resolution, protein solid-state NMR experiments are conducted using spin 1/2 nuclei (^{1}H, ^{13}C,

Simone Furini (ed.), *Potassium Channels: Methods and Protocols*, Methods in Molecular Biology, vol. 2796,
https://doi.org/10.1007/978-1-0716-3818-7_2,

and ^{15}N) that do not have quadrupole moments. The method described here employs ^{1}H-detected experiments, at magic angle spinning (MAS) rates of 40–60 kHz, of $^{15}NH_4^+$ in $^{2}H^{13}C^{15}N$-labelled proteins with exchangeable solvent-exposed sites back-exchanged in H_2O. $^{15}NH_4^+$ ions have previously been used to mimic K^+ binding in solution NMR studies of different biomolecules [1, 2]. Recently, $^{15}NH_4^+$ was also used in a solution NMR study of the K^+-selective ion channel KcsA [3]. In that study, the peaks originating from bound ammonium ions were attributed to the different ion binding sites in the selectivity filter. However, it was not possible to confirm the binding in specific ion binding sites by magnetization transfers between the ions and atoms in the protein. In our recent study, we showed that by using solid-state NMR, it is possible to achieve high-resolution spectra of bound $^{15}NH_4^+$ and that magnetization transfer between the ^{1}H atoms of $^{15}NH_4^+$ and ^{13}C atoms of the selectivity filter enables assignments of ammonium ions to specific ion binding sites [4].

In Subheading 3.1, we describe the expression, purification, and reconstitution into a lipid bilayer for the non-selective ion channel NaK that we have studied in our lab for several years [5, 6]. The protein is expressed in D_2O-based media and purified in H_2O-based buffers. In this process, solvent-exposed exchangeable protons (NH and OH groups) are exchanged from ^{2}H to ^{1}H. Deuteration is important to dilute the proton network, which results in narrow lines in the solid-state NMR spectra of bound $^{15}NH_4^+$ at MAS rates of 40–60 kHz, and reduces potential problems with overlapping peaks that would complicate the assignment process of the bound ions. Subheading 3.2 includes a description of how to efficiently pack the isotope-labelled protein sample into a 1.3 mm solid-state NMR rotor. The final subheading (3.3) contains a description of how to set up and analyze the basic solid-state NMR experiments necessary to identify bound $^{15}NH_4^+$ in the selectivity filter of an ion channel.

2 Materials

2.1 Expression

1. 10× M9 salt in D_2O: 553 mM Na_2HPO_4, 110 mM KH_2PO_4, and 86 mM NaCl, sterile filtered; store at room temperature (*see* **Note 1**).
2. Trace elements in D_2O: 17.1 mM EDTA, 3.3 mM $FeSO_4$, 367 μM $ZnCl_2$, 63 μM $CuSO_4$, pH 7.4, sterile filtered; store at 4 °C.
3. 1.04 M $^{2}H^{13}C$-labelled glucose (D-glucose-$^{13}C_6$,1,2,3,4,5,6,6-d_7) in D_2O, sterile filtered; prepare fresh.
4. 4.59 M ^{15}N-labelled NH_4Cl in D_2O, sterile filtered; prepare fresh.

5. 1 M $MgSO_4$ in D_2O, autoclaved; store at room temperature.
6. 1 M $CaCl_2$ in D_2O, autoclaved; store at room temperature.
7. 409 μM biotin in D_2O, sterile filtered; store at −20 °C.
8. 3 mM thiamine hydrochloride in D_2O, sterile filtered; store at −20 °C.
9. 142 mM carbenicillin disodium salt in D_2O, sterile filtered; store at −20 °C.
10. 1 M IPTG in D_2O, sterile filtered; store at −20 °C.
11. ^{13}C- and ^{15}N-labelled M9 minimal media in D_2O: 55.3 mM Na_2HPO_4, 11 mM KH_2PO_4, 8.6 mM NaCl, 171 μM EDTA, 330 μM $FeSO_4$, 3.7 μM $ZnCl_2$, 0.63 μM $CuSO_4$, 20.7 mM $^2H^{13}C$-labelled glucose, 18 mM ^{15}N-labelled NH_4Cl, 6.1 μM biotin, 4.4 μM thiamine hydrochloride, 142 μM carbenicillin disodium salt, 1 mM $MgSO_4$, and 300 μM $CaCl_2$ (*see* **Note 2**).
12. LB media (autoclaved) with 140 μM carbenicillin disodium (added fresh after autoclaving).
13. *E. coli* NEB Express I^q harboring a pQE-60 vector with the DNA encoding NaKΔ19.

2.2 Purification, Solubilization, and Reconstitution

1. Co^{2+} beads: TALON Superflow beads (GE Healthcare Life Sciences).
2. Lysis buffer: 50 mM Tris–HCl (pH 7.5), 100 mM NaCl, 1 mM $MgCl_2$, protease inhibitor cocktail (cOmplete™, EDTA-free by Roche Diagnostics Germany GmbH), and DNAse (DNase I by AppliChem GmbH, Germany); store at 4 °C (*see* **Note 3**).
3. Solubilization buffer: 50 mM Tris–HCl, 100 mM NaCl, 40 mM DM (n-Decyl-β-maltoside, GLYCON Biochemicals), pH 7.5; store at 4 °C (*see* **Note 4**).
4. Binding buffer: 50 mM Tris–HCl, 100 mM NaCl, 20 mM imidazole, pH 7.5; store at 4 °C.
5. Wash buffer: 20 mM Tris–HCl, 100 mM NaCl, 5 mM DM, 30 mM imidazole, pH 7.5; store at 4 °C.
6. Elution buffer: 20 mM Tris–HCl, 100 mM NaCl, 5 mM DM, 500 mM imidazole, pH 7.5; store at 4 °C.
7. Sample buffer: 20 mM Tris–HCl, 50 mM ^{15}N-labelled NH_4Cl, pH 8; store at 4 °C.
8. Lipid buffer: 20 mM Tris–HCl, 50 mM ^{15}N-labelled NH_4Cl, 5 mM DM, pH 8; store at 4 °C.
9. Lipids: *E. coli* total lipid extract (Avanti Polar Lipids) 10 mg/ml (~14 mM) dissolved in lipid buffer; store at −20 °C.
10. Ultracentrifuges: Beckmann Optima XPN-100 (45TI rotor with 70 mL tubes), Beckmann Optima MAX-XP (TLA 110 rotor with 3.2 mL tubes, TLA 55 rotor with 1.5 mL tubes).

11. Microfluidizer (LM10, Microfluidics).
12. Vivaspin 20 centrifugal concentrator with 10 kDa cutoff (Sartorius).
13. Bio-beads SM2 resin (Bio-Rad).
14. Bradford reagent with 5 mM α-cyclodextrin; prepare fresh.

2.3 Rotor Packing

1. 1.3 mm solid-state NMR rotor with caps (Bruker BioSpin).
2. 1.3 mm rotor packing tools (Bruker BioSpin).
3. Silicone-based glue (LOCTITE SI 5145).
4. K^+ buffer: 20 mM Tris–HCl, 50 mM KCl, pH 8; store at 4 °C.

2.4 Solid-State NMR Experiments

1. NMR spectrometer equipped for solid-state NMR.
2. Three-channel 1.3 mm magic angle spinning (MAS) solid-state NMR probe (Bruker BioSpin), configured to ^{1}H, ^{13}C, and ^{15}N (*see* **Note 5**).

3 Methods

3.1 Protein Expression

1. Day 1. Start culture and D_2O adaption. Inoculate (from a glycerol stock or fresh transformant) 20 mL 50% deuterated medium (10 mL LB + 10 mL deuterated M9 media). Incubate at 37 °C and 150 rpm until the medium turns turbid (OD 0.6, ca 4 h). Add 20 mL deuterated M9 medium. Split into two 100 mL Erlenmeyer flasks (20 mL each), and incubate at 37 °C and 150 rpm until the medium turns turbid (OD 0.6, ca 4 h). Inoculate 270 mL pre-warmed (37 °C) deuterated M9 medium in a 2 L Erlenmeyer flask with both over-day cultures (40 mL). Incubate overnight (ca 16 h) at 37 °C and 150 rpm.
2. Day 2. Divide the cell cultures into eight 50 mL Falcon tubes, and collect the cells by centrifugation (3500 rcf, 20 min at 37 °C).
3. Resuspend each pellet into 15 mL pre-warmed (37 °C) deuterated M9 medium. Inoculate four 2 L Erlenmeyer flasks (containing 470 mL pre-warmed deuterated M9 media each) with 30 mL resuspended cell pellets each.
4. Grow the cells at 37 °C and 150 rpm to an OD of 0.8–1.
5. Lower the temperature to 25 °C, and induce protein expression with 0.4 mM IPTG (200 μL 1 M stock solution per flask). After 6 h, add 0.5 g ^{2}H^{13}C^{15}N-labelled glucose (2.5 mL 1.04 M stock solution) and 0.125 g ^{15}N-labelled NH_4Cl (0.5 mL 4.59 M stock solution) per flask. Incubate for another 18 h (total expression time of 24 h) (*see* **Note 6**).
6. Day 3. Harvest the cells by centrifugation (20 min, 6.000 rcf) at 4 °C (*see* **Note 7**).

7. Weigh the cell pellets, and resuspend them in 10 mL lysis buffer per gram of cell pellet. Add 5 mg DNAse and one protease inhibitor tablet per 100 mL buffer.
8. Lyse the cells by running them five times through a microfluidizer operating at 15,000 psi while keeping the sample cool with ice.
9. Remove the cell debris by centrifugation (20 min, 16,000 rcf at 4 °C). Keep the supernatant.
10. Isolate the membrane fraction from the supernatant by ultracentrifugation (2 h, 150,000 rcf, at 4 °C, rotor 45TI, 70 mL tubes).
11. Resuspend the membrane pellets in 80 mL solubilization buffer, and transfer to a glass beaker. Leave at room temperature for 3 h.
12. Prepare the beads for IMAC. Wash 5 mL Co^{2+} beads with 50 mL 1:1 solubilization buffer/binding buffer. Remove the liquid and add 80 mL binding buffer. Keep the beads at 4 °C.
13. Centrifuge the solubilized membrane pellets (20 min, 16,000 rcf, at 4 °C), and keep the supernatant.
14. Add the supernatant to the Co^{2+} beads and incubate overnight at 4 °C while stirring.
15. Day 4. Purify the protein using IMAC. Add the beads to a disposable chromatography column. Wash with 25 mL wash buffer. Elute the protein in ten times 5 mL elution buffer.
16. Determine the protein concentration using Bradford (*see* **Note 8**).
17. Concentrate the sample to 1 mg/mL using a 10 kDa cutoff centrifugal concentrator. Centrifuge in steps of 10 min at 3000 rcf and room temperature.
18. Centrifuge the concentrated sample for 10 min at 16,000 rcf and 4 °C to remove any precipitated protein.
19. Mix the protein with the solubilized *E. coli* total lipid extract (10 mg/mL in lipid buffer) in a protein/lipid ratio of 2:1 (w/w). Incubate for 1 h at 4 °C while rotating.
20. Put the sample in a dialysis bag with 6–8 kDa cutoff.
21. Dialyze against 2 L sample buffer for 8 days at 4 °C. Exchange the dialysis buffer at least six times. Add 1 g bio-beads to the dialysis buffer in the last two steps (*see* **Note 9**).
22. 8 days later. Ultracentrifuge the sample for 2 h at 300,000 rcf and 4 °C (rotor TLA 110, 3.2 mL tubes). Resuspend the pellets in supernatant, and pool them in a 1.5 mL ultracentrifuge tube. Ultracentrifuge for 2 h at 150,000 rcf and 4 °C (rotor TLA 55, 1.5 mL tubes).

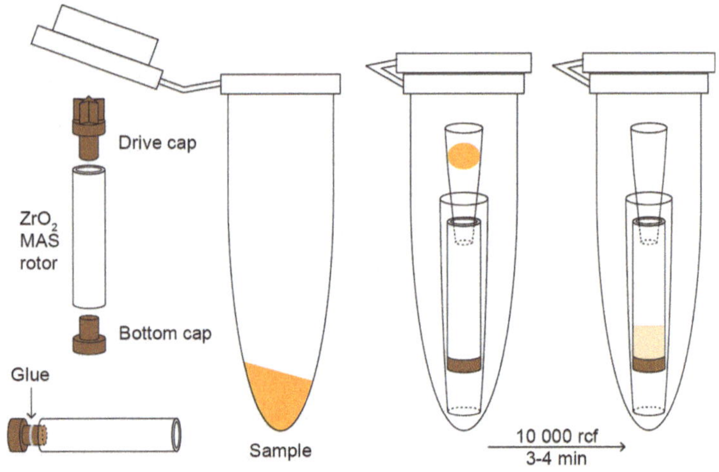

Fig. 1 Scheme of rotor packing
A silicone-based glue is used to prevent water leakage. The sample pellet is transferred into the cutoff pipette tip attached to the rotor, and the rotor is filled by centrifugation

3.2 Solid-State NMR Rotor Packing

1. Remove the supernatant from the ultracentrifuge tube.
2. Prepare two 200 μL pipette tips by cutting them into pieces so that one can be attached into the rotor and the second one can be used to support the rotor when the construction is put into a 1.5 mL Eppendorf tube (*see* Fig. 1).
3. Put the bottom cap on the rotor. Add a thin layer of glue to avoid leakage. Attach a cutoff 200 μL pipette tip.
4. Scoop the pellet into the cutoff pipette tip that is attached to the rotor, and place the rotor in the second cutoff pipette tip. Place the assembled rotor–pipette tip construction into a 1.5 mL Eppendorf tube.
5. Centrifuge in a bench-top centrifuge (10,000 rcf, 4 min). If necessary, repeat until the rotor is full.
6. Put on the drive cap. Add a thin layer of glue to avoid leakage.
7. Carefully wipe the rotor with a laboratory wipe, and use high pressure air to remove any dirt stuck to the rotor and caps. Check the rotor and caps for dirt and damage under a microscope. Make sure that the bottom cap has a mark (for detection of the MAS rate) covering 50% of the circumference.
8. Add the supernatant to the leftover sample pellet and store at 4 °C (*see* **Note 10**).
9. Exchange the ions in the sample from ^{15}N-labelled NH_4^+ to K^+. Ultracentrifuge the leftover sample (1 h, 100,000 rcf, 4 °C, rotor TLA 55, 1.5 mL tubes). Remove the supernatant, and resuspend the pellet in K^+ buffer. Repeat this process three

times with 2–5 days' intervals (for a total washing time of 10–14 days). Leave the sample rotating at 4 °C between buffer exchanges.

10. Repeat the rotor packing process with the sample where the ions have been exchanged from ^{15}N-labelled NH_4^+ to K^+.

3.3 Solid-State NMR Experiments

1. Spin up the rotor (containing ^{15}N-labelled NH_4^+ ions) to 60 kHz while adjusting the sample temperature to ca 20 °C (*see* **Note 11**).
2. Record four 3D experiments for backbone assignments: (H)CANH, (H)CONH, (H)CA(CO)NH, and (H)CO (CA)NH (*see* **Note 12**).
3. Assign each resonance in the spectra to the backbone atoms (HN, N, CA, and CO) in the protein following the backbone walk procedure.
4. Choose a standard ^{1}H-detected 2D (H)NH pulse program with INEPT (insensitive nuclei enhanced by polarization transfer)-based heteronuclear magnetization transfer steps (*see* Fig. 2a for a scheme of the pulse program). To make the experiment selective for $^{15}NH_4^+$ over side-chain and backbone ^{15}NH groups, exchange the 90-degree pulses before and after the water suppression to selective 90-degree pulses (Fig. 2b). Choose Q5 selective 90-degree pulses. Use the shape tool in TopSpin to calculate the pulse length and power level. On a

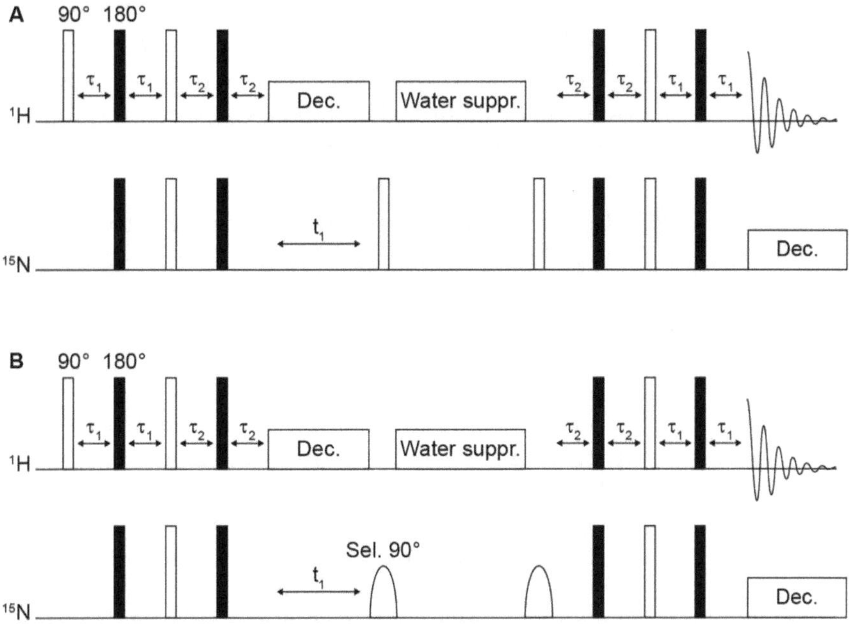

Fig. 2 Scheme of pulse program for INEPT-based ^{1}H-detected ^{1}H–^{15}N correlation spectra
(**a**) Standard pulse program. (**b**) Pulse program with selective 90-degree pulses for bandwidth selective experiments

600 MHz spectrometer, we use a bandwidth of 2 kHz (33 ppm) for the selective pulses, which corresponds to a pulse length of 3091 μs. The power level for the maximum amplitude of the selective pulses corresponds to a nutation frequency of ca 1.5 kHz.

5. Set the INEPT delays to the specific values for $^{15}NH_4^+$: $\tau_1 = 1.05$ ms and $\tau_2 = 2.7$ ms. Set the ^{15}N offset to the expected $^{15}NH_4^+$ region (ca 20 ppm). Optimize the INEPT delays to find the conditions with the highest intensity in 1D (H)(N)H spectra (*see* **Note 13**).
6. Record a 2D (H)NH spectrum with large spectral width (ca 120 ppm, with the ^{15}N offset at 75 ppm) using a standard pulse program for INEPT-based (H)NH experiments (Fig. 2a) with the optimized τ_1 and τ_2 INEPT delays.
7. Set the ^{15}N offset in the middle of the $^{15}NH_4^+$ region (21 ppm for NaK). Record a 2D (H)NH spectrum with small spectral width (6 ppm for NaK) using the INEPT-based pulse program with selective 90-degree pulses on ^{15}N (*see* **Note 14**).
8. Record 2D (H)COH and (H)CXH spectra with 7 ms ^{1}H–^{13}C and ^{13}C–^{1}H CP (cross-polarization) mixing times using a standard ^{1}H-detected (H)CH pulse programs with CP-based heteronuclear magnetization transfer steps. For the (H)COH experiment, set the ^{13}C offset in the middle of the carbonyl region (ca 175 ppm) and the spectral width to ca 14 ppm. For the (H)CXH experiment, set the ^{13}C offset in the middle of the Cα–Cβ region (ca 40 ppm) and the spectral width to ca 70 ppm (*see* **Note 15**).
9. Contacts between $^{15}NH_4^+$ ions and carbon atoms in the protein can be identified in the (H)CH spectra by following the ^{1}H chemical shifts of the $^{15}NH_4^+$ peaks in the (H)NH spectra (*see* Fig. 3b) (*see* **Note 16**).
10. Repeat the NMR experiments with the rotor containing the K^+ bound sample (*see* Fig. 3c for (H)CH spectra).

4 Notes

1. To minimize residual protonation, care has to be taken when preparing the stock solutions for deuterated M9 media. 10× M9 salt and trace elements should be prepared in H_2O and lyophilized and then dissolved in D_2O, lyophilized again, and dissolved in D_2O again. Labelled glucose, labelled NH_4Cl, $MgSO_4$, and $CaCl_2$ can be directly dissolved in D_2O and then lyophilized only once and dissolved in D_2O again. The other components (biotin, thiamine, and carbenicillin) are more fragile and should be directly dissolved in D_2O. The same applies to IPTG.

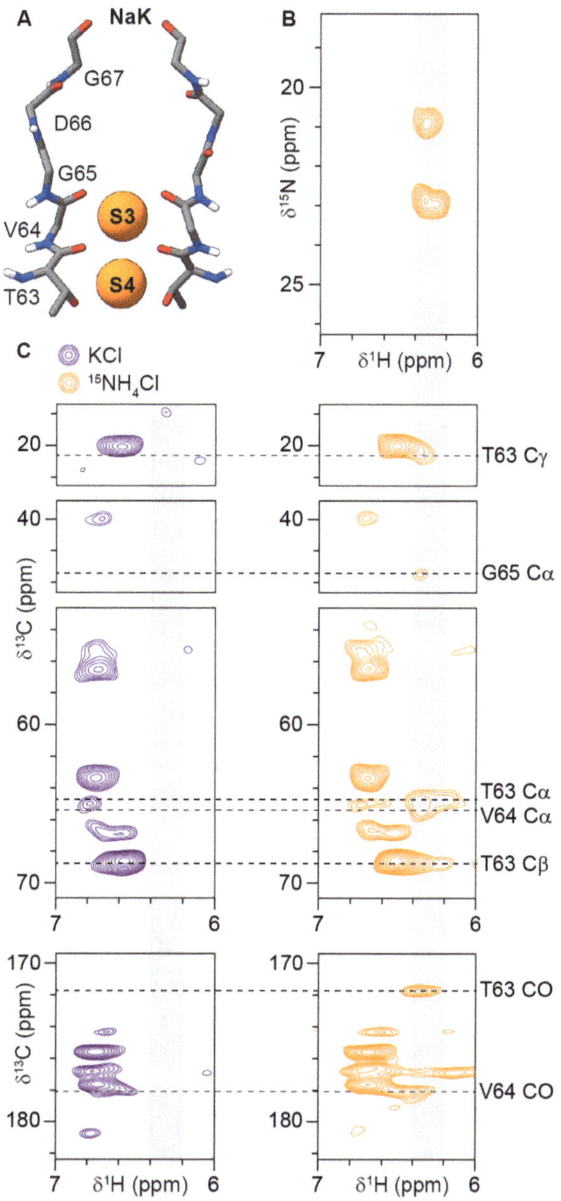

Fig. 3 Detection of bound $^{15}NH_4^+$ ions in the selectivity filter of NaK (**a**) SF structure of NaK, with the ion binding sites S3 and S4 indicated. (**b**) INEPT-based band-selective 2D ^1H–^{15}N correlation spectrum of bound $^{15}NH_4^+$ ions. (**c**) ^1H-detected CP-based ^1H–^{13}C correlation spectra (NaK with K^+ ions in purple and with $^{15}NH_4^+$ ions in orange) with 7 ms CP transfer times. The chemical shifts of carbon atoms in the SF residues T63, V64, and G65 are indicated with dashed lines, and the proton chemical shifts of the $^{15}NH_4^+$ ions are indicated with a gray shade. The spectra at the top were recorded with the carbon carrier frequency in the middle of the aliphatic region ((H)CXH) and the spectra at the bottom in the middle of the carbonyl region ((H)COH). Reproduced from Fig. 2 in Ref. [4], licensed under CC BY-NC-ND; permission for reuse has been given by the original creator (Carl Öster)

2. To prepare 1 L of medium, the following volumes of the stock solutions are mixed (under sterile conditions): 847.2 mL D_2O, 100 mL 10× M9 salt, 10 mL trace elements, 20 mL $^2H^{13}C$-labelled glucose, 4 mL ^{15}N-labelled NH_4Cl, 15 mL biotin, 1.5 mL thiamine, 1 mL carbenicillin, 1 mL $MgSO_4$, and 0.3 mL $CaCl_2$. For 2 L expression, a total of 2.3 L deuterated M9 medium is needed: 300 mL to adapt the cells to D_2O and 2 L for the actual expression. Additionally, 5 mL 1.04 M $^2H^{13}C$-labelled glucose and 1 mL 4.59 M ^{15}N-labelled NH_4Cl are added (per liter of expression medium) after 6 h of expression.
3. Protein inhibitor cocktail should be added to the lysis buffer fresh at the lysis step according to manufacturer information.
4. Buffers containing DM should ideally be prepared on the same day as they will be used. The detergent is not stable in aqueous solution for longer periods of time.
5. The experiments can also be conducted using other solid-state NMR probes dedicated to proton detection if MAS rates of ≥40 kHz can be achieved.
6. Check the glucose level. If there is still glucose left, let the protein expression continue until all glucose has been consumed.
7. The cell pellets can be frozen and kept at −20 °C if the next steps are not started immediately.
8. Prepare the Bradford reagent with 5 mM α-cyclodextrin.
9. Exchange the dialysis buffer more often in the beginning, for example, (1) after 2 h, (2) after 16 h, (3) after 24 h, (4) after 48 h, (5) after 96 h, and (6) after 168 h.
10. 2 L expression of NaK is enough to fill three to four 1.3 mm rotors.
11. Normally the sample temperature can be determined based on the chemical shift of the water peak referenced to DSS (2,2-Dimethyl-2-silapentane-5-sulfonate). For ion channel samples, we don't add DSS to the sample since we want to control the cation concentration and DSS salt contains Na^+ ions. We instead use the lipid peaks (which we have previously referenced to DSS) for referencing. The strong lipid peaks in a ^1H 1D spectrum should be appearing at 0.8 and 1.21 ppm, respectively, in a properly referenced spectrum of a sample containing *E. coli* total lipid extract.
12. A detailed protocol for backbone assignments has been described previously [7]. In the case of NaK, ^{13}C and ^{15}N assignments had been completed earlier using ^{13}C-detected experiments [5]. Therefore, only four 3D experiments were needed. Additional 3D experiments might be needed for de novo assignments and/or for more challenging samples.

13. For NaK, NaK2K, and KcsA-Kv1.3, we observed ^{15}N chemical shifts for $^{15}NH_4^+$ between 18 and 28 ppm [4]. The optimal theoretical values for the INEPT delays, based on a J-coupling strength of 73.5 Hz for NH_4^+ [8], are 1.15 ms (τ_1) and 3.4 ms (τ_2). For NaK, NaK2K, and KcsA-Kv1.3, the optimized values were 1.05–1.15 ms for τ_1 and 2.4–2.7 ms for τ_2 [4].
14. It is not absolutely necessary to use the selective experiment for 2D (H)NH. Due to the much smaller spectral width (6 ppm compared to 120 ppm), fewer points are required in the indirect ^{15}N dimension. However, the selective pulses cause some loss in sensitivity, meaning that more scans are usually needed to obtain the same sensitivity as in the standard experiment. To obtain a good-quality 2D experiment, it is usually necessary to record the experiment overnight, regardless if it's with the standard or selective pulse program. The selective experiment is essential for optimization of the INEPT delays in 1D experiments, since it allows the exclusive detection of NH_4^+ ions. It is most likely also advantageous for investigating contacts between ammonium ions and water molecules in, for example, a 3D H(H)NH spin diffusion experiment since small spectral widths can be used for both indirect dimensions (^1H and ^{15}N). We have, however, not attempted to record these types of 3D spectra without selective pulses, so we have not compared the sensitivities between the experiments.
15. X in (H)CXH represents CH, CH_2, and CH_3 groups that appear with ^{13}C chemical shifts of ca 10–75 ppm in the spectra (Cα, Cβ, and side chains, excluding aromatic CH groups). The (H)COH and (H)CXH experiments are bandwidth selective when low enough power is used on the ^{13}C channel due to the large chemical shift difference between the CO and the CX carbons. With 60 kHz MAS on a 600 MHz spectrometer, we use rectangular pulses with a nutation frequency of ca 15 kHz on the ^{13}C channel and ramped pulses (80–100% for ^1H–^{13}C, 100–80% for ^{13}C–^1H) with maximum amplitudes corresponding to a nutation frequency of ca 50 kHz on the ^1H channel during the cross-polarization transfers. The exact cross-polarization conditions have to be optimized in 1D (H)(C)H experiments for each sample.
16. It can also be helpful to record several (H)CH 2D spectra with cross-polarization transfer times between 1 and 8 ms to see the buildup of magnetization for the ^1H–^{13}C cross-peaks between the ^1H atoms of $^{15}NH_4^+$ and the ^{13}C atoms of the protein backbone and to find the conditions with maximum intensity for these peaks.

Acknowledgments

This work was funded by the Leibniz-Forschungsinstitut für Molekulare Pharmakologie (FMP), the Leibniz Society within the Leibniz Collaborative Excellence funding program (project title: Ion Selectivity and Conduction Mechanism of Cation Channels, project number K305/2020—to A.L.), and the Deutsche Forschungsgemeinschaft (DFG, German Research Foundation), under Germany's Excellence Strategy—EXC 2008/1 (UniSysCat)—390540038 (to A.L.), as well as Project-ID 221545957, SFB 1078/B10 (to A.L.). C. Ö. acknowledges support from the Human Frontier Science Program (LT000303/2019-L).

References

1. Hud NV, Schultze P, Feigon J (1998) Ammonium ion as an NMR probe for monovalent cation coordination sites of DNA quadruplexes. J Am Chem Soc 120:6403–6404. https://doi.org/10.1021/ja9811039
2. Werbeck ND, Kirkpatrick J, Reinstein J, Hansen DF (2014) Using 15 N-ammonium to characterise and map potassium binding sites in proteins by NMR spectroscopy. Chembiochem 15: 543–548. https://doi.org/10.1002/cbic.201300700
3. Eichmann C, Frey L, Maslennikov I, Riek R (2019) Probing ion binding in the selectivity filter of the KcsA potassium channel. J Am Chem Soc 141:7391–7398. https://doi.org/10.1021/jacs.9b01092
4. Öster C, Tekwani Movellan K, Goold B et al (2022) Direct detection of bound ammonium ions in the selectivity filter of ion channels by solid-state NMR. J Am Chem Soc 144:4147–4157. https://doi.org/10.1021/jacs.1c13247
5. Shi C, He Y, Hendriks K et al (2018) A single NaK channel conformation is not enough for non-selective ion conduction. Nat Commun 9: 717. https://doi.org/10.1038/s41467-018-03179-y
6. Hendriks K, Öster C, Shi C et al (2021) Sodium ions do not stabilize the selectivity filter of a potassium channel. J Mol Biol 433:167091. https://doi.org/10.1016/j.jmb.2021.167091
7. Fricke P, Chevelkov V, Zinke M et al (2017) Backbone assignment of perdeuterated proteins by solid-state NMR using proton detection and ultrafast magic-angle spinning. Nat Protoc 12: 764–782. https://doi.org/10.1038/nprot.2016.190
8. Sanders JKM, Hunter BK, Jameson CJ, Romeo G (1988) Isotope effects on proton chemical shifts and coupling constants in the ammonium ions 15, 14 NH4-nDn+. Chem Phys Lett 143: 471–476. https://doi.org/10.1016/0009-2614(88)87398-6

Chapter 3

Conformational Dynamic Studies of Prokaryotic Potassium Channels Explored by Homo-FRET Methodologies

Ana Coutinho, José Antonio Poveda, and María Lourdes Renart

Abstract

Fluorescence techniques have been widely used to shed light over the structure–function relationship of potassium channels for the last 40–50 years. In this chapter, we describe how a Förster resonance energy transfer between identical fluorophores (homo-FRET) approach can be applied to study the gating behavior of the prokaryotic channel KcsA. Two different gates have been described to control the K^+ flux across the channel's pore, the helix-bundle crossing and the selectivity filter, located at the opposite sides of the channel transmembrane section. Both gates can be studied individually or by using a double-reporter system. Due to its homotetrameric structural arrangement, KcsA presents a high degree of symmetry that fulfills the first requisite to calculate intersubunit distances through this technique. The results obtained through this work have helped to uncover the conformational plasticity of the selectivity filter under different experimental conditions and the importance of its allosteric coupling to the opening of the activation (inner) gate. This biophysical approach usually requires low protein concentration and presents high sensitivity and reproducibility, complementing the high-resolution structural information provided by X-ray crystallography, cryo-EM, and NMR studies.

Key words Potassium channels, KcsA, Intrinsic fluorescence, Fluorescent dyes, Gating dynamics, Conformational plasticity, Energy transfer, Homo-FRET, Fluorescence spectroscopy, Time-resolved fluorescence

1 Introduction

Potassium channels are integral membrane proteins (usually homo-oligomeric entities) that play their function through an ensemble of dynamic conformations, far away from behaving as a single static structure. The modulation of their functional properties, i.e., selectivity, high-rate K^+ permeation, blockade, and inactivation, is orchestrated by multiple ion–protein, protein–protein, and lipid–protein interactions, among other factors [1–4]. The unraveling of the molecular basis of such physiological processes requires the ability to precisely define the structure and dynamics of these proteins in different experimental conditions. High-resolution

Simone Furini (ed.), *Potassium Channels: Methods and Protocols*, Methods in Molecular Biology, vol. 2796,
https://doi.org/10.1007/978-1-0716-3818-7_3,

biophysical techniques such as X-ray crystallography, single-particle cryo-electron microscopy (cryo-EM), and nuclear magnetic resonance (NMR), among others, have provided extremely powerful and insightful information about the structure of K^+ channels, usually representing detailed snapshots of a particular state [5]. Nevertheless, complementary information gathered from techniques performed under milder/more physiological conditions is also needed, especially to define the conformational dynamics of these membrane proteins. In this chapter, we illustrate how to use Förster resonance energy transfer (FRET) between identical fluorophores (homo-FRET) as a powerful tool to map conformational transitions in homo-multimeric potassium channels like KcsA, a prokaryotic channel from *Streptomyces lividans*. The formalisms reviewed here to determine intersubunit distances can be applied to any homotetrameric system.

1.1 Fundamental Principles of FRET: Hetero- and Homo-FRET

In general terms, FRET is a non-radiative, long-range dipole–dipole coupling mechanism in which energy is transferred from an excited donor fluorophore (D) to a nearby ground-state acceptor (A) with an absorption spectrum overlapping the emission spectrum of the donor. Typically, the donor and acceptor pair consists of two chemically distinct molecules (hetero-FRET), and this process is usually monitored by changes in the fluorescence intensity (I_D) or lifetime (τ_D) of the donor [6, 7].

According to the Förster theory of resonance energy transfer, the energy transfer rate constant, k_T, between a single D and A pair separated by a distance R is given by:

$$k_T = \frac{1}{\tau_D} \times \left(\frac{R_0}{R}\right)^6 \tag{1}$$

where τ_D is the fluorescence lifetime of the donor in the absence of the acceptor and R_0 is the Förster or critical radius [6, 7]. Energy transfer is often described by its efficiency, E, defined as the probability that an excited donor returns to the ground state by non-radiatively transferring its energy to a nearby acceptor. This can be expressed in terms of a competition between the deactivation pathways available to the excited donor:

$$E = \frac{k_T}{k_f + k_{nr} + k_T} = \frac{k_T}{{\tau_D}^{-1} + k_T} \tag{2}$$

where k_f and k_{nr} are the radiative and non-radiative rate constants of the donor in the absence of an acceptor, respectively. The combination of Eqs. (1) and (2) yields:

$$E = \frac{R_0^6}{R_0^6 + R^6} \tag{3}$$

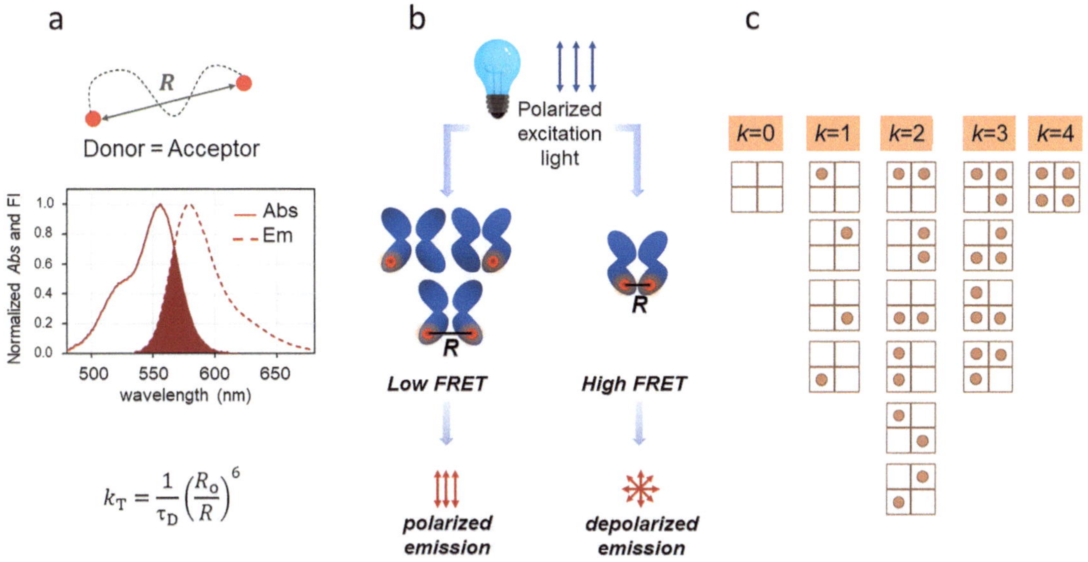

Fig. 1 FRET. (**a**) Highlights of a homo-FRET process between identical fluorophores separated by a distance *R*. This process takes place when the absorption (excitation) spectrum partially overlaps the emission spectrum of the fluorophore (small Stokes shift). The energy transfer rate constant (κ_T) is described by Förster's theory to inversely depend on the sixth power of the distance between the dyes (*R*). (**b**) When a fluorescently labeled sample is excited with vertically polarized light, the efficiency of energy migration among the identical fluorophores will depend (among other factors) on the final degree of labeling and the *R* value. If the probes are distant ($R >> R_0$) or the degree of protein labeling is very low, there will be a low homo-FRET efficiency, and the emitted fluorescence remains highly polarized (high fluorescence anisotropy). On the contrary, when there is a high degree of protein labeling and a reduced inter-fluorophore distance, the efficiency of energy migration among the conjugated dyes will be higher, and the emitted fluorescence will be more depolarized (low fluorescence anisotropy). (**c**) Considering the random labeling of a homotetrameric protein (with one labeling site per subunit), the sample will consist of a heterogeneous population of fluorescently labeled species with a variable number of dyes, *k*, conjugated to each protein: unlabeled proteins ($k = 0$), single-labeled proteins ($k = 1$), double-labeled proteins ($k = 2$), triple-labeled proteins ($k = 3$), and fully labeled proteins ($k = 4$). Upon increasing the final D/P molar ratio, the dominant species progressively shift from being the unlabeled proteins to a more complex solution enriched in multiply labeled species

According to this equation, the Förster radius, R_0, can be interpreted as the distance at which the energy transfer efficiency between an isolated *D* and *A* pair is 50%. In addition, the dependence of the energy transfer efficiency with the inverse of the sixth power of the distance, *R*, between *D* and *A* confers a steep sensitivity of the FRET signal to the donor–acceptor distances around R_0, effectively allowing its use as a "spectroscopic ruler" (*see* Figs. 1a and 2a) [8].

FRET can also occur between two or more identical molecules (homo-FRET or donor–donor energy migration) when their absorption and emission spectra display a small Stokes shift [6–9] (*see* Fig. 1a). Each energy transfer step within a cluster of identical fluorophores leads to a FRET-induced angular displacement

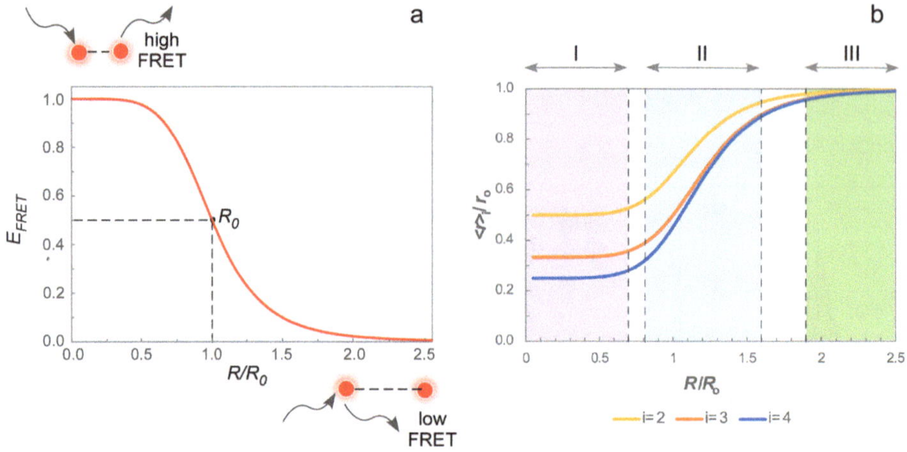

Fig. 2 Energy transfer. (**a**) Dependence of the energy transfer efficiency (E_{FRET}) with the ratio between the distance (R) and Förster radius (R_0) for an isolated donor–acceptor pair. Ideally, the FRET system will detect changes in distances when these fall between 0.5 $R_0 < R <$ 1.5 R_0. For a donor and acceptor pair in very close proximity (very high FRET) or very distant from each other (very low or null FRET), the FRET efficiency becomes independent of the D–A distance. (**b**) The fluorescence anisotropy of a set of "i" fluorophores undergoing fully reversible isotropic homo-FRET within a cyclic molecular or supramolecular organization is determined by the specific geometry of the system. Steady-state fluorescence anisotropy, $<r>/r_0$, for a pair of fluorophores ($i = 2$) and a triad ($i = 3$) or a quartet ($i = 4$) of fluorophores arranged in an equilateral triangle or square geometry, respectively. r_0 is the initial (fundamental) anisotropy of the fluorophore. Regions I ($R/R_0 < 0.7$), II ($0.8 < R/R_0 < 1.6$), and III ($R/R_0 > 1.9$) are described in the text. The overall rotational tumbling of each system in solution was neglected

(depolarization) of the emitted fluorescence, so this homo-FRET process can be tracked by performing steady-state and/or time-resolved fluorescence anisotropy measurements [7, 9]. In fact, and in contrast to the hetero-FRET phenomenon, a fully reversible homo-FRET process (i.e., when the fluorophores experience the same local environment) is not accompanied by any changes in the fluorescence intensity/lifetime of the participating fluorophores.

1.2 Fluorescence Anisotropy

When a molecule is electronically excited with linear ultraviolet/visible polarized light, the probability of its excitation is proportional to $\cos^2 \theta_A$, where θ_A is the angle between the electric vector of the incident polarized light and the absorption transition dipole moment of the chromophore. This implies that an anisotropic distribution of excited molecules can be produced from an isotropic distribution of ground-state fluorophores (i.e., when there is a random orientation of the dipole moment of each molecule present in the ensemble) because those molecules whose transition moments have an orientation closer to the electric vector of the incident polarized light are preferentially excited (photoselection effect) [6, 7]. Fluorescence anisotropy is a parameter that describes the extent of polarization of the emitted light:

$$<r> = \frac{I_{\parallel} - I_{\perp}}{I_{\parallel} + 2\, I_{\perp}} \tag{4}$$

where $I_{\parallel}$ and $I_{\perp}$ are the parallel and perpendicular fluorescence intensities of the molecule relative to the polarized excitation light, respectively.

There are three main causes responsible for fluorescence depolarization, the first one being the existence of noncollinear absorption and emission transition dipole moments in the fluorophore. The fundamental anisotropy, r_0, is the fluorescence anisotropy displayed by a fluorophore in the absence of any rotational diffusion. For an isotropic distribution of fluorophores is determined by:

$$r_0 = \frac{2}{5}\left(\frac{3\,\cos^2\theta_{AE} - 1}{2}\right) \tag{5}$$

where θ_{AE} is the angle between the absorption and emission transition moments. For a randomly oriented population of fluorophores under frozen/high viscosity conditions, r_0 reaches a maximum value of 0.4 when $\theta_{AE} = 0°$ (parallel transition dipole moments). Secondly, an angular displacement of the emission transition dipole moment due to rotational diffusion of the fluorophore during the time it spends in its excited state gives rise to an additional depolarization of the emitted light. Thus, fluorescence anisotropy measurements provide useful information about the molecular mobility and hydrodynamic volume of the emitting species. Finally, the transfer of the excitation energy through a FRET mechanism to a nearby fluorophore that has a distinct dipole orientation (homo-FRET) is responsible for an additional depolarization route of the emitted fluorescence [6, 7, 10].

Fluorescence anisotropy is usually measured under steady-state conditions, $<r>$ (i.e., under continuous illumination of the sample), providing a time-averaged value of this parameter. However, time-resolved measurements are much more informative because the fluorescence anisotropy decays, $r(t)$, which describe the fluorescence depolarization kinetics of a macromolecule, allow to dissect different modes of its dynamics that reflect the local/segmental and global rotational motions of the fluorescent reporter group [10–13]. Usually, the fluorescence depolarization kinetics of a fluorescently labeled protein is described by a sum of two exponentials:

$$r(t) = r_0 \bullet (\beta_{local} \bullet \exp(-t/\varphi_{local}) + \beta_{global} \bullet \exp(-t/\varphi_{global})) \tag{6}$$

where β_i are the amplitudes associated with ϕ_{local}, the rotational correlation time related to the fast local motion of the covalently linked fluorophore, and ϕ_{global}, the rotational correlation time linked to the slow overall tumbling of the conjugated protein in solution, respectively [6, 11].

1.3 Using Homo-FRET Measurements as a Molecular Ruler

Recently, Cardoso and Berberan-Santos derived a set of analytical solutions describing the time evolution of fluorescence depolarization due to energy hopping among several fluorophores symmetrically disposed in a multichromophoric system [14]. By integrating the solutions obtained over time, it is possible to compute the variation of the reduced steady-state fluorescence anisotropy, $<r>_i/r_0$, as a function of the relative lateral inter-fluorophore distance in the absence of any rotational motion [15, 16]. As exemplified in Fig. 2b for a dimer ($i = 2$), trimer ($i = 3$), and tetramer ($i = 4$), two limiting situations can be easily identified in this representation. For $R/R_0 < 0.7$ (Region I), migration of the excitation energy is so efficient that it becomes evenly distributed among all the fluorophores, and the steady-state fluorescence anisotropy reaches a limiting value that is inversely proportional to the number of fluorescent molecules present in the multichromophoric system under consideration. This region has been extensively exploited in the literature to evaluate the average number of fluorescently labeled peptides/proteins in a cluster [17–19] and the degree of oligomerization of proteins involved in a cooperative partition equilibrium [20]. The other extreme corresponds to $R/R_0 > 1.9$ (Region III), that is, the lateral inter-fluorophore distances are much larger than the Förster radius, and so the fluorescence emission arises primarily from the initially excited fluorophore due to the absence of homo-FRET, i.e., there is no depolarization of the fluorescence emitted. The central region of this plot, corresponding to $0.8 < R/R_0 < 1.6$ (Region II), is the most relevant to our studies because, under this regime, the fluorescence anisotropy strongly varies with the interchromophore distance and so homo-FRET studies can be also exploited as a molecular ruler to measure inter-fluorophore distances and detect conformational changes in proteins (*see* Figs. 1 and 2) [9, 21–24].

1.4 Experimental Approaches to Study the Structural Dynamics of KcsA by Homo-FRET Measurements

In this chapter, we focus on homo-FRET protocols used to investigate the conformational dynamics of the prototypical bacterial KcsA potassium channel as a case study. KcsA was cloned from *Streptomyces lividans* in 1995 [25], and its high-resolution structure was first solved in 1998 [26]. Its high homology to several eukaryotic potassium channels, together with its simpler structure and the ability of being purified in milligrams–grams quantity, has made it the most commonly used ion channel model. Thus, several biophysical techniques have been applied to KcsA to understand the molecular basis of how ion channels work, mainly how selectivity, ion conduction, and inactivation processes take place. This homotetrameric potassium channel presents two different but allosterically communicating gates: the intracellular (inner) gate that opens the inner mouth of the protein when the pH drops from neutral to 4–5 and the outer gate that corresponds to the selectivity filter (SF) domain (*see* Fig. 3) [27–29]. The SF is constituted by the

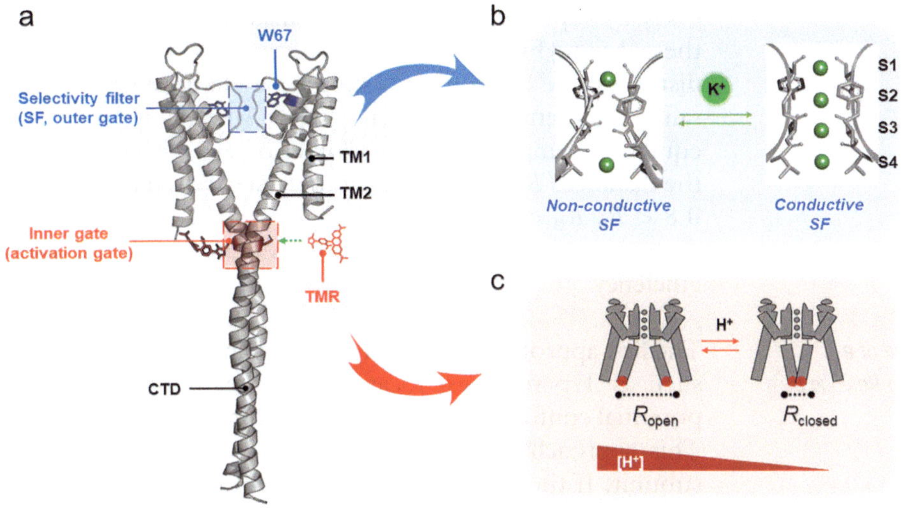

Fig. 3 General structure of the homotetrameric potassium channel KcsA and its two gates. (**a**) Full-length structure of KcsA in the closed state (PDB: 3EFF). Two of the four monomers are depicted for clarity. Each monomer consists of two transmembrane helices (TM1 and TM2); a short, tilted pore helix; the selectivity filter (SF); and the C-terminal domain (CTD). For the homo-FRET experiments, W67 was selected as reporter of the disorder-to-order transitions of the SF (outer gate), and tetramethylrhodamine-5-maleimide (TMR-M) was selected as reporter of the conformational changes undergone by the inner gate, near the pH sensor of the channel. (**b**) Conformational equilibrium of the SF: at low [K^+], only S1 and S4 binding sites are occupied, and the center of this domain is "collapsed" (non-conductive) (PDB: 1K4D). An increase of the [K^+] leads to a structural change at the center of the SF associated with fully occupied binding sites (S1–S4). This latter structure represents the conductive state (PDB: 1K4C). The rearrangements at the SF in the presence of different type/concentration of cations are also translated to a molecular rearrangement of the pore helices, where the W67 reporter is located. (**c**) The conformational transition of the inner gate between the open and closed states is modulated by the [H^+] in solution. After gating (pH < 5.0), the bundle crossing is released, and the accompanying opening of this gate can be monitored by an increase in the TMR fluorescence anisotropy in a multiply TMR-labeled G116C KcsA channel

signature sequence of amino acids $T_{75}VGYG_{79}$, which delimit four consecutive binding sites (S1–S4) able to interact with different permeant and blocking cations. The SF present a certain degree of dynamics, adopting non-conductive and conductive conformations that highly contribute to the selectivity, inactivation, and permeation processes (*see* Fig. 3a) [16, 27, 30, 31]. As in many eukaryotic channels, few milliseconds after gating, KcsA evolves from a conductive to a C-type inactivated state, which is characterized by an open intracellular gate and a non-conductive SF conformation [32, 33]. This slow inactivation process is important for the regulation of the activity of the channels and hence to modulate the excitability of different types of cells [34–37].

Two main strategies were used to study the conformational dynamics of KcsA potassium channel through homo-FRET measurements that can eventually be applied to any homotetrameric potassium channel (or any other homotetrameric protein). In both cases, a couple of experimental requisites are mandatory to perform these homo-FRET studies and eventually calculate intersubunit distances: the fluorophores should be identical and experience the same microenvironment (i.e., they should be spectroscopically equivalent, implying the possibility of fully reversible FRET), and the distance (R) among these molecules should be within the range $0.8 < R/R_0 < 1.6$, so that changes in the lateral inter-fluorophore distance result in a pronounced variation of the homo-FRET efficiency.

1.4.1 Use of a Tryptophan Residue as a Reporter

The first approach consists in using intrinsic fluorescent amino acids such as Trp residues, located at strategic positions, to monitor potential conformational changes within different protein domains. This approach requires the presence of a single Trp residue per subunit. If there is more than one residue in the native protein, a mutagenesis strategy should be carried out to change the non-required intrinsic fluorescent amino acid to a non-fluorescent residue (often a Trp to Phe substitution). It is important to remark that after the mutagenesis procedure is completed, a thorough biophysical characterization (including electrophysiological measurements) must be performed to check for the preservation of the wild-type structure and function.

Previous studies performed in our laboratory described that the intrinsic fluorescence emitted by the five Trp residues of KcsA (W26 and W113 in the intracellular mouth and W67, W68, and W87 near the SF at the top of the transmembrane region) responds to the opening of the inner gate and also to the concentration and types of cations within the pore [1, 38, 39]. By performing site-directed mutagenesis, the single Trp mutant W26,68,87,113F was developed (W67 KcsA hereafter), which presents, according to previous crystallographic data, a lateral distance (Cα–Cα) circa 15–18 Å. Since R_0 values for a Trp-Trp pair usually fall among 4–16 Å [6], W67 was initially considered to be a possible good homo-FRET reporter of any potential conformational change in the pore helix and surroundings areas (e.g., the SF) (*see* Figs. 3 and 4). In addition, the W67 residue is part of a modulatory network of the so-called inactivation triad (E71–D80–W67) that plays an important role on the slow inactivation process in this prokaryotic channel. Electrophysiological recordings performed on the asolectin-reconstituted protein confirmed that the W67 channel behaves as a wild-type-like protein that gates at acidic pH, conducts K^+ at high rates, is blocked by Na^+, and enters into an inactivated state responsible for the low open probability detected during the electrophysiological recordings [15].

By working with the detergent-solubilized channel, both steady-state and time-resolved fluorescence characterizations were performed on this single Trp W67 KcsA mutant. Considering that four W67 residues display a fourfold symmetry, a formal analysis of

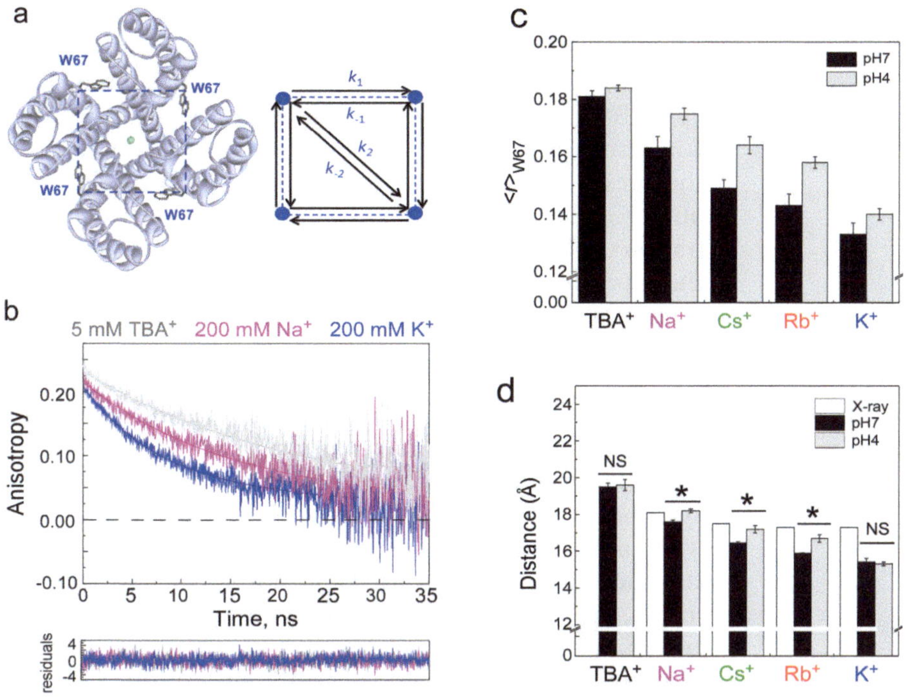

Fig. 4 Energy migration in the W67 KcsA channel. (**a**) Top view of the KcsA channel highlighting the positions of the W67 residues (one in each monomer) and the square geometry delimited by their position. The k_i and k_{-i} parameters are the Förster's transfer rates for this residue (considering that the initially excited W67 residue is at subunit 1). (**b**) Representative time-resolved anisotropy decays of 6 μM W67 KcsA in 20 mM HEPES and 5 mM DDM, pH 7.0, supplemented with TBA^+, Na^+, or K^+. Solid lines represent the best fits of Eq. 7 to the data. (**c**) Dependence of the steady-state anisotropy ($<r>$, λ_{ex}, 300 nm; λ_{em}, 345 nm) of the W67 KcsA protein on the cation present in the buffer at both pH 7.0 (20 mM HEPES, 5 mM DDM) and pH 4.0 (10 mM succinic acid, 5 mM DDM). (**d**) W67–W67 intersubunit (lateral) distances determined from the time-resolved anisotropy decays of W67 KcsA in different experimental conditions and comparison to the distances determined from available X-ray crystallography data. Both $<r>_{W67}$ and distance values show a clear dependence on the type of cation bound to the SF, a phenomenon that is not so clearly defined by the high-resolution X-ray crystallographic structures available so far probably to the use of a co-adjuvant crystallization Fab fragment

the anisotropy decays allowed for the calculation of the W67–W67 intersubunit (lateral) distances, using the following model:

$$r(t) = \frac{r(0)}{4}\left[1 + \exp(-4k_1 t) + 2\exp\left(-\frac{9}{4}k_1 t\right)\right] \cdot \exp\left(-t/\phi_g\right) \tag{7}$$

where $r(0)$ is the initial anisotropy, k_1 is the rate constant for FRET (between neighboring tryptophan residues in a square geometry) (*see* Fig. 4a), and ϕ_g is the global rotational correlation time of the KcsA–DDM complex (43 ns). This way, the inter-tryptophan lateral distance R can be directly calculated via Eq. 1 using the fitted k_1.

In summary, both steady-state ($<r>$) and time-resolved anisotropy measurements performed on the W67 KcsA channel in the presence of different permeant (K^+, Rb^+, Cs^+) and blocking (Na^+, Li^+, TBA^+, TEA^+) cations were found to be directly related to the conformation of the SF, being more efficient (shorter distances) in the presence of the permeant species [15] (*see* Fig. 4). We also observed that the $<r>$ and R values were sensitive to the concentration of the cations bound to the SF, being thus able to calculate the apparent dissociation constant (K_D) of the ion binding equilibrium when the inner gate was at the closed (pH 7.0) or open (pH 4.0) states [40].

1.4.2 Use of an Extrinsic Fluorescent Dye as a Reporter

The second approach relies on fluorescent labeling the protein by using a natural or an engineered cysteine residue in the amino acid sequence by site-directed mutagenesis. These natural/engineered cysteines are the targets for covalently attaching an extrinsic label with an adequate highly reactive moiety, like a maleimide or iodoacetamide group [41–43]. In contrast to the hetero-FRET measurements, the use of a single extrinsic fluorophore greatly simplifies the labeling of the samples. However, one must bear in mind that an incomplete random labeling of a multimeric protein, even with only one labeling site/subunit, always results in various subpopulations of fluorescently labeled channels [44]. In the case of a tetrameric protein, these populations correspond to unlabeled, single-, double-, triple-, and fully labeled proteins (*see* Fig. 1c). The relative proportion of each subpopulation is controlled by the final average degree of protein labeling (dye-to-protein (D/P) molar ratio), which, in turn, is critically dependent on the initial D/P molar ratio used in the labelling reaction. Therefore, the application of steady-state and time-resolved fluorescence measurements to the characterization of these systems requires performing a comparative study of two samples, prepared with a low and a high D/P, respectively. The first case is used to characterize the fluorescence properties of a predominantly singly labeled sample, i.e., when essentially no homo-FRET occurs. On the other hand, samples prepared with a higher degree of labeling are enriched in multiply labeled species and are used to characterize the changes in the efficiency of the homo-FRET process within the fluorescently labeled proteins associated with the conformational change we are monitoring [45].

Thus, to monitor the conformational landscape of the inner gate of the KcsA channel, a cysteine residue should be introduced near the pH sensor and then chemically modified by an extrinsic dye through the maleimide chemistry. Previous EPR experiments have shown that a good candidate is position G116 and mutation to cysteine did not affect the wild-type-like behavior [46–48]. Tetramethylrhodamine-5-maleimide (TMR-M) was chosen as an extrinsic dye to label KcsA at this C116 residue, firstly used by

Francisco Bezanilla and collaborators to detect the pH-induced opening of the inner gate of KcsA using fluorescence lifetime measurements in the frequency domain [49]. A similar approach was applied in voltage-clamp fluorometry studies to monitor conformational modifications of the voltage sensor domain by monitoring changes in the fluorescence intensity associated with channel gating [50, 51].

As mentioned above, the homo-FRET studies of the W67 G116C KcsA channel conjugated to TMR required analyzing the influence of the degree of protein labeling on this process. An under-labeled sample, predominantly consisting of singly labeled proteins, allows for the characterization of the rotational dynamics of the conjugated dye/protein and to study the influence of pH on its fluorescence properties. On the other hand, samples prepared with a higher degree of labeling, which are enriched in multiply labeled protein species, give information about the extent of the homo-FRET process among the fluorescently labeled subunits within each tetrameric channel, which, in turn, is sensitive to the conformational transition between an open and closed state of the inner activation gate (*see* Fig. 4b). Using the latter condition, we were able to determine the activation pK_a of KcsA in different experimental conditions and found that the occupation of the S2 site at the SF (by K^+ or Ba^{2+}) modified this value by one pH unit when compared to the value obtained in the presence of Na^+ or Rb^+ at the SF [45], thus demonstrating a bidirectional cross talk between both the inner and outer gates (*see* Fig. 4). The occupation of this S2 binding site has been claimed by several groups as a key factor to modulate the rate of the slow inactivation process in KcsA [52–54].

In the following sections, the reader will find a step-by-step guide on how to purify, label, prepare, and measure W67 KcsA and W67 G116C TMR KcsA samples in different experimental conditions and how to perform the formal analysis of the data to calculate the W67–W67 intersubunit distances and the activation pK_a of the channel. A labeling procedure to conjugate *N*-(1-pyrenyl)-maleimide to L90C KcsA is also described to measure the global rotational correlation time of the KcsA–detergent complex.

By choosing the proper position of the single Trp residue, these homo-FRET studies have already been translated to the analysis of conformational dynamics of the selectivity filter from other tetrameric potassium channels, such as the NaK2K mutant channel from *Bacillus cereus* [55]).

Both homo-FRET approaches, using intrinsic or extrinsic dyes, have proven to be a valuable tool to monitor the conformation of the two channel gates and thus could potentially be used to characterize the interaction of potassium channels with different types of modulators such as diverse types of natural or synthetic lipids, anesthetics, etc.

2 Materials

Prepare all solutions using ultrapure water (resistivity of 18 MΩ.cm at 25 °C) and analytical grade reagents. Store all solutions at 4 °C (unless indicated otherwise). Diligently follow all waste disposal regulations when disposing waste materials. Wear individual protection equipment (IPE) when necessary.

2.1 Protein Expression and Purification

1. pQE30 expression plasmid containing the KcsA gene (ampicillin resistance selection, N-terminal 6xHis tag) (*see* **Note 1**).
2. Competent *E. coli* M15 cells (pRep4; kanamycin resistance selection).
3. Solid culture medium: A Luria–Bertani agar plate supplemented with ampicillin (100 μg/mL, final concentration) and kanamycin (25 μg/mL, final concentration).
4. Pre-culture medium: 100 mL of autoclaved Luria–Bertani (LB) medium supplemented with ampicillin (100 μg/mL, final concentration) and kanamycin (25 μg/mL, final concentration).
5. Culture medium: 1 L of autoclaved 2xYT medium supplemented with ampicillin (100 μg/mL), kanamycin (25 μg/mL), and glucose (0.04% w/v).
6. 1 M isopropyl-β-D-thiogalactopyranoside (IPTG; dioxane-free, high purity) stock solution: Dissolve 238.3 mg of IPTG in 1 mL of ultrapure H_2O with subsequent sterile filtration (PVDF membrane filters, 0.22 μm pore size) of the solution. Keep at 4 °C for immediate use or freeze at −20 °C for posterior practice.
7. Resuspension buffer: 20 mM Tris (Tris(hydroxymethyl)-aminomethane), 0.45 M sucrose, pH 8.0 supplemented with an EDTA-Free Protease Inhibitor Cocktail (one tablet/ 200 mL buffer) and egg lysozyme (0.4 mg/mL) (*see* **Note 2**).
8. Solubilization buffer: 20 mM HEPES, 200 mM KCl, 10 mM n-dodecyl-β-D-maltoside (DDM; Ultrol grade), 40 mM imidazole, pH 7.5 (*see* **Note 3**).
9. Washing buffer: 20 mM HEPES, 100 mM KCl (or NaCl), 5 mM DDM, 40 mM imidazole, pH 7.5 (*see* **Note 4**).
10. Elution buffer: 20 mM HEPES, 100 mM KCl (or NaCl), 5 mM DDM, 500 mM imidazole, pH 7.0. Use diluted HCl stock to adjust the pH.
11. Dialysis buffer: 20 mM HEPES, 100 mM KCl (or NaCl), 5 mM DDM, pH 7.0. Use a concentrated 1 M N-methyl-D-glucamine (NMDG) stock to adjust the final pH.
12. Immobilized metal affinity chromatography (IMAC) resin: Ni^{2+}-Sepharose Fast Flow, equilibrated in solubilization buffer.

2.2 Protein Labeling with Extrinsic Fluorophores

1. TMRM stock solution: Dissolve tetramethylrhodamine-5-maleimide in anhydrous dimethyl sulfoxide (DMSO) to 10–30 mM (5–15 mg/mL) final concentration. Quantification of the solution can be performed by dilution in methanol (spectroscopic grade) and using a molar extinction coefficient of 91,000 M^{-1} cm^{-1} ($\lambda = 541$ nm) (*see* **Note 5**).
2. Pyrene stock solution: Dissolve *N*-(1-pyrenyl)-maleimide in DMSO to 10–30 mM (3–9 mg/mL) final concentration. Quantification of the solution can be performed by dilution in methanol (spectroscopic grade) and using a molar extinction coefficient of 40,000 M^{-1} cm^{-1} ($\lambda = 338$ nm) (*see* **Note 6**).
3. Tris(2-carboxyethyl)phosphine (TCEP) stock solution: Dissolve TCEP in dialysis (sample) buffer to 10–30 mM final concentration (3–9 mg/mL). TCEP is soluble in water up to 310 mg/mL (1.08 M).

2.3 Sample Preparation

1. Influence of ionic environment: 20 mM HEPES, 5 mM DDM, pH 7.0 buffer or 10 mM succinic acid, 5 mM DDM, pH 4.0 buffers, supplemented with high concentrations (usually 200 mM) of NaCl, LiCl, $BaCl_2$, RbCl, CsCl, or KCl. 5–10 mM tetrabutylammonium (TBA) chloride or 100 μM tetraoctylammonium (TOA) chloride salts can also be tested. pH adjustment is performed by using a 1 M N-methyl-D-glucamine (NMDG) stock (this solution has a pH ~12). To facilitate the procedure, 10× concentrated stocks of pH 7.0 and pH 4.0 buffers (prior to the supplementation with the selected salt) can also be prepared and stored at +4 °C until use. Upon dilution, check the final pH of the solution and adjust the pH if necessary.
2. Influence of pH: 10 mM succinic acid, 5 mM DDM, pH 3.5, 4.0, 4.5, 5.0, 5.5, 6.0, or 6.5. pH adjustment should be performed by adding an adequate volume of 1 M NMDG. Succinic buffers can be then supplemented with the desired concentration of blocking or permeant cations (NaCl, LiCl, $BaCl_2$, RbCl, CsCl, KCl, etc.). To facilitate the procedure, 10× concentrated stocks of all buffers (without supplementation with the selected salt) can also be prepared and stored at +4 °C until use. Upon dilution, check the final pH of the solution and adjust the pH if necessary.
3. Dialysis tubing cellulose membrane MWCO 14 kDa or D-tube dialyzers MWCO 6–8 kDa (Mini (10–250 μL) or Midi (50–800 μL) capacity; Merck-Sigma-Aldrich, catalog references 71,504 and 71,507, respectively).

2.4 Reference Solutions

1. 10 μM *N*-acetyl-*L*-tryptophanamide (NATA) in water.
2. 1 μM rhodamine 101 in ethanol (spectroscopic grade).
3. 0.1% Ludox® (colloidal silica).

2.5 Instrumentation

1. SDS-PAGE kit.
2. Double-beam UV-vis spectrophotometer.
3. Spectrofluorometer equipped with double-grating monochromators and polarizers at both the excitation and emission light paths (L format). A temperature-controlled cell holder is highly recommended.
4. Single photon timing system for performing time-resolved fluorescence measurements with pulsed excitation light source. An appropriate pulsed laser, laser diode, or light-emitting diode can be used as the pulsed light source, depending on the excitation wavelength of the sample and available equipment. A temperature-controlled cell holder allowing for magnetic stirring of the samples is highly recommended.
5. Quartz (QS-Suprasil) absorbance cuvettes with 1 cm of optical path (with reduced volume if possible).
6. Quartz (QS-Suprasil) fluorescence cuvettes with a reduced volume format are highly recommended (usually 0.5 cm versus 0.5 cm of optical path is used).

2.6 Software

1. TRFA Data Processing Package version 1.4 of the Scientific Software Technologies Center (Belarusian State University).
2. Excel spreadsheet prepared by Professor Mario Nuno Berberan-Santos that allows for the calculation of W–W homo-FRET rate constant (κ_1) and the W–W intersubunit distances (this spreadsheet can be made available upon request).

3 Methods

3.1 Selecting the Position to Insert the Fluorophores: Intrinsically Fluorescent Amino Acids and Extrinsic Fluorescent Probes

3.1.1 Intrinsic Fluorescence Studies

The selected Trp residue should ideally be in a solvent-shielded position in order to present a high quantum yield and a high spectral overlap between its fluorescence emission and absorption spectra, leading to a high Förster radius. In addition, the local rotational freedom of this amino acid residue should be restricted in order to display a high steady-state fluorescence anisotropy at low cation concentration. In this way, the sensitivity of this fluorescence parameter to the conformational rearrangements undergone by the selectivity filter is maximized.

3.1.2 Extrinsic Fluorescence Studies

The cysteine residues (one in each monomer, native or introduced by mutagenesis) selected for the determination of the rotational correlation time by their conjugation to an extrinsic long-lived fluorescent probe should be located in a position that does not interfere with the overall conformational dynamics of the protein and is easily reached by the dye (solvent exposed). In the case of KcsA, the introduction of the L90C mutation allowed for the characterization of the high-resolution X-ray crystallographic structure without interfering with the wild-type behavior of the channel [26, 27]. This residue is located at the extracellular loop, far away from the domains which conformational dynamics are going to be characterized.

On the other hand, the position chosen to report on the dynamics of the inner gate should be close to the protein domain involved in this gating process and cause a minimal interference with the WT functional properties of the channel. The residue G116 (near the inner bundle crossing that stabilizes the closed conformation of KcsA activation gate; *see* Fig. 3) was described as a good position to mutate and was subsequently labeled with an extrinsic fluorescent probe [49].

Regarding the selection of the adequate fluorescent probe, *N*-(1-pyrenyl)-maleimide was selected to measure the slow rotational correlation time since it typically presents a very long fluorescence lifetime (70–80 ns in the labeled KcsA channel), which allows to increase the time window available to monitor the overall tumbling of the detergent-solubilized pyrene-labeled protein in solution (~43 ns). As a rule of thumb, rotational correlation times can be measured with high precision within the range $0.1 \langle\tau\rangle < \phi < 10 \langle\tau\rangle$ [56].

To determine the conformational landscape of the inner gate of KcsA, tetramethylrhodamine-5-maleimide presents a combination of desirable photophysical characteristics: red–orange fluorescence with high photostability, relative pH insensitivity, strong absorptivity of the thiol conjugates, high fluorescence quantum yields [57], a short linker length (minimization of the local fluorescence depolarization), and a Förster radius within the range of the expected distance changes between the open and closed conformations of the activation gate [58]. In addition, this probe has a propensity to form ground-state dimers [59, 60], which was an additional valuable characteristic in our study [45].

3.2 Protein Expression and Purification

1. Pick a single colony of the *E. coli* M15 (pRep4) host strain previously transformed with pQE30 plasmid carrying the KcsA gene (following standard heat shock procedure) and selected by double antibiotic resistance (ampicillin and kanamycin) [61]. For the purpose of the following experiments, W67, W67 G116C, and L90C KcsA mutant channels are selected for purification.

2. Inoculate 100 mL of pre-culture medium (LB + ampicillin + kanamycin) with the single colony and grow overnight at 30 °C in an orbital incubator with 200 rpm rotation.
3. Inoculate 1 L of the culture medium (2xYT + ampicillin + kanamycin + glucose) with 50 mL of the pre-culture (*see* **Note 7**). Grow at 37 °C in an orbital incubator at 200 rpm until a final optical density of the culture at 600 nm, OD_{600} ~ 0.6–0.8, is reached (log phase).
4. Add 550 μL of 1 M IPTG to induce protein expression and grow for 2–3 h at 30 °C with uniform rotation at 200 rpm.
5. Harvest the cells by centrifugation at 6353 × g at 4 °C for 15 min.
6. Discard the supernatant and freeze the pellets for at least 2 h at −20 °C, typically overnight.
7. Thaw the pellets and add 10 mL of resuspension buffer/g of pelleted cells. Resuspend the pellets with a sterile disposable pipette while working on an ice bath.
8. Lyse the resuspended cells by sonication (ultrasonic processer) with increasing power in 45 s sonication on/1 min off cycle in an ice bath. Usually, 5–6 rounds are sufficient to break all the cell membranes. High-pressure homogenizers can also be used to disrupt the cells.
9. Transfer the suspension to polycarbonate ultracentrifuge tubes and centrifuge the cell lysate at 100,000 × g for 45 min at 4 °C.
10. Collect the membrane fraction (pellet) and resuspend it in solubilization buffer to extract the expressed membrane protein (~5 mL/g of cell pellet). After the resuspension with a pipette, the use of a homogenizer (Polytron®) is highly recommended since it reduces the particle size, favoring the solubilization process. Proceed always in an ice bath.
11. Incubate the solution during 2 h at room temperature or at 4 °C overnight, with gentle stirring.
12. Remove any insoluble debris by centrifugating the solution at 100,000 × g for 45 min at 4 °C.
13. Prepare a Ni^{2+}-Sepharose Fast Flow IMAC column by equilibrating the resin with 5–10 column volumes of solubilization buffer containing 40 mM of imidazole (*see* **Note 8**). Due to the high binding capacity of this medium, use short, wide columns (typically 5–15 cm bed height) for rapid protein purification, even at low flow velocity. Equilibrate all materials to the temperature at which the separation will be performed.
14. Gently load the supernatant obtained after the last centrifugation step onto the IMAC column prepared to purify the protein containing the 6xHis tag.

15. Proceed to the washing step of the column by gently adding washing buffer containing 40 mM imidazole and until the absorbance at 280 nm of the flow-through is circa zero (using the washing buffer as a blank). Usually, it takes 50–100 mL of washing buffer per liter of initial bacterial culture to effectively wash the column.
16. Elute the KcsA mutant channel by slowly adding the elution buffer containing 500 mM of imidazole to the column. Collect 10–12 of 1.5 mL fractions in sterile microtubes.
17. Check for the protein-containing fractions by performing SDS-PAGE electrophoresis (4% stacking–13.5% resolving gel, non-reducing conditions). This is also helpful to check for the channel final purity and integrity (usually, 90–95% of the total protein in the tetrameric state), which is resistant to SDS-induced denaturation [62].
18. Dialyze the protein-containing fractions against a 200–500-fold volume of dialysis buffer using a 14 kDa MWCO tubing membranes at 4 °C. You can pool all the fractions or dialyze only the most concentrated ones (according to the planned experiments). Typically, removal of all the imidazole usually requires two or three changes of dialysis buffer (1 L each) for 10–15 mL of collected fractions (*see* **Note 9**).
19. Collect the dialyzed fractions and proceed with the spectrophotometric quantification of the protein concentration as explained in Sect. 3.5.1.
20. Store the purified protein stock solutions in sterile tubes at 4 °C until posterior use.
21. Typically, the purification yield for W67 and W67 G116C KcsA mutant proteins is ~0.25 mg/L cell culture, whereas 1 mg/L cell culture is usually obtained in the case of the L90C mutant channel.

3.3 Protein Labeling with Extrinsic Fluorophores

1. Equilibrate a single-cysteine mutant channel (L90C or W67 G116C KcsA) solution (typically 20–50 μM, monomer based in 20 mM HEPES, 5 mM DDM, and 100 mM NaCl (or KCl), pH 7.0) at room temperature for 15–30 min. Bubble through an inert gas (nitrogen or argon) to prevent oxidation of the thiol groups.
2. Add a tenfold molar excess of TCEP dissolved in the same buffer of the sample. Reduction of disulfide bonds in the protein is best carried out at this stage. Incubate at room temperature for 30 min (*see* **Note 10**). For example, we are labeling 1 mL of a 40 μM protein stock solution. In this step, we add 4 μL of a 100 mM TCEP stock solution in sample buffer (400 μM final concentration).
3. Prepare a 10–30 mM stock solution of the fluorescent dye (TMRM or *N*-(1-pyrene)-maleimide) in DMSO immediately

prior to use. From now on, protect all stock solutions from light by wrapping containers in aluminum foil or using opaque tubes (*see* **Note 11**).

4. The final labeling yield of the sample can be controlled by varying the fluorescent dye/protein (D/P) molar ratio used in the conjugation reaction. To maximize the labeling yield, add a 10- to 20-fold molar excess of the fluorescent dye with respect to the single-cysteine mutant protein stock solution. To minimize the labeling yield, an equimolar TMRM/W67 G116C KcsA ratio can be used. In both cases, add the reagent dropwise to the protein solution as it is gently stirred. The final volume of organic solvent must be kept below 1% (v/v) of the total sample volume (*see* **Note 12**). For example, in this step, we add 13 μL of a 30 mM stock solution of TMRM in DMSO (~400 μM final concentration) to 1 mL of a 40 μM protein stock solution.
5. Allow the reaction to proceed for 2 h at room temperature or overnight at 4 °C under gently mixing in a rotator. Always protect the reaction mixture from light as much as possible. Upon completion of the reaction with the protein, an excess of glutathione, 2-mercaptoethanol, or other soluble low molecular weight thiol can be added to consume excess thiol-reactive reagent, thus ensuring that no reactive species are present during the purification step.
6. Remove the free dye from the conjugated protein by affinity (IMAC) chromatography (*see* **Note 13**). To do so, equilibrate a Ni^{2+}-Sepharose Fast Flow or an equivalent IMAC column with a 20 mM HEPES, 5 mM DDM, 100 mM NaCl (or KCl), or pH 7.0 buffer as previously described. Load the mixture of fluorescently labeled protein and free dye on the column, and wash it with 10 column volumes of washing buffer. To test if the washing procedure is complete, the absorbance of the eluate must be checked at the maximum absorption wavelength of the free dye ($\lambda = 340$ nm for pyrene, $\lambda = 550$ nm for TMR). When the absorbance at these wavelengths is circa zero, the conjugated protein can thus be eluted by adding the elution buffer (20 mM HEPES, 5 mM DDM, 100 mM NaCl (or KCl), 500 mM imidazole, pH 7.0). Check for the fluorescently labeled protein-containing fractions by monitoring their absorbance values at the maximum absorption wavelength of the conjugated dye. Remove the imidazole from the purified samples by dialyzing against the appropriate buffer (e.g., 20 mM HEPES, 5 mM DDM, 100 mM NaCl, pH 7.0). This dialysis step further ensures the removal of any traces of free dye.
7. Calculate the concentration of the conjugated dye by spectrophotometric measurements as described in Sect. 3.5.2.

3.4 Sample Preparation for Absorbance and Fluorescence Recordings

3.4.1 Sample Preparation from Diluted Protein Stocks

1. Get your diluted protein stock (6–30 μM) purified in 20 mM HEPES, 5 mM DDM, and 100 mM NaCl, pH 7.0, from the refrigerator (T = 4 °C). Pick an aliquot of W67 KcsA, TMR-labeled W67 G116C, or Pyr-labeled L90C KcsA stock solution according to the planned experiments. Remember Trp measurements will require around 6 μM protein (final concentration), whereas the pyrene- or TMR-labeled proteins can be measured at 1–2 μM.
2. Charge a dialysis tubing membrane or a D-Tube dialyzer previously equilibrated with dialysis buffer. For the study of the influence of the ionic media on the SF conformation, use 20 mM HEPES and 5 mM DDM, pH 7.0 (inner gate closed), or 10 mM succinic acid and 5 mM DDM, pH 4.0 (inner gate open), supplemented with saturating amounts of blocking or permeant cations. These different ionic environments include 200 mM LiCl, NaCl, CsCl, RbCl, or KCl or 10 mM TBA^+. In the case of the study of the activation pK_a of the inner gate and its connection to the SF conformation, prepare buffers with a 0.5 pH unit difference from 3.5 to 7.0 using 10 mM succinic acid and 5 mM DDM solutions and adjusting the pH with a 1 M NMDG solution. Use 200 mM NaCl, KCl, or RbCl as ionic background to study the influence of the SF conformation on the pK_a value.
3. Tightly seal the device and place it into a beaker with dialysis buffer. For an optimal procedure, the total volume of dialysis buffer should be at least 200-fold the sample volume (can be split in several times). Dialysis should be performed at 4 °C with gentle stirring.
4. Change the dialysis buffer every 3 h. The last stage of dialysis can be run overnight.
5. Collect the samples from the dialysis bag/D-Tube dialyzer.
6. Centrifuge the sample at 14,000 × g at 4 °C for 5 min to spin down any insoluble material.
7. After centrifuging, pick the supernatant and store the sample at 4 °C until use.

3.4.2 Sample Preparation from Concentrated Protein Stocks

1. If a highly concentrated stock of protein (50–100 μM in 20 mM HEPES, 5 mM DDM, and 100 mM NaCl, pH 7.0) is obtained after the purification process, the dialysis steps can be skipped. For example, to prepare 350 μL of 6 μM W67 KcsA in 10 mM succinic acid, 5 mM DDM, 200 mM RbCl, and pH 4.0 buffer, mix 30 μL of a 70 μM protein stock, 35 μL of a 10× buffer 10 mM succinic acid pH 4.0, 70 μL of a 1 M stock of RbCl (in H_2O), and 215 μL of ultrapure H_2O. The presence of a low Na^+ concentration in the background (5–10 mM) will not have an impact on the final results since it is easily replaced by other cations at the SF level.

Table 1
Example of the preparation of KcsA samples (6 μM) to test the binding of K^+ to the channel at pH 4.0, using a concentrated stock (70 μM) of protein purified in 20 mM HEPES, 5 mM DDM, and 100 mM NaCl, pH 7.0

$[K^+]$ final (mM)	Stock of W67 KcsA (μM)	[KcsA] final (μM)	Volume of protein stock (μL)	10× 10 mM succinic acid, 5 mM DDM, pH 4.0 buffer (μL)	50 mM KCl stock (μL)	H_2O (μL)	Final volume (μL)
1	70	6	30	35	7	278	350
2.5	70	6	30	35	17.5	267.5	350
5	70	6	30	35	35	250	350
10	70	6	30	35	70	215	350

There is no particular order during the mixing of the solutions except for the addition of the protein stock that should be added as the last one

2. The same direct dilution procedure (without dialysis) can be used to prepare a series of samples with increasing concentrations of a particular cation to subsequently determine the apparent dissociation constant (K_D) of the binding process (*see* **Note 14**). For example, prepare the samples as in Table 1.

3.5 UV-Vis Spectrophotometric Measurements

3.5.1 Protein and Extrinsic Dye Quantification

1. Quantification of KcsA concentration is usually performed by measuring the absorbance spectra (blank corrected) of the sample and calculating the concentration of the protein (in monomeric units) using the Lambert–Beer equation:

$$[\text{protein}] = A_{280} \times \varepsilon_{280}^{P} \times l \qquad (8)$$

being ε_{280}^{P} the molar extinction coefficient of the mutant KcsA at 280 nm (ε_{280}^{P} = 12,950 M^{-1} cm^{-1} for W67 and W67 G116C KcsA; ε_{280}^{P} = 34,950 M^{-1} cm^{-1} for L90C KcsA) and l the optical pathlength (usually, 1 cm). If the absorbance at 340 nm (A_{340}) is >0, then use the (A_{280}–A_{340}) difference to calculate the protein concentration (scattering correction) (*see* **Note 15**).

2. In cysteine mutants labeled with TMR or pyrene, the concentration of the conjugated dye is calculated by spectrophotometric measurements as:

$$[\text{dye}] = \frac{A_{\max}}{\varepsilon_{\max}^{D} \times l} \qquad (9)$$

The $A_{\max}$ are the absorbance values registered at the maximum absorption wavelength of the fluorescent dye (339 and 552 nm for pyrene and TMR, respectively). The corresponding molar extinction coefficients are ε_{339}^{D} = 40,000 M^{-1} cm^{-1} and ε_{552}^{D} = 83,871 M^{-1} cm^{-1} for the pyrene- and TMR-labeled KcsA channels, respectively. The optical path, l, is usually 1 cm.

3. Protein quantification in the presence of conjugated probes requires a first step of estimation of the theoretical absorbance of the dye at 280 nm as:

$$A'_{280} = [\text{dye}] \times \varepsilon^{\text{D}}_{280} \times l \quad (10)$$

At this wavelength, the estimated values for the molar extinction coefficients of the dye, $\varepsilon^{\text{D}}_{280}$, are 20,000 and 24,900 M^{-1} cm^{-1} for Pyr-labeled KcsA and TMR-labeled KcsA, respectively.

Then, the corrected protein concentration is calculated as:

$$[\text{KcsA}] = \frac{[A_{280} - A'_{280}] - A_{340}}{\varepsilon^{\text{P}}_{280} \times l} \quad (11)$$

Eventually, protein concentration can also be determined using a colorimetric assay (e.g., DC Protein Assay, Bio-Rad).

3.5.2 Determination of the Degree of Labeling

1. Calculate the efficiency of labeling by determining the final dye/KcsA (D/P) molar ratio:

$$\text{D/P} = \frac{[\text{dye}]}{[\text{KcsA}]} \quad (12)$$

Typically, the final D/P is between 0.3 and 0.5 in the case of the pyrene conjugates (Pyr L90C KcsA). For the TMR-labeled channels (TMR G116C KcsA), a final D/P of approximately 0.12 and 0.45 is obtained when using an equimolar and a tenfold molar excess of fluorescent dye in the labeling reaction, respectively (*see* **Note 16**).

3.5.3 Detecting the Formation of TMR H-Type Dimers

1. Record the absorption spectra of the TMR-labeled W67 G116C between 475 and 620 nm to characterize the influence of the degree of protein labeling and/or pH buffer on the spectroscopic properties of the covalently linked TMR dye.
2. Evaluate the $\text{Abs}_{554}/\text{Abs}_{520}$ ratio to detect the potential formation of ground-state H-type dimers between the conjugated rhodamine dyes. When the extent of H-dimer formation is relevant, the absorption spectra are characterized by the appearance of a shoulder at ~520 nm.

3.6 Steady-State Fluorescence Measurements

3.6.1 Spectroscopic Characterization of the Samples

1. Characterize the fluorescence emission properties of each sample by measuring its fluorescence emission spectrum under the different experimental conditions being tested (i.e., variable monovalent salt type/concentration or pH of the sample buffer). Typically, the single-Trp-containing proteins (~6 μM) are excited at 295 nm to avoid any significant contribution from the tyrosine residues, and the emission is scanned from 305 up to 410 nm. For the pyrene- and TMR-labeled samples (~1–2 μM), the excitation wavelengths used are 335 nm and

490 nm, respectively, and the emission scans are acquired from 350 to 450 nm and from 550 to 650 nm, respectively. The fluorescence measurements are temperature-sensitive, so these recordings must be performed under temperature-controlled conditions (typically, the temperature is set at 25 °C) (*see* **Note 17**).

2. The emission spectrum of each sample must be corrected by subtracting the emission contribution of the corresponding blank (buffer) under the same experimental conditions (*see* **Note 18**).
3. For each blank-corrected emission spectrum, calculate its fluorescence spectral center of mass (intensity-weighted average emission wavelength, $<\lambda>$) according to:

$$<\lambda> = \sum_i \frac{\lambda_i \bullet I(\lambda_i)}{\sum_i I(\lambda_i)} \qquad (13)$$

where $I(\lambda_i)$ is the fluorescence intensity at the wavelength λ_i. The $<\lambda>$ parameter quantifies the position and shape of the emission spectrum, which are indeed sensitive to local structural changes affecting (directly or indirectly) the fluorescent properties of the fluorophore. So, this analysis is important to gain site-specific insights about the conformational changes undergone by the potassium channel at the level of either of its gates that are caused by a change in the medium conditions (ionic environment or pH).

3.6.2 Quantum Yield Determination

1. Dilute the selected protein stock solution with the adequate sample buffer to reach an absorbance value of 0.05 (<0.1) at 280 nm (for W67 KcsA) or 550 nm (for TMR-labeled W67 G116C KcsA).
2. Register the absorbance value at the excitation wavelength of each fluorophore.
3. Register the complete emission spectra of the fluorophore at the selected excitation wavelength and the corresponding blank.
4. Integrate the blank-corrected emission spectra to calculate the area under each curve.
5. Repeat the previous steps with the chosen fluorescent reference: NATA in the case of Trp residues ($Q_R = 0.14$ in water) [63] and rhodamine 101 for TMR measurements ($Q_R = 0.915$ in ethanol) [64].
6. Calculate the quantum yield of the fluorophore as:

$$Q_F = Q_R \frac{IA_F}{IA_R} \cdot \frac{Abs_R}{Abs_F} \cdot \frac{n_F^2}{n_R^2} \qquad (14)$$

where Q_i is the quantum yield, IA_i is the integrated area of the emission spectrum, Abs_i is the absorbance at the excitation wavelength, and n is the refractive index of the solution (1.33 for water, 1.36 for ethanol, 1.6 for a Trp residue buried in the protein core [65], and 1.4 for the detergent-solubilized TMR-labeled protein). The subscripts F and R refer to the sample (W67 or TMR) and reference fluorophores, respectively. The quantum yield of Trp67 residues in W67 KcsA is 0.29 in 20 mM HEPES, 5 mM DDM, and 200 mM KCl [15] and 0.21 for the TMR-conjugated protein (D/P = 0.05) in the same buffer [45].

3.6.3 Steady-State Fluorescence Anisotropy Measurements

1. Fill a fluorescence cuvette with the proper volume of the sample to be studied. Check for the absence of air bubbles or high scattering level (turbidity) of the sample. A final concentration of 5–6 μM (monomer based) of KcsA gives optimum Trp registers. For the TMR or pyrene measurements, more diluted samples can be used (1–2 μM) (*see* **Note 19**).
2. Set the measurement temperature at 25 °C.
3. Set up the equipment to obtain a good signal/noise ratio. Use the following excitation/emission conditions: Trp residues, λ_{ex} = 300 nm and λ_{em} = 345 nm (*see* **Note 20**); TMR, λ_{ex} = 490 nm and λ_{em} = 580 nm; and pyrene, λ_{ex} = 335 nm and λ_{em} = 400 nm.
4. Register the I_{VV}, I_{VH}, I_{HV}, and I_{HH} polarization components of each sample five to ten times where I_{VV} and I_{VH} are the fluorescence intensities of the vertically and horizontally polarized emission, when the sample is excited with vertically polarized light, and I_{HV} and I_{HH} are the fluorescence intensities of the vertically and horizontally polarized emission, when the sample is excited with horizontally polarized light, respectively.
5. Repeat the same measurements for a blank sample.
6. For each polarization component, subtract the average value of the blank to the average value of the respective component presented by the sample. Use the average blank-corrected intensities to calculate the steady-state fluorescence anisotropy $<r>$ as follows:

$$<r> = \frac{I_{VV} - G \cdot I_{VH}}{I_{VV} + 2\ G \cdot I_{VH}} \qquad (15)$$

If your equipment does not automatically calculate the G factor (grating factor, instrument correction factor; *see* **Note 21**), you can calculate it as:

$$G = \frac{I_{HV}}{I_{HH}} \qquad (16)$$

7. At least three independent samples should be used to calculate average steady-state anisotropy values (± standard deviation, SD) at each experimental condition tested.
8. The steady-state fluorescence anisotropy measurements provide a time-averaged value of the depolarization kinetics of the fluorophore that can be used to characterize the equilibrium binding of different cations to the selectivity filter (Sect. 3.8.5) and to determine the transition midpoint (inner gate activation pK_a) from its sigmoidal dependence with the pH (Sect. 3.8.6).

3.7 Time-Resolved Fluorescence Measurements

Time-resolved fluorescence measurements are performed according to the single photon timing technique [6, 7].

3.7.1 Fluorescence Intensity Decays

1. Turn on the equipment and set the temperature of the cell holder to 25 °C.
2. Measure the fluorescence intensity decay of your protein sample using magic angle conditions, i.e., by positioning the polarizers at the magic angle (vertically polarized excitation light, 54.7° polarized emission light). In our studies, an excitation beam produced by a frequency-doubled rhodamine 6G laser was used to excite the W67 residues of the KcsA channels at 300 nm [66]. On the other hand, the TMR dye in the fluorescently labeled KcsA channels was excited at 488 nm using a BDS-SM-488FBE pulsed picosecond diode laser from Becker and Hickl (*see* **Note 22**).
3. Acquire the instrument response function (IRF) (*see* **Note 23**) of the instrument using a very diluted scattering solution such as colloidal silica (Ludox®). In this case, the emission wavelength should be set at the same wavelength used to excite the sample. In order to better define this rather narrow function, 50,000 counts are usually accumulated at its peak channel.

3.7.2 Fluorescence Anisotropy Decays

1. Turn on the equipment and set the temperature of the cell holder to 25 °C.
2. Acquire the polarized fluorescence intensity decays of your protein sample by alternately recording the parallel and perpendicular polarized components of the emitted fluorescence ($I_{VV}(t)$ and $I_{VH}(t)$, respectively) relatively to the vertical plane of polarization of the excitation beam using identical accumulation times (*see* **Note 24**).
3. Time-resolved anisotropy measurements are registered to determine the global rotational correlation time of the KcsA–DDM complex (for ulterior calculation of Trp-Trp intersubunit distances) using the Pyr-labeled potassium channel [15], to

evaluate the homo-FRET rate constant in the W67 mutant and its dependence on the type/concentration of cations bound to the SF [15], and also to characterize the splaying of the C-terminal domain on the TMR-conjugated KcsA mutants under different pH conditions [45].

3.8 Data Analysis

3.8.1 Calculation of the Förster Radius

1. Register the absorbance of the sample at its excitation wavelength. The absorbance should be around 0.05 ($A < 0.1$).
2. Register the corrected emission spectra of the sample at the selected excitation wavelength.
3. Calculate the spectral overlap ($J(\lambda)$) as follows:

$$J(\lambda) = \int_0^{\infty} F_D(\lambda)\tilde{n}\varepsilon_A(\lambda)\tilde{n}\lambda^4\tilde{n}d\lambda / \int_0^{\infty} F_D(\lambda)\tilde{n}d\lambda \qquad (17)$$

where $F_D(\lambda)$ is the corrected emission spectrum of the donor and $\varepsilon_A(\lambda)$ is the absorption spectra of the acceptor in M^{-1} cm^{-1}. A practical guide to calculate the spectral overlap can be downloaded from [67].

4. Calculate the Förster radius of the sample as follows:

$$R_0 = 0.2108 \cdot \sqrt[6]{\kappa^2 \cdot n^{-4} \cdot Q_F \cdot J(\lambda)} \qquad (18)$$

where the orientation factor, κ^2, and the refractive index of the medium, n, were assumed to be 2/3 (i.e., the dynamical isotropic limit value [7]) and 1.6 [10] or 1.4 [68], respectively. In this equation, using nm units for the calculation of the overlap integral $J(\lambda)$, R_0 is obtained in Å.

3.8.2 Analysis of the Fluorescence Intensity Decays

1. The fluorescence intensity decay obtained for each sample is analyzed by fitting a sum of exponentials:

$$i_m(t) = \sum_{i=1}^{n} \alpha_i \cdot \exp(-t/\tau_i) \qquad (19)$$

where α_i and τ_i are the normalized preexponential factor and lifetime of the ith decay component, respectively.

2. The goodness of the fits is judged from obtaining a reduced $\chi^2 < 1.3$ and a random distribution of the weighted residuals and autocorrelation plots [22] (*see* Fig. 5).
3. In our laboratory, the TRFA Data Processor Advanced version 1.4 was used for this analysis, but any software that allows performing an iterative reconvolution can be employed instead.
4. Calculate the amplitude-weighted mean fluorescence lifetime, $<\tau>_1$, as follows:

$$<\tau>_1 = \sum_{i=1}^{n} \alpha_i \cdot \tau_i \qquad (20)$$

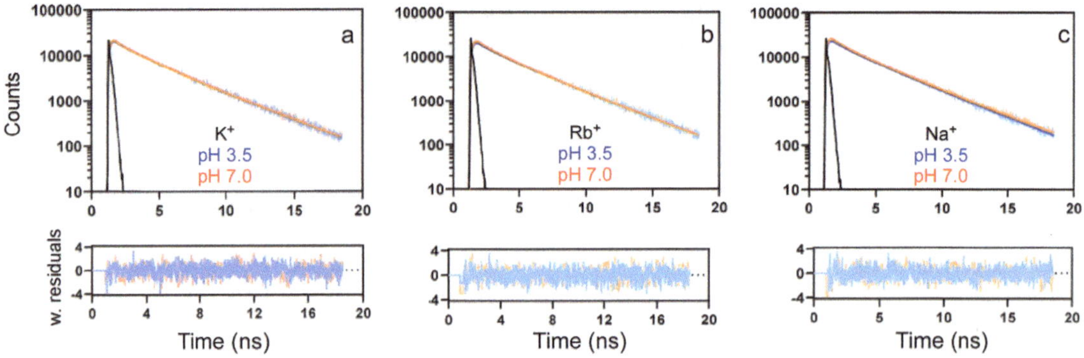

Fig. 5 Representative time-resolved fluorescence intensity decays of TMR–W67 G116C KcsA registered in the presence of 200 mM (**a**) KCl, (**b**) RbCl, and (**c**) NaCl, all at pH 3.5 (inner gate open) and 7.0 (inner gate closed). In the three samples, the D/P ratio was >0.4. Solid lines represent the best fit of Eq. 19 to the data using the TRFA Data Processing Package. The average lifetime $<\tau>_2$ was ~3.4 ns in all cases

and the intensity-weighted mean fluorescence lifetime, $<\tau>_2$, according to:

$$<\tau>_2 = \frac{\sum_{i=1}^{n} \alpha_i \cdot \tau_i^2}{\sum_{i=1}^{n} \alpha_i \cdot \tau_i} \tag{21}$$

The amplitude-weighted mean fluorescence lifetime, $<\tau>_1$, is a parameter proportional to the area under the decay curve, whereas the intensity-weighted mean fluorescence lifetime, $<\tau>_2$, is the average time that a fluorophore spends in its excited state.

5. The influence of the ionic environment and/or pH on the mean fluorescence lifetimes of the fluorophores provides information about the changes in the polarity of its immediate environment and/or the occurrence of dynamic quenching effects [15].

3.8.3 Analysis of the Fluorescence Anisotropy Decays: KcsA Variants Conjugated to an Extrinsic Fluorophore

1. Perform a global analysis of the obtained anisotropy decays [69] by a two-step procedure using an adequate software (TRFA Data Processing Package version 1.4 in our case).
2. The fluorescence intensity decay parameters are first obtained by iterative reconvolution of a sum of exponentials, $i_m(t)$ (Eq. 19), with the IRF, which is then fitted to the experimentally recorded data, $I_m(t)$, calculated as:

$$I_m(t) = I_{VV}(t) + 2\, I_{VH}(t) \tag{22}$$

using a nonlinear least-squares regression method. The instrumental G factor for our setup is 1 because each polarized emitted component was fully depolarized before detection by using a polarization scrambler.

3. The anisotropy decay parameters are then obtained by simultaneous iterative reconvolution of $i_{VV}(t)$ and $i_{VH}(t)$:

$$i_{VV}(t) = \frac{i_m(t)}{3}[1 + 2\, r(t)] \tag{23}$$

$$i_{VH}(t) = \frac{i_m(t)}{3}[1 - r(t)] \tag{24}$$

with the IRF and globally fitting to the experimentally acquired parallel and perpendicular polarized components of the fluorescence intensity ($I_{VV}(t)$ and $I_{VH}(t)$, respectively) [69], after fixing in this analysis the fluorescence decay parameters to the values obtained after the first step of the fitting procedure.

4. The fluorescence anisotropy decay curves, $r(t)$, are described by a sum of discrete exponential terms:

$$r(t) = \sum_{i=1}^{n} \beta_i \exp(-t/\phi_i) \tag{25}$$

where β_i and ϕ_i are the amplitude and the rotational correlation time of the *i*th decay component of the anisotropy, respectively (*see* **Note 25**).

5. The usual statistical criteria, namely, a reduced $\chi^2 < 1.3$ and a random distribution of weighted residuals and autocorrelation plots, are used to evaluate the goodness of the fits (*see* Fig. 6).

3.8.4 Analysis of the Fluorescence Anisotropy Decays: Calculation of Trp-Trp Intersubunit Distances and Comparison to the Crystallographic Data

1. The Excel spreadsheet is used to fit Eq. 7 to the experimental fluorescence anisotropy decays obtained for W67 KcsA under different experimental conditions. Only two parameters are allowed to vary in this fitting procedure, namely, $r(0)$, the fluorescence anisotropy at time zero, and k_1, the energy transfer rate constant among the neighboring Trp residues. The global rotational correlation time of the detergent-solubilized protein is determined independently using a pyrene-labeled KcsA channel (Sect. 3.8.3) and kept fixed in this analysis.

2. The lateral Trp-Trp distances are then calculated using Eq. 1 where τ is now the intensity-weighted mean fluorescence lifetime of W67 KcsA in each tested experimental condition and R_0 is the corresponding Förster radius (*see* **Note 26**).

3. Compare the distances recovered using different experimental conditions (e.g., blocking against permeant cations, closed/open inner gate, addition of modulating compounds, etc.) with the $C(\delta_2)$–$C(\varepsilon_2)$ inter-tryptophan lateral distances obtained from X-ray crystallographic data obtained for KcsA and its variants (*see* Fig. 4).

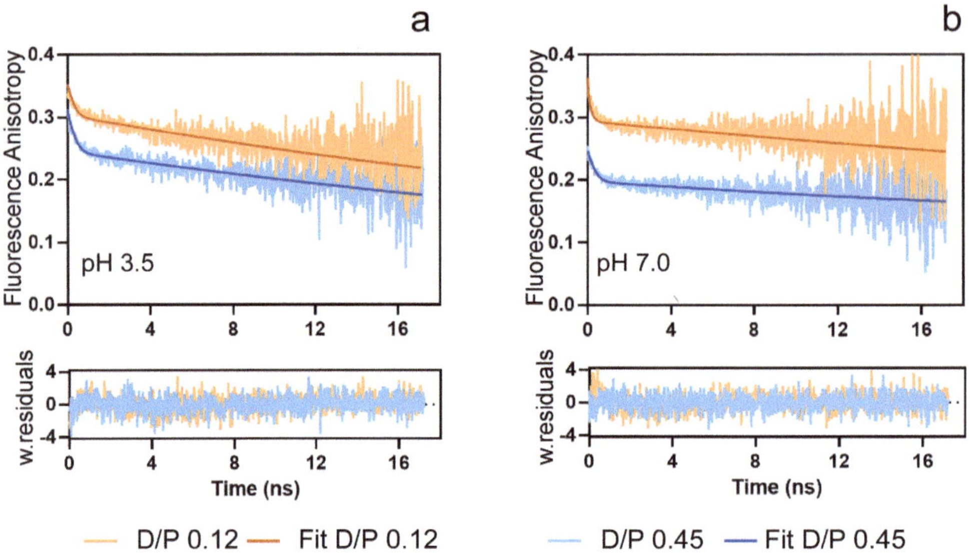

Fig. 6 Representative time-resolved anisotropy decays of the TMR–W67 G116C KcsA channels (λ_{ex}, 488 nM; λ_{em}, 575 nM) measured at pH 3.5 (**a**) and 7.0 (**b**) and at low (0.12) and high (0.45) labeling ratio (D/P). Whereas the decays of the under-labeled protein at a D/P of 0.12 (orange data) present no significant differences between acidic and neutral pH, the increase of the D/P to 0.45 (blue data) increases the efficiency of the homo-FRET process among the TMR dyes conjugated to each protein subunit, particularly at pH 7.0 (the initial much faster decay of the anisotropy detected in these conditions is the spectroscopic signature for the occurrence of very fast energy migration in a multichromophoric system)

3.8.5 Determination of the Cation Binding Affinity to the SF

The changes of the $<r>$ W67 values according to the concentration of, for instance, a permeant cation, can be used to determine the apparent dissociation constant (K_D) of the binding equilibrium by using the following model:

$$<r> = <r>_{NC} + (Q<r>_{C/I} - <r>_{NC}) \cdot \frac{[X^+]^h}{K_D{}^h + [X^+]^h} \quad (26)$$

where $<r>_{NC}$ and $<r>_{C/I}$ are the steady-state fluorescence anisotropy values of the non-conductive (collapsed) and conductive (pH 7.0) or inactivated (pH 4.0) conformations of the SF, respectively, Q is the relative change of W67 KcsA quantum yield upon cation binding ($\Phi_{C/I}/\Phi_{NC}$), $[X^+]$ is the concentration of cation (in M or mM), and h is the Hill coefficient. The parameter Q was fixed at 0.95 in the experiments performed at pH 7.0 and 0.8 at pH 4.0 [40] (*see* Fig. 7).

3.8.6 Determination of the Apparent Activation pK_a of the Inner Gate and Correlation to the SF Occupancy

1. Measure the steady-state fluorescence anisotropy of TMR-labeled KcsA samples from pH 3.5 to pH 7.0.
2. Plot the changes of the $<r>_{TMR}$ values against the pH values (*see* Fig. 8).
3. Determine the apparent activation pK_a using a Hill equation as:

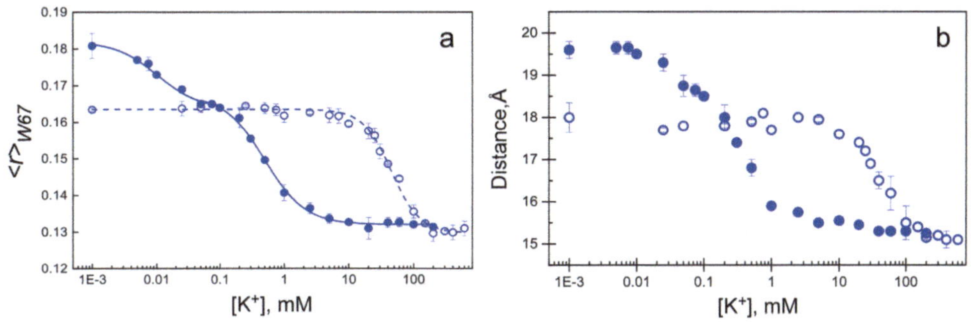

Fig. 7 Characterization of the K^+ binding behavior of W67 KcsA monitored by the changes in the $\langle r \rangle$W67 ($\lambda_{ex} = 300$ nm; $\lambda_{em} = 345$ nm) (**a**) and the W67–W67 intersubunit (lateral) distances (calculated from the time-resolved anisotropy decay analysis of each sample) (**b**). (**a**) Two consecutive binding events were detected at pH 7.0 (filled blue circle) with K_D values in the order of μM and low mM (~10 μM and 0.5 mM, respectively), which were accompanied by a pronounced approximation of the pore helices, as depicted in (**b**). On the other hand, the opening of the inner gate at pH 4.0 (empty circles) leads to a pronounced decrease in the affinity for K^+ (K_D ~ 45 mM) and a concomitant decrease in the conformational dynamics of the pore helices at low K^+ concentrations (the first binding event is no longer detected). Full (pH 7.0) and dashed (pH 4.0) lines represent the best fits of Eq. 26 to the fluorescence anisotropy data

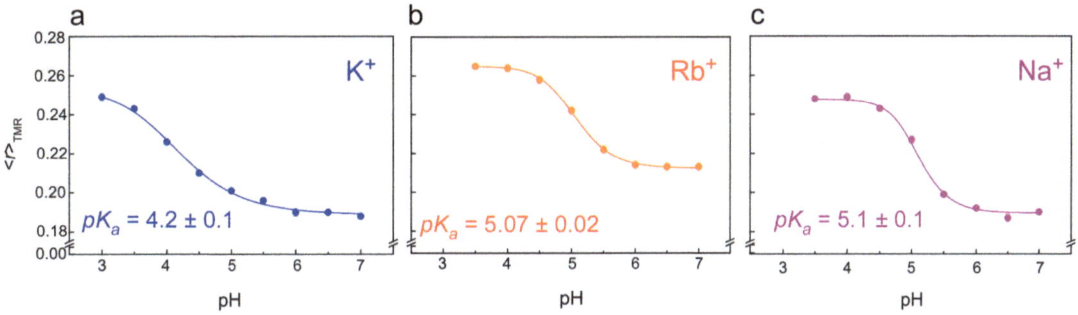

Fig. 8 Determination of the pK_a of the inner gate of KcsA and its dependence on the SF ion occupancy. The steady-state fluorescence anisotropy of the conjugated TMR–G116C W67 KcsA varies with the pH in a sigmoidal manner that was empirically fitted using Eq. 27. The experiments were conducted in the presence of (**a**) 200 mM (D/P 0.45), (**b**) 200 mM RbCl (D/P 0.38), and (**c**) 200 mM NaCl (D/P 0.45). The measuring conditions were $\lambda_{ex} = 490$ nm, $\lambda_{em} = 580$ nm, and T = 25 °C

$$< r >_{\mathrm{TMR}} = < r >_{\mathrm{i}} + (< r >_{\mathrm{f}} - < r >_{\mathrm{i}}) \times \frac{\mathrm{pH}^{\mathrm{n}}}{\mathrm{p}K_{\mathrm{a}}^{\mathrm{n}} + \mathrm{pH}^{\mathrm{n}}} \quad (27)$$

Here, $<r>_{\mathrm{I}}$ and $<r>_{\mathrm{f}}$ are the steady-state TMR fluorescence anisotropy values at pH 3.5 and 7.0, respectively, and n are the number of cooperative sites (*see* **Note 27**).

4 Notes

1. kcsa gene inserted in the pQE30 plasmid vector does not have a tag removal sequence to selectively cut out the tag from the expressed protein. Even though several biophysical studies have shown that the presence of such His tag does not affect the function and structure of KcsA [70], another expression vector should be considered if the tag must be removed. There is also the possibility of locating the His tag next to the C-terminal domain of the protein.
2. Protease inhibitor cocktail can be replaced by a mix of phenylmethylsulfonyl fluoride (PMSF) and benzamidine. *Optional*: you can add 10 μL of DNAase/100 mL resuspension buffer in order to reduce the suspension viscosity. In our hands, if the resuspended pellet is extremely viscous, the protein purification yield will be very low.
3. DDM is a very expensive detergent. After reception from the supplier, it is advisable to divide the total amount into 10 g aliquots and freeze at −20 °C. Using fluorescent impurity-free reagents, including the detergent, is mandatory.
4. If you prepare a 10× stock, white precipitate could be found when stored at 4 °C. Gently warm the 10× stock to re-dissolve the components and then proceed to prepare the 1× solution.
5. Whenever possible, you should prepare and use dye stock solutions on the same day. However, if you need to make up stock solutions in advance, we recommend that you store the solution as aliquots in tightly sealed vials at −20 °C. Maleimides do not react with histidine or methionine, and the optimum pH for the reaction is near 7.0.
6. Whenever possible, you should prepare and use solutions on the same day. Pyrene derivatives exhibit structured spectra. The absorption maximum is usually about 340 nm with a subsidiary peak at about 325 nM. There are also strong absorption peaks below 300 nm. The emission maximum is usually about 376 nm with a subsidiary peak at 396 nm. Excimer emission at about 470 nm may be observed at high concentrations.
7. Expression levels of W67 and W67 G116C channels are relatively low (0.25 mg/L culture), and larger volume of culture medium is needed to obtain proper amounts of protein. L90C KcsA mutant can be purified with a 1–2 mg/L culture yield.
8. During the purification step, you can either use a pre-packed Ni^{2+}-Sepharose column or the user can pack its own column by buying the slurry and the empty vessel. The Ni^{2+}-Sepharose Fast Flow slurry is typically sold in 20% ethanol and has a dynamic binding capacity of 40 mg of a histidine-tagged

protein/mL medium. Estimate the amount of medium required to bind the sample of interest and use five times this amount to pack the column. Since KcsA purification yield is typically 1–2 mg protein/L culture, low volumes of Ni^{2+}-Sepharose slurry can be charged into the column (i.e., 250 μL Ni^{2+}-Sepharose slurry/L culture).

9. Imidazole can also be removed by using gel filtration chromatography. This step will dilute the sample but increase the overall purity of the channel. In the case of KcsA, the purity after the IMAC step is usually >95%.

10. Using TCEP instead of DTT avoids a dialysis step during conjugation with maleimides (or iodoacetamides). It may be advisable to carry out thiol modifications in an oxygen-free environment because thiols can be oxidized to disulfides. In contrast to dithiothreitol or β-mercaptoethanol, TCEP has the advantages of being odorless, a more powerful reducing agent, an irreversible reducing agent, more hydrophilic, and more resistant to oxidation in air. It also does not reduce metals used in immobilized metal affinity chromatography. It gives a selective, complete and quantitative, and quick reduction (in less than 5 min) [71].

11. If a frozen stock of TMR-M in DMSO is going to be used, we recommend that you allow your product to equilibrate to room temperature for at least 1 h before opening the vial.

12. Decomposition of maleimide derivatives to maleamic acids above pH 8.0 can also take place as a competing reaction, so check the pH of the solution carefully. If you want to measure the fluorescence properties of the protein conjugates under non-saturating conditions, i.e., using an under-labeled protein, then you should test several labeling conditions where the D/P ratio is varied during the labeling process (e.g., use of an equimolar or lower D/P ratio).

13. The elimination of the free dye from the conjugated protein can also be performed by gel filtration chromatography or by extensive dialysis.

14. In the case you are preparing samples with a final pH different from the protein stock, first prepare a control by adding all the reagents but replacing the protein stock solution with the same volume of its buffer (protein-free), and check if the final pH is the selected for the experiment. If not, an aliquot of the protein stock should be dialyzed to a more suitable condition. You can also prepare a cation binding experiment using a very diluted protein stock, but a previous dialysis step is required to lower the Na^+ concentration so that its final concentration in the samples falls between 5 and 10 mM.

15. If you have an absorbance value higher than 0.5 at 280 nm, it is advisable to dilute the sample and measure the absorbance again in order to avoid high scattering artifacts.
16. In the case of the pyrene-labeled KcsA mutant, a D/P ratio < 0.5 minimized the possibility of pyrene excimer formation, which present different spectroscopic characteristics that can lead to an erratic interpretation of the results.
17. To obtain a good signal-to-noise ratio (S/R) for the fluorescence measurements, the excitation and emission slits of the spectrofluorometer can be adjusted. However, to minimize the interference of photobleaching during the fluorescence recordings, particularly for the single-Trp-containing proteins, the use of narrow excitation slits (nominal bandpass of no more than 5 nm) is highly recommended. For the most critical cases, the overall irradiation time must be kept to a minimum while the samples are magnetically stirred during data acquisition. Stirring the sample helps renovating the fresh molecules that are excited by the spectrofluorometer lamp along its excitation pathway. Finally, the saturation limits of the detector present in the equipment must always be obeyed to prevent spectral distortions and damaging this important component of the spectrofluorometer.
18. Blank correction is particularly important for weak emitting fluorescent samples to avoid potential artifacts caused by the contribution of Raman scattering to the signal detected.
19. Samples presenting a high turbidity can cause scattering of both the incident light and the emitted photons prior to their detection. Each scattering event causes a significant depolarization of the scattered photon, ultimately producing an artifactually pronounced decrease in the fluorescence anisotropy measured for the sample [6, 12]. To avoid this interference, it is particularly important to use fluorescence cuvettes with a short pathlength for this set of measurements and to make sure that any insoluble material is removed by a spin down of the sample prior to the measurements.
20. The near-UV electronic absorption of tryptophan has contributions from two low-lying singlet excited states, which are designated as ^1L_a and ^1L_b. These excited states have nearly perpendicular transition moments [72], and their relative contribution to the absorption spectrum of tryptophan varies with the excitation wavelength [72]. Due to the occurrence of an ultrafast conversion between these singlet excited states, the emissive state of tryptophan residues in proteins is always the ^1L_a state. Accordingly, the fundamental anisotropy, r_0, of tryptophan residues strongly varies along its long-wavelength absorption band. By exciting the single-Trp-containing

proteins at 300 nm, essentially only the ^1L_a state absorbs, and therefore, the fundamental anisotropy of tryptophan is very high (r_0 (300 nm) > 0.3 compared to, e.g., r_0 (295 nm) ~ 0.19 [72]). This working condition further allows to maximize the dynamic range of the subsequent time-resolved anisotropy measurements of these samples.

21. The *G* factor is an instrumental factor that accounts for the variable sensitivity of the detection system for horizontal and vertically polarized light. For a given setup, this factor should be constant for each pair of excitation and emission wavelengths used [6].
22. In order to avoid pulse pile-up effects, the average count rate of the detection system should be limited to approximately 1% of the repetition rate of the pulsed excitation light source used in the equipment. If necessary, the sample count rate at the detector can be adjusted by placing an appropriate neutral density filter along the emission pathlength and/or by narrowing the slits of the emission monochromator. An adequate cutoff filter is also introduced between the sample and the detection system to effectively prevent any excitation light from reaching the detector. Ideally, the time scale (ps/channel) selected for data acquisition should allow for a decrease of two orders of magnitude in the fluorescence counts relative to the peak channel of the intensity decay within the time window available (corresponding to 1024 channels in our multichannel analyzer). To obtain a good S/N ratio, at least 10,000 counts should be accumulated at the peak channel of the fluorescence intensity decay obtained for the sample.
23. The IRF reflects the distribution of photons from the excitation pulse (width of the excitation pulse) and is used as an input in the subsequent data analysis that requires its iterative reconvolution with the fitted function (usually, a sum of exponentials).
24. The possible interference of variable photobleaching effects on each polarized intensity decay curve during data acquisition must be minimized by magnetically stirring the samples. Also, the implementation of an automated alternate acquisition of each polarized intensity for short time periods (15–30 s each) is highly recommended. Usually, at least 20,000 counts are accumulated at the peak channel of the vertically polarized fluorescence intensity decay obtained for the sample, $I_{VV}(t)$.
25. In the case of the analysis of the time-resolved anisotropy decay obtained for Pyr-labeled KcsA to determine the global rotational correlation time of the KcsA–DDM complex, firstly analyze the initial time range of 0–60 ns, then fix the obtained rotational correlation times, and proceed to fit the complete decay.

26. The R_0(Trp-Trp) value is calculated using $k^2 = 2/3$ (dynamic condition, that is, fast isotropic rotational motion of the fluorophores is admitted), which does not strictly hold for W67 KcsA. However, as discussed in detail in [15], the mixed polarization of the Trp absorption spectrum (*see* **Note 20** above) and the fast internal reorientations of Trp residues during their excited-state lifetime (revealed by $r(0) < r_0$) both contribute to reduce the orientational constraints of the system.
27. Apparent activation pK_a can also be calculated by using changes in the TMR emission spectra (spectral center of mass) or by the absorbance ratio A_{554}/A_{520} [45]. In addition to the spectral alterations detected in the absorption spectra of the fluorescently labeled protein, the formation of non-fluorescent H-dimers by a fluorescent dye conjugated to a Cys-containing protein usually results in a pronounced decrease in its overall steady-state fluorescence intensity, while its mean fluorescence lifetime remains invariant (static quenching) [59, 60].

Acknowledgments

We thank Aleksander Fedorov, Manuel Prieto, and Mário Nuno Berberan-Santos from iBB (Portugal) and Clara Díaz-García, Ana Marcela Giudici, José Manuel González-Ros, and Eva Martínez from IDiBE (Spain) for all the extensive and wise contributions to the work described here. This work was partly supported by grants PGC2018-093505-B-I00 from the Spanish "Ministerio de Ciencia e Innovación"/FEDER, U.E., and national funds from FCT Fundação para a Ciência e a Tecnologia, I.P., under the scope of the projects UIDB/04565/2020 and UIDP/04565/2020 of the Research Unit Institute for Bioengineering and Biosciences (iBB) and the project LA/P/0140/2020 of the Associate Laboratory Institute for Health and Bioeconomy (i4HB).

References

1. Renart ML, Barrera FN, Molina ML et al (2006) Effects of conducting and blocking ions on the structure and stability of the potassium channel KcsA. J Biol Chem 281:29905–29915. https://doi.org/10.1074/jbc.M602636200
2. Renart ML, Giudici AM, Díaz-García C et al (2020) Modulation of function, structure and clustering of K+ channels by lipids: lessons learnt from KcsA. Int J Mol Sci 21:2554. https://doi.org/10.3390/ijms21072554
3. Molina ML, Barrera FN, Fernández AM et al (2006) Clustering and coupled gating modulate the activity in KcsA, a potassium channel model. J Biol Chem 281:18837–18848. https://doi.org/10.1074/jbc.M600342200
4. Krishnan MN, Bingham J-P, Lee SH et al (2005) Functional role and affinity of inorganic cations in stabilizing the tetrameric structure of the KcsA K+ channel. J Gen Physiol 126:271–283. https://doi.org/10.1085/jgp.200509323

5. Lerner E, Barth A, Hendrix J et al (2021) FRET-based dynamic structural biology: challenges, perspectives and an appeal for open-science practices. elife 10:e60416. https://doi.org/10.7554/eLife.60416
6. Lakowicz JR (2006) Principles of fluorescence spectroscopy. Springer, Boston
7. Valeur B, Berberan-Santos MN (2012) Molecular fluorescence: principles and applications, 2nd edn. Wiley, London
8. Stryer L, Haugland RP (1967) Energy transfer: a spectroscopic ruler. Proc Natl Acad Sci 58: 719–726. https://doi.org/10.1073/pnas.58.2.719
9. Kalinin S, Johansson LB-Å (2004) Utility and considerations of donor–donor energy migration as a fluorescence method for exploring protein structure-function. J Fluoresc 14: 681–691. https://doi.org/10.1023/B:JOFL.0000047218.51768.59
10. Thaler C, Blank P, Koushik S, Vogel S (2009) Time-resolved fluorescence anisotropy. In: Flim microscopy in biology and medicine. Chapman and Hall/CRC, pp. 245–320
11. Majumdar A, Mukhopadhyay S (2018) Fluorescence depolarization kinetics to study the conformational preference, structural plasticity, binding, and assembly of intrinsically disordered proteins. Methods Enzymol 611: 347–381. https://doi.org/10.1016/bs.mie.2018.09.031
12. Ameloot M, van de Ven M, Acuña AU, Valeur B (2013) Fluorescence anisotropy measurements in solution: methods and reference materials (IUPAC technical report). Pure Appl Chem 85:589–608. https://doi.org/10.1351/PAC-REP-11-11-12
13. Millar DP (2000) Time-resolved fluorescence methods for analysis of DNA-protein interactions. Methods Enzymol 323:442–459. https://doi.org/10.1016/S0076-6879(00)23377-6
14. Cardoso S, Berberan-Santos MN (2021) Reversible electronic energy transfer (homo-FRET) in cyclic molecular and supramolecular systems: fluorescence anisotropy decays for the isotropic interaction. J Phys Chem A 125: 8476–8481. https://doi.org/10.1021/acs.jpca.1c04975
15. Renart ML, Giudici AM, Poveda JA et al (2019) Conformational plasticity in the KcsA potassium channel pore helix revealed by homo-FRET studies. Sci Rep 9:6215. https://doi.org/10.1038/s41598-019-42405-5
16. Coutinho A, Díaz-García C, Giudici AM, Renart ML (2022) Insights into the conformational dynamics of potassium channels using homo-FRET approaches, pp 443–478. In R. Šachl, M. Amaro (eds.), Fluorescence Spectroscopy and Microscopy in Biology, Springer Ser Fluoresc (2023). https://doi.org/10.1007/4243_2022_24
17. Runnels LW, Scarlata SF (1995) Theory and application of fluorescence homotransfer to melittin oligomerization. Biophys J 69:1569–1583. https://doi.org/10.1016/S0006-3495(95)80030-5
18. Bader AN, Hoetzl S, Hofman EG et al (2011) Homo-FRET imaging as a tool to quantify protein and lipid clustering. ChemPhysChem 12:475–483. https://doi.org/10.1002/cphc.201000801
19. Blackman SM, Piston DW, Beth AH (1998) Oligomeric state of human erythrocyte band 3 measured by fluorescence resonance energy homotransfer. Biophys J 75:1117–1130. https://doi.org/10.1016/S0006-3495(98)77601-5
20. Melo AM, Fedorov A, Prieto M, Coutinho A (2014) Exploring homo-FRET to quantify the oligomer stoichiometry of membrane-bound proteins involved in a cooperative partition equilibrium. Phys Chem Chem Phys 16: 18105–18117. https://doi.org/10.1039/C4CP00060A
21. Bergström F, Hägglöf P, Karolin J et al (1999) The use of site-directed fluorophore labeling and donor–donor energy migration to investigate solution structure and dynamics in proteins. Proc Natl Acad Sci 96:12477–12481. https://doi.org/10.1073/pnas.96.22.12477
22. Lillo MP, Cañadas O, RobertE D, Acuña AU (2002) Location and properties of the taxol binding center in microtubules: a picosecond laser study with fluorescent Taxoids. Biochemistry 41:12436–12449. https://doi.org/10.1021/bi0261793
23. Thaler C, Koushik SV, Puhl HL et al (2009) Structural rearrangement of CaMKIIα catalytic domains encodes activation. Proc Natl Acad Sci 106:6369–6374. https://doi.org/10.1073/pnas.0901913106
24. Kayser V, Turton DA, Aggeli A et al (2004) Energy migration in novel pH-triggered self-assembled β-sheet ribbons. J Am Chem Soc 126:336–343. https://doi.org/10.1021/ja035340+
25. Schrempf H, Schmidt O, Kümmerlen R et al (1995) A prokaryotic potassium ion channel with two predicted transmembrane segments from Streptomyces lividans. EMBO J 14: 5170–5178. https://doi.org/10.1002/j.1460-2075.1995.tb00201.x

26. Doyle DA, Cabral JM, Pfuetzner RA et al (1998) The structure of the potassium channel: molecular basis of K^+ conduction and selectivity. Science 280:69–77. https://doi.org/10.1126/science.280.5360.69
27. Zhou Y, Morais-Cabral JH, Kaufman A, MacKinnon R (2001) Chemistry of ion coordination and hydration revealed by a K+ channel–Fab complex at 2.0 Å resolution. Nature 414:43–48. https://doi.org/10.1038/35102009
28. LeMasurier M, Heginbotham L, Miller C (2001) KcsA: It's a Potassium Channel. J Gen Physiol 118:303–314. https://doi.org/10.1085/jgp.118.3.303
29. Uysal S, Vásquez V, Tereshko V et al (2009) Crystal structure of full-length KcsA in its closed conformation. Proc Natl Acad Sci 106:6644–6649. https://doi.org/10.1073/pnas.0810663106
30. Morais-Cabral JH, Zhou Y, MacKinnon R (2001) Energetic optimization of ion conduction rate by the K+ selectivity filter. Nature 414:37–42. https://doi.org/10.1038/35102000
31. Lockless SW, Zhou M, MacKinnon R (2007) Structural and thermodynamic properties of selective ion binding in a K+ channel. PLoS Biol 5:e121. https://doi.org/10.1371/journal.pbio.0050121
32. Gao L, Mi X, Paajanen V et al (2005) Activation-coupled inactivation in the bacterial potassium channel KcsA. Proc Natl Acad Sci 102:17630–17635. https://doi.org/10.1073/pnas.0505158102
33. Cordero-Morales JF, Jogini V, Chakrapani S, Perozo E (2011) A multipoint hydrogen-bond network underlying KcsA C-type inactivation. Biophys J 100:2387–2393. https://doi.org/10.1016/j.bpj.2011.01.073
34. Xu Y, McDermott AE (2019) Inactivation in the potassium channel KcsA. J Struct Biol X 3:100009. https://doi.org/10.1016/j.yjsbx.2019.100009
35. Kiss L, Korn SJ (1998) Modulation of C-type inactivation by K^+ at the potassium channel selectivity filter. Biophys J 74:1840–1849. https://doi.org/10.1016/S0006-3495(98)77894-4
36. Ogielska EM, Aldrich RW (1999) Functional consequences of a decreased potassium affinity in a potassium channel pore. J Gen Physiol 113:347–358. https://doi.org/10.1085/jgp.113.2.347
37. Hsu H, Huang E, Yang XC et al (1993) Slow and incomplete inactivations of voltage-gated channels dominate encoding in synthetic neurons. Biophys J 65:1196–1206. https://doi.org/10.1016/S0006-3495(93)81153-6
38. Montoya E, Lourdes Renart M, Marcela Giudici A et al (2017) Differential binding of monovalent cations to KcsA: deciphering the mechanisms of potassium channel selectivity. Biochim Biophys Acta Biomembr 1859:779–788. https://doi.org/10.1016/j.bbamem.2017.01.014
39. Renart ML, Montoya E, Fernández AM et al (2012) Contribution of ion binding affinity to ion selectivity and permeation in KcsA, a model potassium channel. Biochemistry 51:3891–3900. https://doi.org/10.1021/bi201497n
40. Giudici AM, Díaz-García C, Renart ML et al (2021) Tetraoctylammonium, a long chain quaternary ammonium blocker, promotes a noncollapsed, resting-like inactivated state in KcsA. Int J Mol Sci 22:490. https://doi.org/10.3390/ijms22020490
41. Ahern CA, Pless SA (2015) Novel chemical tools to study ion channel biology. Springer, New York
42. Nanda JS, Lorsch JR (2014) Labeling of a protein with fluorophores using maleimide derivitization. In: Methods in enzymology, pp 79–86. https://doi.org/10.1016/B978-0-12-420070-8.00007-6
43. Akabas MH (2015) Cysteine modification: probing channel structure, function and conformational change. In: Advances in experimental medicine and biology, pp 25–54. https://doi.org/10.1007/978-1-4939-2845-3_3
44. Sadler EE, Kapanidis AN, Tucker SJ (2016) Solution-based single-molecule FRET studies of K+ channel gating in a lipid bilayer. Biophys J 110:2663–2670. https://doi.org/10.1016/j.bpj.2016.05.020
45. Díaz-García C, Renart ML, Poveda JA et al (2021) Probing the structural dynamics of the activation gate of KcsA using homo-FRET measurements. Int J Mol Sci 22:11954. https://doi.org/10.3390/ijms222111954
46. Perozo E, Cortes DM, Cuello LG (1998) Three-dimensional architecture and gating mechanism of a K+ channel studied by EPR spectroscopy. Nat Struct Biol 5:459–469. https://doi.org/10.1038/nsb0698-459
47. Tilegenova C, Cortes DM, Cuello LG (2017) Hysteresis of KcsA potassium channel's activation– deactivation gating is caused by structural changes at the channel's selectivity filter. Proc Natl Acad Sci 114:3234–3239. https://doi.org/10.1073/pnas.1618101114
48. Perozo E, Marien D, Cortes, Cuello LG (1999) Structural rearrangements underlying

K^+ -channel activation gating. Science 285:73–78. https://doi.org/10.1126/science.285.5424.73

49. Blunck R, Cordero-Morales JF, Cuello LG et al (2006) Detection of the opening of the bundle crossing in KcsA with fluorescence lifetime spectroscopy reveals the existence of two gates for ion conduction. J Gen Physiol 128:569–581. https://doi.org/10.1085/jgp.200609638
50. Gupta K, Toombes GE, Swartz KJ (2019) Exploring structural dynamics of a membrane protein by combining bioorthogonal chemistry and cysteine mutagenesis. elife 8. https://doi.org/10.7554/eLife.50776
51. Faure É, Starek G, McGuire H et al (2012) A limited 4 Å radial displacement of the S4-S5 linker is sufficient for internal gate closing in Kv channels. J Biol Chem 287:40091–40098. https://doi.org/10.1074/jbc.M112.415497
52. Matulef K, Komarov AG, Costantino CA, Valiyaveetil FI (2013) Using protein backbone mutagenesis to dissect the link between ion occupancy and C-type inactivation in K^+ channels. Proc Natl Acad Sci 110:17886–17891. https://doi.org/10.1073/pnas.1314356110
53. Devaraneni PK, Komarov AG, Costantino CA et al (2013) Semisynthetic K^+ channels show that the constricted conformation of the selectivity filter is not the C-type inactivated state. Proc Natl Acad Sci 110:15698–15703. https://doi.org/10.1073/pnas.1308699110
54. Matulef K, Annen AW, Nix JC, Valiyaveetil FI (2016) Individual ion binding sites in the K+ channel play distinct roles in C-type inactivation and in recovery from inactivation. Structure 24:750–761. https://doi.org/10.1016/j.str.2016.02.021
55. Giudici AM, Renart ML, Coutinho A et al (2022) Molecular events behind the selectivity and inactivation properties of model NaK-derived ion channels. Int J Mol Sci 23:9246. https://doi.org/10.3390/ijms23169246
56. Wahl P (1979) Analysis of fluorescence anisotropy decays by a least square method. Biophys Chem 10:91–104. https://doi.org/10.1016/0301-4622(79)80009-5
57. Johnson I, Spence MTZ (2010) Molecular probes handbook, a guide to fluorescent probes and labeling technologies, 11th edn. Life Technologies
58. Pantazis A, Westerberg K, Althoff T et al (2018) Harnessing photoinduced electron transfer to optically determine protein sub-nanoscale atomic distances. Nat Commun 9:4738. https://doi.org/10.1038/s41467-018-07218-6
59. Ogawa M, Kosaka N, Choyke PL, Kobayashi H (2009) H-type dimer formation of fluorophores: a mechanism for activatable, in vivo optical molecular imaging. ACS Chem Biol 4:535–546. https://doi.org/10.1021/cb900089j
60. Donaphon B, Bloom LB, Levitus M (2018) Photophysical characterization of interchromophoric interactions between rhodamine dyes conjugated to proteins. Methods Appl Fluoresc 6:045004. https://doi.org/10.1088/2050-6120/aad20f
61. Molina ML, Encinar JA, Barrera FN et al (2004) Influence of C-terminal protein domains and protein-lipid interactions on tetramerization and stability of the potassium channel KcsA. Biochemistry 43:14924–14931. https://doi.org/10.1021/bi048889+
62. Giudici AM, Molina ML, Ayala JL, et al (2013) Detergent-labile, supramolecular assemblies of KcsA: relative abundance and interactions involved Biochim Biophys Acta Biomembr 1828:193–200. https://doi.org/10.1016/j.bbamem.2012.09.020
63. Stryer L (1978) Fluorescence energy transfer as a spectroscopic ruler. Annu Rev Biochem 47:819–846. https://doi.org/10.1146/annurev.bi.47.070178.004131
64. Würth C, Grabolle M, Pauli J et al (2013) Relative and absolute determination of fluorescence quantum yields of transparent samples. Nat Protoc 8:1535–1550. https://doi.org/10.1038/nprot.2013.087
65. Babul J, Stellwagen E (1969) Measurement of protein concentration with interferences optics. Anal Biochem 28:216–221. https://doi.org/10.1016/0003-2697(69)90172-9
66. Poveda JA, Prieto M, Encinar JA et al (2003) Intrinsic tyrosine fluorescence as a tool to study the interaction of the shaker B "ball" peptide with anionic membranes. Biochemistry 42:7124–7132. https://doi.org/10.1021/bi027183h
67. Visser A, Vysotski ES, Lee J. http://photobiology.info/Experiments/Biolum-Expt.html
68. Strop P, Brunger AT (2005) Refractive index-based determination of detergent concentration and its application to the study of membrane proteins. Protein Sci 14:2207–2211. https://doi.org/10.1110/ps.051543805
69. Cross AJ, Fleming GR (1984) Analysis of time-resolved fluorescence anisotropy decays. Biophys J 46:45–56. https://doi.org/10.1016/S0006-3495(84)83997-1
70. Cortes DM, Cuello LG, Perozo E (2001) Molecular architecture of full-length KcsA. J Gen Physiol 117:165–180. https://doi.org/10.1085/jgp.117.2.165

71. Han JC, Han GY (1994) A procedure for quantitative determination of Tris(2-Carboxyethyl)phosphine, an odorless reducing agent more stable and effective than dithiothreitol. Anal Biochem 220:5–10. https://doi.org/10.1006/abio.1994.1290

72. Valeur B, Weber G (1977) Resolution Of The Fluorescence Excitation Spectrum Of Indole into the ^1L_a and ^1L_b excitation bands. Photochem Photobiol 25:441–444. https://doi.org/10.1111/j.1751-1097.1977.tb09168.x

Chapter 4

Purification of Potassium Ion Channels Using Styrene–Maleic Acid Copolymers

Grigory Glukhov, Maria Karlova, Ekaterina Kravchuk, Anna Glukhova, Elizaveta Trifonova, and Olga S. Sokolova

Abstract

Structural studies require the production of target proteins in large quantities and with a high degree of purity. For membrane proteins, the bottleneck in determining their structure is the extraction of the target protein from the cell membranes. A detergent that improperly mimics the hydrophobic environment of the protein of interest can also significantly alter its structure. Recently, using lipodiscs with styrene–maleic acid (SMA), copolymers became a promising strategy for the purification of membrane proteins. Here, we describe in detail the one-step affinity purification of potassium ion channels solubilized in SMA and sample preparation for future structural studies.

Key words Human Kv channels, SMALP, Affinity purification, Electron microscopy, DLS

1 Introduction

Voltage-dependent (VD) ion channels play a major role in conducting the electrical impulse that de- or repolarizes membranes in excitable cells in response to changes in membrane action potential. Potassium Kv channels comprise four identical subunits consisting of six transmembrane segments (S1–S6). Among these, the S1–S4 segments form the characteristic voltage sensor domain (VSD), which contains a number of positively charged amino acid residues, and the S5–S6 segments form the channel pore and are connected to the S4 segment through a S4–S5 linker [1]. Mutations, interactions with numerous drugs, and post-translational modifications can be the reason for changes in the primary and secondary structure of ion channels. These changes can lead to disturbances in the function of the ion channel, including the action potential

Grigory Glukhov and Maria Karlova have equally contributed to this chapter.

Simone Furini (ed.), *Potassium Channels: Methods and Protocols*, Methods in Molecular Biology, vol. 2796, https://doi.org/10.1007/978-1-0716-3818-7_4,

[2, 3]. For example, alterations in the Kv7.1 channel activity lead to the development of a prolonged QT interval (LQT) on the electrocardiogram (ECG) [4]. *KCNH2* gene mutations with enhanced function result in accelerated repolarization and are characterized by abnormally short QT intervals on ECGs [5].

A detailed understanding of the structure of ion channels is important for the interpreting of structural and functional effects of mutations and may help in drug development. However, for structural studies, it is necessary to obtain large amounts of protein with a high degree of purity. In this regard, sample preparation becomes one of the limiting factors. The presence of lipids around the membrane protein is important for the extraction procedure. Detergents used in solubilization protocols can significantly alter the protein structure [6, 7]. To preserve the lipid environment of the protein, various membrane mimetics are extensively used. Among them are nanodiscs—disc-shaped phospholipid bilayers originally based on the use of human apolipoprotein ApoA1, surrounded by molecules of the amphipathic α-helical protein MSP, which limits the size of the disc [8–11].

The most recent method of protein isolation in a natural lipid environment is based on the use of a styrene and maleic acid copolymer (SMA). SMA is an amphiphilic copolymer, due to the hydrophilic part of the maleic acid and the hydrophobic part of the styrene. When the SMA polymer is added to a protein-containing membrane, protein molecules become enclosed in the forming discs; the protein is cut out of the membrane, while preserving a certain layer of lipids surrounding it. These polymer-bound lipoprotein particles are called "lipodiscs" [12–14]. Membrane proteins solubilized within lipodiscs are highly stable and suitable for further structural investigation using a wide range of methods [3, 15]. In recent years, a large variety of amphiphilic copolymers have been suggested, including SMA and its derivatives (SMAnh), as well as the precursor of SMA, and diverse non-SMA-based polymers [16]. A significant advantage of using SMA is the complete absence of detergent in the ion channel purification protocol. As a result, SMA-solubilized proteins can be extracted together with their natural lipid environment and ligands.

2 Materials

2.1 Ion Channel Protein Expression in Eukaryotic Cells

2.1.1 Cell Line

1. COS-1 cell line (see also **Note 1**).

2.1.2 Equipment

1. Class II Biological Safety Cabinet.
2. Cell culture CO_2 incubator.
3. Inverted microscope.
4. Hemocytometer (or automated cell counter as an alternative).
5. Table-top centrifuge.
6. Ice bath.
7. Autoclave.
8. Ultra-low temperature (ULT) freezer (optional).

2.1.3 Chemicals

1. Dulbecco's Modified Eagle Medium (DMEM) with glutamine, 4.5 g/L glucose content cell culture media with 10% fetal bovine serum (FBS), sterile filtered.
2. Trypsin solution 0.25%, sterile.
3. Phosphate-buffered saline (PBS), sterile.
4. Transfection reagent (*see* **Note 1**).
5. Plasmid DNA, sterile and preferably endotoxin-free.
6. Trypan Blue solution.
7. Cryopreservation solution: 20% glycerol in PBS supplemented with protease inhibitor cocktail (PIC).
8. Antibiotics mix for cell culture (optional) (100 IU/mL penicillin and 100 μg/mL streptomycin).

2.2 Ion Channel Protein Solubilization and Purification

2.2.1 Equipment

1. Centrifuge with swinging-bucket rotor.
2. End-over-end rotator.
3. Refrigerated benchtop high-speed centrifuge.
4. Ultracentrifuge capable of >100,000 g.
5. Ultrasonic sonicator.
6. SDS-PAGE and immunoblotting equipment.

2.2.2 SMA Stock Solution

1. SMA buffer: 10 mM Tris–HCl, 150 mM NaCl, pH 8 (*see* **Notes 2–4**).
2. Centrifugal filter units with 10 kDa cutoff.
3. SMA copolymer: styrene–maleic acid (SMA) copolymer with styrene-to-maleic acid molar ratio of 3:1 (MW 9500 Da, supplied as an aqueous sodium salt solution SMA 3000 HNa). The 5% (w/v) solution of SMA should be extensively dialyzed against the SMA buffer with a centrifugal filter unit. The dialyzed SMA stock solution is used for the preparation of SMALPs (*see* **Note 5**).

2.2.3 Preparation of Protein-Containing SMALPs and Affinity Chromatography

1. SMA stock solution.
2. Affinity resin:
   (a) Anti-rho-1D4 affinity resin (NHS-activated Sepharose conjugated with monoclonal anti-rho-1D4 antibody).
   (b) Anti-V5-tag pAb-agarose.
   (c) Any other appropriate affinity resin.
3. Rho-1D4 peptide, lyophilized, or V5 peptide, lyophilized in PBS, or any appropriate peptide.
4. Buffer A: 10 mM Tris–HCl, 150 mM NaCl, 2 mM DTT, 1 mM EDTA, protease inhibitor cocktail (PIC), pH 8.
5. Buffer B: buffer A, supplemented with 330 mM NaCl, pH 8.
6. Elution buffer for affinity resin:
   (a) Rho-1D4-EB elution buffer for anti-rho-1D4 affinity resin: buffer B with 0.2 mg/mL rho-1D4 peptide.
   (b) V5-EB elution buffer for anti-V5-tag pAb-agarose: lyophilized in PBS V5 peptide, reconstituted with deionized water to the final concentration of 2 mg/mL with addition of 150 mM NaCl, 40 mM KCl, 2 mM DTT, 1 mM EDTA, and protease inhibitor cocktail (PIC), pH 8.

2.3 SMALPs Validation and Verification

2.3.1 SDS-PAGE Electrophoresis

1. SDS-PAGE cell.
2. Precast or handcast gels.
3. Trans blot system.
4. Primary antibodies directed against the protein of interest and appropriate labeled secondary antibodies.

2.3.2 Dynamic Light Scattering (DLS)

1. DLS instrument.
2. Microcuvette for DLS measurements.
3. Syringe filter with cutoff of 0.22 μm.

2.3.3 Electron Microscopy

1. Transmission electron microscope with LaB6 electron source and 100–200 kV accelerated voltage.
2. CCD camera incorporated into microscope.
3. Glow-discharge apparatus (or plasma cleaner).
4. TEM copper grids 300–400 mesh, formvar/carbon-coated.
5. Tweezers with fine tips.
6. Negative stain solution (e.g., 1% UA solution or any uranium-less EM stain).
7. Ash-free blotting paper.
8. Parafilm M.
9. Glass slide.

10. Grid storage box.
11. SerialEM ver. 3.8 [17].
12. CryoSPARC [18].

3 Methods

3.1 Protein Production in Eukaryotic Cells

1. All manipulations with cell cultures should be carried out under sterile conditions, using sterile instruments.
2. The COS-1 cell line should be maintained in Dulbecco's Modified Eagle Medium (DMEM), supplemented with 10% of fetal bovine serum at 5% CO_2 and 37 °C in a humidified incubator. Antibiotics (100 IU/ml penicillin and 100 μg/mL streptomycin) use is optional, depending on the practical necessity and the cell culture facility rules.
3. For transient transfection, cells should be seeded 24 h before the procedure.
4. For cell passaging, remove the culture medium, wash the cells with sterile PBS (or fresh culture medium without serum), and detach with trypsin solution. Sediment by centrifugation, discard supernatant. Resuspend cell pellet in fresh complete culture medium.
5. Take an aliquot of cell suspension, add 1:1 (v/v) trypan blue solution, and count the number of viable cells using a hemocytometer or automated cell counter.
6. Seed an appropriate amount of cells into dishes to achieve confluence of around 70–80% after 24 h of cultivation. A quantity of 2.5*10^5 cells for the COS-1 line for the 35 mm plate can be used as a starting point, but the actual growth curve should be determined experimentally.
7. Cell visual confluency should be about 70–80% on the day of transfection for better protein expression levels.
8. Mix sterile plasmid DNA with an appropriate transfection reagent in sterile conditions according to the manufacturer's recommendations, incubate for 15 min, and drop gently into cell plates. Swirl the plate very gently to distribute the reagent. The exact quantity of plasmid and transfection reagent is to be determined experimentally. The excess of the transfection reagent could be toxic to cells. For example, 2 μg of plasmid DNA per 4 μL of Metafectene® Pro works well for the 35 mm cell plate of COS-1 cells with 80% of visual confluence. It is possible to use other methods of transient transfection (*see* **Note 6**).
9. Incubate cells for 48 h.

10. Wash the cells with ice-cold PBS twice.
11. Add ice-cold PBS with protease inhibitors. Harvest cells with a cell scraper, transfer suspension to microtubes, and centrifuge at 1000 g for 3 min at 4 °C. Discard the supernatant. Use cell pellet for protein extraction.
12. If the cells will not be used immediately for protein extraction, add 100 μL PBS supplemented with 20% of glycerol, and protease inhibitors before freezing. Resuspend and flash-freeze in liquid nitrogen. Keep at −70 °C until use.

3.2 Protein Solubilization and Purification

3.2.1 Preparation of Protein-Containing SMALPs

1. Thoroughly resuspend cells expressing ion channel proteins into buffer A. Use 100 μL of buffer A per 1*10^6 cells (cell monolayer from a 35 mm cell plate) (*see* **Note 7**).
2. Add the appropriate volume of the SMA stock solution so that the final concentration of the SMA copolymer is 2.5% (w/v).
3. Incubate for 30 min at 4 °C with gentle end-over-end mixing.
4. Sonicate cell lysate with a thin ultrasound probe at low power pulses on ice for 15 s (*see* **Note 8**).
5. Incubate for an additional 30 min at 4 °C with gentle end-over-end mixing.
6. Centrifuge cell lysate for 15 min at 200,000 g in an ultra-centrifuge.
7. Take aliquots of pellet and supernatant for further SDS-PAGE and immunoblotting analysis (*see* Fig. 1).
8. Check supernatant for the presence of nanoparticles in the range of 12 nm using DLS (*see* **step 3.3**).

3.2.2 Affinity Chromatography

1. 20 μL aliquots of affinity resin were used for protein purification (the quantity of resin depends on its binding capacity and the estimated amount of protein).
2. Resuspend the agarose slurry in a storage buffer, and transfer a small portion of suspension into a clean microcentrifuge tube. Allow the resin to settle out by gravity, remove the supernatant, and estimate the volume of the resin.
3. Equilibrate the resin with 30 column volumes (CV) of buffer A. Add 10 CV of buffer A and gently resuspend the slurry. Allow the resin to settle out by gravity, and remove the supernatant. Repeat these steps three times.
4. Add the clarified cell lysate to the appropriate affinity resin, pre-equilibrated with buffer A.
   - Add clarified SMA-solubilized hKCNQ1 and hKCNH5 solutions to the anti-rho-1D4 affinity resin.
   - Add SMA-solubilized hKCNE1–hKCNQ1 solution to the anti-V5-tag pAb-agarose.

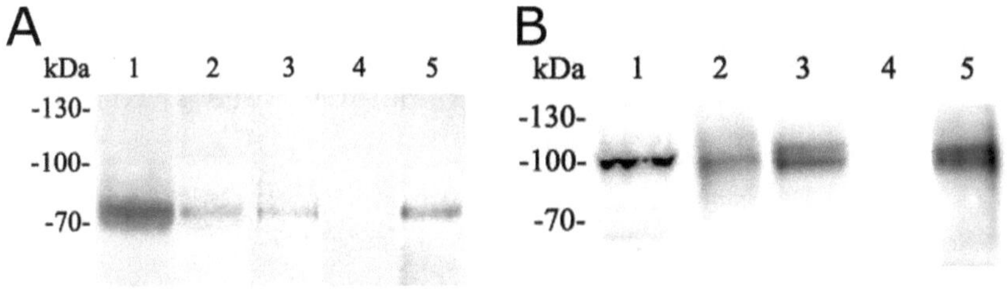

Fig. 1 The solubilization and purification of hKCNQ1 and hKCNE1–hKCNQ1 by SMA copolymer, immunoblot. (**a**) hKCNQ1 protein expression, solubilization with SMA, and purification on anti-rho-1D4 affinity resin. Line 1, COS-1 cells expressing the hKCNQ1 protein; 2, solubilization with 2.5% of SMA copolymer—supernatant; 3, pellet; 4, column wash; and 5, elution fraction. (**b**) hKCNE1–hKCNQ1 fusion protein expression, solubilized with SMA copolymer and purified on anti-V5 affinity resin. Line 1, COS-1 cells expressing the hKCNE1–hKCNQ1 fusion protein; 2, solubilization with 2.5% of SMA copolymer—supernatant; 3, pellet; 4, column wash; and 5, elution fraction. (Adapted from [15])

5. Incubate with gentle end-over-end mixing:
   - Overnight at 4 °C for anti-rho-1D4 affinity resin.
   - 1 h at 4 °C for anti-V5-tag pAb-agarose.
6. Transfer the binding suspension to a disposable spin column with a capped bottom outlet. Allow the resin to settle out by gravity. Use the supernatant to rinse the binding tube, and remove resin adhered to the wall.
7. Remove the bottom cap of the column and collect the flow-through. Take an aliquot of the column flow-through for further SDS-PAGE and immunoblotting (flow-through fraction).
8. Wash the resin with 30 CV of buffer B (at least three times in 10 CV portions). For the anti-V5-tag pAb-agarose: additionally wash the resin with 5 CV of PBS. Take aliquots for SDS-PAGE and immunoblotting (wash fractions).
9. Elute the protein with appropriate elution buffer:
   - For the anti-rho-1D4 resin: add 5 CV of the rho-1D4-EB buffer, close and rotate the column for 1 h at 4 °C. Remove the top and bottom cap of the column, and collect the eluate. Repeat elution steps at least three times (*see* **Note 9**).
   - For the anti-V5 resin: add 4 CV of V5-EB buffer, close the column, and incubate for 5 min RT with gentle shaking. Centrifuge for 1 min and collect the supernatant. Repeat elution steps three times.
10. Proceed to **step 3.3**.

3.3 SMALPs Visualization and Size Estimation

In order to substantiate the results obtained using SMAPLs, one needs to demonstrate (i) the presence and concentration of purified proteins and (ii) the size of purified proteins. To achieve this, the following steps are recommended:

1. Use SDS-PAGE and immunoblotting to analyze elution fractions for the presence of protein.
2. Analyze elutions by DLS for the presence of nanoparticles.
3. Analyze elutions in TEM for the size and shape of nanoparticles.

3.3.1 SDS-PAGE and Immunoblotting

Electrophoresis and blotting can be performed routinely. There are a few points to keep in mind:

1. Determination of protein concentration in samples taken for SDS-PAGE. Be very careful when choosing this method, as SMA derivatives have their own strong absorbance and fluorescence and can interfere severely with measurements.
2. After mixing the samples with the appropriate amount of loading buffer, it is not recommended to boil them, as boiling often leads to irreversible aggregation of membrane proteins and results in a band shift on the gel or multiple mass bands. Instead, incubate samples at 46 °C for 30 min.
3. Since SMALPs are very stable structures and the SMA copolymer at its physiological pH has a strong negative charge, SMA-solubilized proteins can be analyzed in native PAGE without the addition of detergent or other charge-shifting agents [19].

3.3.2 Dynamic Light Scattering (DLS)

1. The buffer solution should be filtered through 0.22 μm membrane filters.
2. The cuvette should be washed with a filtered buffer prior to measurements to avoid dust contamination.
3. DLS experiments were carried out on a Brookhaven 90 Plus instrument, in a thermostated cell at 20 °C.
4. The scattered light was recorded at an angle of $\theta = 90°$, and the accumulation time of the signal was 1 min. The measurements should be repeated 3–5 times and averaged.
5. The mathematical processing of the experimentally recorded autocorrelation functions of the scattered light should be carried out using a package of programs provided by the manufacturer.
6. Results are to be presented as a DLS curve (*see* Fig. 2).

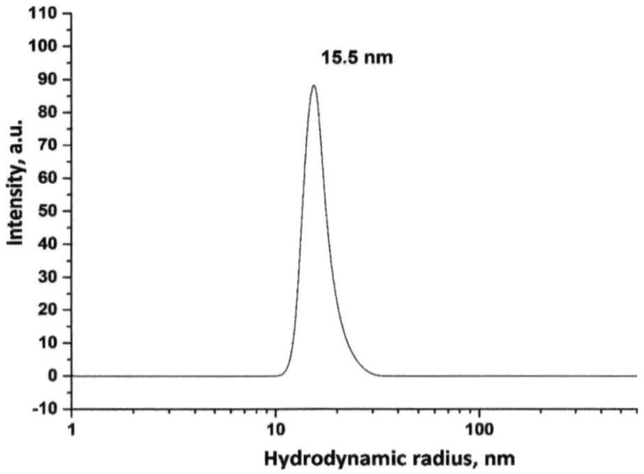

Fig. 2 DLS curve of the purified hKCNH5 protein, solubilized with SMA. Major peak corresponds to protein-containing SMALPs with an average hydrodynamic radius of 15 nm

3.3.3 Electron Microscopy Sample Preparation

1. Treat the copper TEM grids in an appropriate plasma cleaner (the suggested regime for EasyGlow is −20 mA, 45 s).
2. Briefly spin column-purified SMALPs in a table-top centrifuge, and apply 3 μL of solution to the glow-discharged TEM grid (carbon side). Incubate for 30 s. Carefully remove excess of liquid with ash-free blot paper.
3. Immediately put the grid onto a drop of 1% UA solution trying to minimize the time the grid spends in the air. The less the liquid layer dries out on the grid surface, the less the protein will be disturbed by the air/liquid interface.
4. Incubate for 30 s. Carefully remove excess of liquid with ash-free blot paper.
5. Repeat **steps 3–4** once again.
6. Air-dry the grid and put it in a grid box. The grids can be additionally vacuum dried if electron microscopy is planned for the same day.

3.3.4 Electron Microscopy and Image Processing

Micrographs should be acquired in a transmission electron microscope (here we used Jem-2100 (JEOL, Japan) equipped with a 2Kx2K CCD camera Ultrascan 1000XP (Gatan, USA)). The microscope was operated at 200 kV in a low dose mode, with a magnification of x40000 and a defocus of 0.5–1.9 μm.

Micrographs are acquired with SerialEM or any appropriate software of the electron microscope (*see* Fig. 3).

We recommend using CryoSPARC [18] to further process the data. Collected single particles can be classified, and 3D reconstruction can be obtained for ion channel in lipodiscs (*see* **Note 10**).

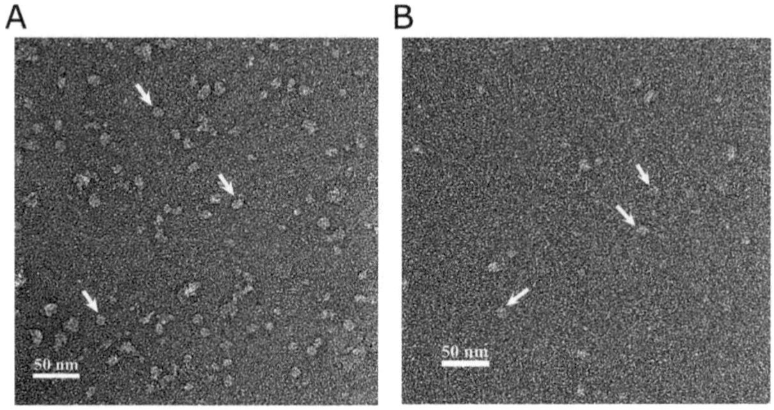

Fig. 3 EM micrographs of potassium ion channel proteins (**a**) hKCNQ1 and (**b**) hKCNE1–hKCNQ1 solubilized with a SMA copolymer and affinity purified. Samples stained with 1% UA. Arrows indicate channel's single particles

4 Notes

1. The choice of the transfection method depends on the cell type and the number of cells to be transfected. In our experiments we tried liposomal transfection reagents and electroporation. Though this protocol describes the Metafectene® Pro transfection, the calcium phosphate method is cost-effective and less toxic to cells. On the other hand, few cell lines are suitable for this method, HEK293 cells being one of them.
2. While using various SMA for membrane protein purification, factors affecting lipodisc size should be considered, especially the lipid/polymer ratio. A higher concentration of polymers with respect to lipids yields smaller lipodiscs [20–22]. The size also depends on the molecular weight of the amphiphilic polymers [23]. Low-molecular-weight zwitterionic zSMA polymers can interact with the membrane and solubilize it more efficiently [24]. On the other hand, large-weight polymer solubilization is ineffective, due to the tendency to aggregate, as well as steric hindrance [24]. Thus, the optimal molecular weight of SMA for solubilization lies in the range from 1.2–3 kDa [25, 26] to 10 kDa [27, 28].
3. The concentration of monovalent and divalent ions in the buffer affects the dynamics and conformation of lipodiscs [27, 29, 30]. An increase in the concentration of monovalent ions, particularly NaCl, enhances polymer formation, raises the protein extraction yield, and increases the lipid transfer rate [16, 31]. Divalent cations tend to be chelated by carboxylate groups and at some concentrations lead to the precipitation of carboxyl-containing polymers/lipodiscs. Varying the

concentration of bivalent ions and using modified polymer derivatives is helpful to avoid precipitation [20, 29, 32].

4. Changes in pH lead to the protonation of acidic carboxyl/carboxylate groups and to a change in the total charge of the polymer, which affects its solubility and structure. Therefore, the threshold has a pH value of about 5.5, and, at values lower than pKa, SMA forms aggregates and loses its ability to solubilize [33]. The impossibility of using this polymer in biochemical assays at low pH is the reason for developing other alternative SMA derivatives with a wider working pH spectrum [34].
5. Since the development of this protocol, SMA solutions that do not require dialysis have become commercially available. In addition, a wide variety of polymer compositions appeared—with different polymer mass, monomer ratio, and chemical features. Ready-to-use formulations in different buffers can also be found.
6. An alternative HEK 293 Expi suspension cell culture can be used for protein expression in big amounts of cells. For PEI transfection, one must (i) seed an appropriate amount of HEK 293 Expi cells into a sterile flask and (ii) prepare two mixes:
   (a) Mix A: Dilute a PEI MAX transfection reagent in the FreeStyle 293 Expression Medium (+) GlutaMAX. The stock solution of PEI MAX (1 mg/mL) is a 250-fold concentrate. Mix well.
   (b) Mix B: Dilute sterile plasmid DNA in the same volume (as in Mix A) of FreeStyle 293 Expression Medium (+) GlutaMAX. Mix well.
   (c) Incubate for 30 min at room temperature, and add the Master mix (1:1 ratio of Mix A to Mix B) to the cell flask. Mix well. 5 μg of plasmid DNA per 40 μL of PEI MAX (1 mg/mL stock) works well for the 10 mL cell culture of HEK 293 Expi cells with 90% of visual confluence.
   (d) Incubate cells for 72 h in the cell culture CO_2 shaking incubator.
7. This protocol suggests the solubilization of whole cells. An alternative option would be to obtain a membrane fraction of the cells beforehand. This significantly reduces the amount of SMA required, but also adds an extra manipulation step. Moreover, it is necessary to have tools for cell homogenization.
8. The addition of SMA results in immediate membrane lysis and DNA release, and the cell suspension turns into a very viscous substance. For further work, the mixture can be sonicated with a thin ultrasound probe at low power pulses and on ice. An alternative option is to use benzonase to disrupt nucleic acids.

From our observations, at least for the Kv7 channel protein, neither of these methods gives an advantage in terms of protein yield.

9. Affinity resin after protein elution should be analyzed by SDS-PAGE and immunoblotting. Sometimes it is necessary to perform more than three elutions or to adjust the peptide concentration. As for the concentration of rho-1D4 peptide: it is very important to control the pH of the elution buffer, as the acidic peptide destroys the SMA–lipid complexes by shifting the pH.
10. Note that the SMA in electron microscopical images does not form an extra density around the membrane part of the channel, unlike nanodiscs [3, 35].

Acknowledgments

Authors thank Dr. Natalia Voskoboynikova for fruitful discussions and Cray Valley (Exton, PA, USA) for gifting the SMA copolymer. AG and GG are part of the "Innovation drug development team based on structural biology and bioinformatics" of Shenzhen MSU-BIT University, Guangdong province, P.R.C. This work was supported by RSF (22-14-00088) to OSS.

References

1. Whicher JR, MacKinnon R (2016) Structure of the voltage-gated K^+ channel Eag1 reveals an alternative voltage sensing mechanism. Science 353(6300):664–669. https://doi.org/10.1126/science.aaf8070
2. Hishigaki H, Kuhara S (2011) hERGAPDbase: a database documenting hERG channel inhibitory potentials and APD-prolongation activities of chemical compounds. Database (Oxford). 2011:bar017. https://doi.org/10.1093/database/bar017
3. Kacher YG, Karlova MG, Glukhov GS et al (2021) The integrative approach to study of the structure and functions of cardiac voltage-dependent ion channels. Crystallogr Rep 66:711–725. https://doi.org/10.1134/S1063774521050072
4. Duggal P, Vesely MR, Wattanasirichaigoon D et al (1998) Mutation of the gene for IsK associated with both Jervell and Lange-Nielsen and Romano-Ward forms of long-QT syndrome. Circulation 97(2):142–146. https://doi.org/10.1161/01.cir.97.2.142
5. Hu D, Li Y, Zhang J et al (2017) The phenotypic spectrum of a mutation hotspot responsible for the short QT syndrome. JACC Clin Electrophysiol 3(7):727–743. https://doi.org/10.1016/j.jacep.2016.11.013
6. Marius P, Alvis SJ, East JM et al (2005) The interfacial lipid binding site on the potassium channel KcsA is specific for anionic phospholipids. Biophys J 89(6):4081–4089. https://doi.org/10.1529/biophysj.105.070755
7. Loussouarn G, Park KH, Bellocq C et al (2003) Phosphatidylinositol-4,5-bisphosphate, PIP2, controls KCNQ1/KCNE1 voltage-gated potassium channels: a functional homology between voltage-gated and inward rectifier K+ channels. EMBO J 22(20):5412–5421. https://doi.org/10.1093/emboj/cdg526
8. Peters-Libeu CA, Newhouse Y, Hall SC et al (2007) Apolipoprotein E*dipalmitoylphosphatidylcholine particles are ellipsoidal in solution. J Lipid Res 48(5):1035–1044. https://doi.org/10.1194/jlr.M600545-JLR200
9. Johansen NT, Tidemand FG, Nguyen TTTN et al (2019) Circularized and solubility-enhanced MSPs facilitate simple and

high-yield production of stable nanodiscs for studies of membrane proteins in solution. FEBS J 286(9):1734–1751. https://doi.org/10.1111/febs.14766

10. Nasr ML, Baptista D, Strauss M et al (2017) Covalently circularized nanodiscs for studying membrane proteins and viral entry. Nat Methods 14(1):49–52. https://doi.org/10.1038/nmeth.4079
11. Zhao Z, Zhang M, Hogle JM et al (2018) DNA-corralled Nanodiscs for the structural and functional characterization of membrane proteins and viral entry. J Am Chem Soc 140(34):10639–10643. https://doi.org/10.1021/jacs.8b04638
12. Xue M, Cheng L, Faustino I et al (2018) Molecular mechanism of lipid Nanodisk formation by styrene-maleic acid copolymers. Biophys J 115(3):494–502. https://doi.org/10.1016/j.bpj.2018.06.018
13. Gulamhussein AA, Uddin R, Tighe BJ et al (2020) A comparison of SMA (styrene maleic acid) and DIBMA (di-isobutylene maleic acid) for membrane protein purification. Biochim Biophys Acta Biomembr 1862(7):183281. https://doi.org/10.1016/j.bbamem.2020.183281
14. Hardin NZ, Ravula T, Mauro GD, Ramamoorthy A (2019) Hydrophobic functionalization of polyacrylic acid as a versatile platform for the development of polymer lipid Nanodisks. Small 15(9):e1804813. https://doi.org/10.1002/smll.201804813
15. Karlova MG, Voskoboynikova N, Gluhov GS et al (2019) Detergent-free solubilization of human Kv channels expressed in mammalian cells. Chem Phys Lipids 219:50–57. https://doi.org/10.1016/j.chemphyslip.2019.01.013
16. Orekhov PS, Bozdaganyan ME, Voskoboynikova N et al (2022) Mechanisms of formation, structure, and dynamics of lipoprotein discs stabilized by amphiphilic copolymers: a comprehensive review. Nanomaterials (Basel) 12(3):361. https://doi.org/10.3390/nano12030361
17. Mastronarde DN (2005) Automated electron microscope tomography using robust prediction of specimen movements. J Struct Biol 152(1):36–51. https://doi.org/10.1016/j.jsb.2005.07.007
18. Punjani A, Rubinstein JL, Fleet DJ, Brubaker MA (2017) cryoSPARC: algorithms for rapid unsupervised cryo-EM structure determination. Nat Methods 14(3):290–296. https://doi.org/10.1038/nmeth.4169
19. Pollock NL, Rai M, Simon KS et al (2019) SMA-PAGE: a new method to examine complexes of membrane proteins using SMALP nano-encapsulation and native gel electrophoresis. Biochim Biophys Acta Biomembr 1861(8):1437–1445. https://doi.org/10.1016/j.bbamem.2019.05.011
20. Voskoboynikova N, Margheritis EG, Kodde F et al (2021) Evaluation of DIBMA nanoparticles of variable size and anionic lipid content as tools for the structural and functional study of membrane proteins. Biochim Biophys Acta Biomembr 1863(6):183588. https://doi.org/10.1016/j.bbamem.2021.183588
21. Oluwole AO, Danielczak B, Meister A et al (2017) Solubilization of membrane proteins into functional lipid-bilayer nanodiscs using a diisobutylene/maleic acid copolymer. Angew Chem Int Ed Engl 56(7):1919–1924. https://doi.org/10.1002/anie.201610778
22. Esmaili M, Brown CJ, Shaykhutdinov R et al (2020) Homogeneous nanodiscs of native membranes formed by stilbene-maleic-acid copolymers. Nanoscale 12(32):16705–16709. https://doi.org/10.1039/d0nr03435e
23. Ravula T, Hardin NZ, Di Mauro GM, Ramamoorthy A (2018) Styrene maleic acid derivates to enhance the applications of bio-inspired polymer based lipid-nanodiscs. Eur Polym J 108:597–602. https://doi.org/10.1016/j.eurpolymj.2018.09.048
24. Fiori MC, Jiang Y, Altenberg GA, Liang H (2017) Polymer-encased nanodiscs with improved buffer compatibility. Sci Rep 7(1):7432. https://doi.org/10.1038/s41598-017-07110-1
25. Kopf AH, Lijding O, Elenbaas BOW et al (2022) Synthesis and evaluation of a library of alternating amphipathic copolymers to solubilize and study membrane proteins. Biomacromolecules 23(3):743–759. https://doi.org/10.1021/acs.biomac.1c01166
26. Ball LE, Riley LJ, Hadasha W et al (2021) Influence of DIBMA polymer length on lipid nanodisc formation and membrane protein extraction. Biomacromolecules 22(2):763–772. https://doi.org/10.1021/acs.biomac.0c01538
27. Morrison KA, Akram A, Mathews A et al (2016) Membrane protein extraction and purification using styrene-maleic acid (SMA) copolymer: effect of variations in polymer structure. Biochem J 473(23):4349–4360. https://doi.org/10.1042/BCJ20160723
28. Swainsbury DJK, Scheidelaar S, Foster N et al (2017) The effectiveness of styrene-maleic acid (SMA) copolymers for solubilisation of integral membrane proteins from SMA-accessible and SMA-resistant membranes. Biochim Biophys Acta Biomembr 1859(10):2133–2143.

https://doi.org/10.1016/j.bbamem.2017.07.011

29. Hawkins OP, Jahromi CPT, Gulamhussein AA et al (2021) Membrane protein extraction and purification using partially-esterified SMA polymers. Biochim Biophys Acta Biomembr 1863(12):183758. https://doi.org/10.1016/j.bbamem.2021.183758
30. Pollock NL, Lee SC, Patel JH, Gulamhussein AA, Rothnie AJ (2018) Structure and function of membrane proteins encapsulated in a polymer-bound lipid bilayer. Biochim Biophys Acta Biomembr 1860(4):809–817. https://doi.org/10.1016/j.bbamem.2017.08.012
31. Scheidelaar S, Koorengevel MC, van Walree CA et al (2016) Effect of polymer composition and pH on membrane Solubilization by styrene-maleic acid copolymers. Biophys J 111(9):1974–1986. https://doi.org/10.1016/j.bpj.2016.09.025
32. Esmaili M, Acevedo-Morantes C, Wille H, Overduin M (2020) The effect of hydrophobic alkyl sidechains on size and solution behaviors of nanodiscs formed by alternating styrene maleamic copolymer. Biochim Biophys Acta Biomembr 1862(10):183360. https://doi.org/10.1016/j.bbamem.2020.183360
33. Dörr JM, Scheidelaar S, Koorengevel MC et al (2016) The styrene-maleic acid copolymer: a versatile tool in membrane research. Eur Biophys J 45(1):3–21. https://doi.org/10.1007/s00249-015-1093-y
34. Burridge KM, Harding BD, Sahu ID et al (2020) Simple derivatization of RAFT-synthesized styrene-maleic anhydride copolymers for lipid disk formulations. Biomacromolecules 21(3):1274–1284. https://doi.org/10.1021/acs.biomac.0c00041
35. Yoder N, Gouaux E (2020) The his-Gly motif of acid-sensing ion channels resides in a reentrant 'loop' implicated in gating and ion selectivity. elife 9:e56527. https://doi.org/10.7554/eLife.56527

Chapter 5

Ready-to-Record Cells for Kinetic Screening of VGICs

Emmanuelle Logette and Rajnish Ranjan

Abstract

Voltage-gated ion channels (VGICs) are integral membrane proteins crucial for transmitting electrical signals in excitable cells. Understanding the kinetics of these ion channels requires conducting patch-clamp experiments using genetically modified cell lines that express a single type of ion channel gene. However, this process relies on the continuous maintenance of cell lines to ensure an adequate supply of sample cells for patch-clamp experiments. Advancements in automated patch-clamp methods have enabled researchers to significantly increase the number of patch-clamped cells per experiment, from just a few cells to as many as 384 cells. Despite this progress, the manual task of preparing the cell samples remains a significant bottleneck in the kinetic screening of VGICs. Here we describe a method to address this challenge by generating ready-to-record (RTR) VGIC-expressing cells that can be frozen and stored separately from patch-clamp experiments. This decoupling of the cell sample preparation process from the patch-clamp experiments offers a streamlined approach to studying VGICs on manual or an automated patch-clamp system.

Key words CHO cells, Patch-clamp, Voltage-clamp, Automated patch-clamp, Kinetic screening

1 Introduction

Voltage-gated ion channels (VGICs) are crucial integral membrane proteins that facilitate the passage of ions across cell membranes, playing a pivotal role in generating and transmitting electrical signals in excitable cells, such as neurons and muscle cells. To characterize the kinetics of VGICs, researchers primarily rely on patch-clamp experiments, particularly voltage-clamp, performed on cell lines that have been genetically modified to express a single type of ion channel gene [1].

To achieve efficient kinetic screening of VGICs, a high-throughput patch-clamp method and easily accessible ion channel cell lines are essential. Advancements in automated patch-clamp techniques have made it possible to record data from as many as 384 cells in a single experiment [2, 3]. However, the cell line maintenance and cell preparation process for high-throughput

Simone Furini (ed.), *Potassium Channels: Methods and Protocols*, Methods in Molecular Biology, vol. 2796,
https://doi.org/10.1007/978-1-0716-3818-7_5,

patch-clamping remains a manual task and represents the true limiting factor for the kinetic screening of VGICs. Additionally, standardized cell line generation is essential to ensure meaningful and reliable data for the screening process.

Here we describe steps for producing ready-to-record (RTR) VGIC-expressing cells under standardized conditions, facilitating electrophysiological recordings. The foundation of this approach relies on the Flp-In™-T-Rex™ expression system to create a library of isogenic expressing cell lines. Furthermore, we develop a process to prepare standardized frozen "ready-to-record" vials, suitable for both manual and automated patch-clamp recordings. Traditionally, patch-clamp experiments rely on continuously maintained cell lines for sample production, but proposed method separates these tasks through the generation of RTR cells that can be produced independently and utilized on-demand with manual or automated patch-clamp systems. This approach promises to streamline and enhance the efficiency of patch-clamp experiments for ion channel kinetic characterization and kinetic screening.

The Flp-In™-T-Rex™ expression system was selected as the ideal system for ion channel heterologous expression and electrical recording. Detailed methods for the generation of the CHO Flp-In™–T-Rex™ ion channel-expressing cell lines (i.e., gene amplification, cloning, host cell transfection, selection, and validation) are available in our previous study [4]. However, the key principles of this system are explained below:

The *Flp-In™ system* [5] is a genetic tool for site-specific integration of an exogenous gene in mammalian cells. It is based on the Flp recombinase enzyme, which recognizes specific DNA sequences called Flp recognition target (FRT) sites. The Flp-In system allows stable integration of a gene of interest into a predetermined genomic locus in the host cell's genome. This leads to the generation of isogenic cells preventing any off-target effects of exogenous gene insertion, hence allowing high standardization in gene expression.

The *T-Rex™ system* [6] is a tetracycline-regulated expression system for mammalian cells. Briefly, the expression of the gene of interest is repressed in the absence of tetracycline and induced, on-demand, in the presence of tetracycline. This "on-demand" expression system has two main advantages; it avoids potential toxicity of the constitutive expression of some ion channels, and it allows a standardization in the gene expression, even so the expression may vary from one gene to another, which cannot be controlled (half-life of the gene or the protein for example).

Multiple host cells, such as HEK, CV1, and HeLa, are commonly used for studying ion channels. However, Chinese hamster ovary (CHO) cells offer several distinct advantages for these studies. CHO cells are highly efficient for expressing heterologous proteins, making them ideal for ion channel research.

Additionally, CHO cells are highly compatible for whole cell patch-clamp techniques and exhibit a low background of voltage-gated current, enhancing the accuracy of ion channel measurements in these cells.

Moreover, ion channel genes of interest were cloned in their native form, including start and stop codons, without the addition of any tags. This approach was chosen to avoid potential alterations in the protein's conformation and activity that could result from the presence of tags.

RTR cells: to simplify the maintenance of a large number of ion channel cell lines for kinetic screening, we have developed a method to generate RTR cells. In this approach, recombinant Flp-In™–T-Rex™ are induced with tetracycline for ion channel expression and then frozen in a manner that allows for rapid preparation within 10 min before performing patch-clamp experiments. The following sections provide a comprehensive guide on the preparation of these on-demand RTR cell vials and their utilization for electrophysiological recording. This method offers a practical solution, ensuring the immediate availability of cells for kinetic screening without the complexities associated with maintaining multiple live cell lines.

2 Materials

2.1 Consumables

1. 2.0 mL cryotubes.
2. 60 cm^2 tissue culture dishes.
3. 14 mL conical PP centrifuge tube.
4. 50 mL conical PP centrifuge tube.
5. CoolCell freezing container.
6. Pasteur glass pipette.
7. Brown tubes 1.5 mL (autoclaved beforehand).
8. Black tubes 0.5 mL (autoclaved beforehand).

2.2 Chemical Products and Antibiotics

1. DMSO: liquid, ready to use; keep at room temperature.
2. 50 mg/mL hygromycin B: liquid, ready to use; keep at 4 °C and protect from light.
3. 50 mg blasticidin S HCl: To create a 10 mg/mL stock concentration, first, resuspend the powder in a tube containing 1 mL of nanopure water. Vortex the tube and let it rest at room temperature for 30 min. Vortex the tube again before filling it up to 5 mL with nanopure water, using sterile tips. Finally, aliquot the solution into 1.5 mL tubes, each containing 500 μL of the solution. To preserve the solution, keep it at −20 °C. It is crucial to work in a fume hood due to toxicity, and when

opening the vial, take care to prevent any powder from escaping the bottle.

4. 5 g doxycycline hyclate: To prepare a working solution of doxycycline at 1 μg/μL, first, take 0.5 g of doxycycline and resuspend it in 10 mL of nanopure water. Mix the solution well and prepare aliquots of 250 μL into 1.5 mL brown tubes. Keep the 50 mg/mL stock solution aliquots at −20 °C. To make the 1 μg/μL working solution, add 25 μL of the 50 mg/mL stock solution to 1225 μL of nanopure water and mix well. Next, aliquot 50 μL of the working solution into 0.5 mL black tubes and store them at −20 °C, protected from light. When ready to use, dilute 10 μL of the working solution in 10 mL of free culture medium to get a final concentration of 1 μg/mL.

2.3 Culture Media

1. DPBS 1× (without $CaCl_2$ and $MgCl_2$): store at 4 °C. However, once opened, it can be stored at room temperature.
2. RPMI-1640: keep at 4 °C.
3. Fetal calf serum (FCS): store at −20 °C. Aliquot in 50 mL Falcon tube and keep aliquots at −20 °C; once opened, they can be stored at 4 °C.
4. 0.05% trypsin–EDTA (1×): store at −20 °C. Aliquot in 50 mL Falcon tube and keep aliquots at −20 °C; once opened, they can be stored at 4 °C.
5. CHO complete medium (for CHO-FT-IC cell lines maintenance): RPMI-1640 + 5% FCS + 10 μg/mL blasticidin S HCl + 500 μg/mL hygromycin B, to be conserved at 4 °C.
6. CHO free medium (for CHO-FT-IC cell lines induction): RPMI-1640 + 5% FCS, to be conserved at 4 °C.
7. RTR freezing medium: RPMI-1640 + 5% FCS + 10% DMSO, to be freshly prepared and kept cold before use.
8. Diluted trypsin: when indicated, trypsin is used diluted 1:1 in DPBS 1x (the same volume of DPBS 1x and trypsin is added in a 50 mL Falcon tube and kept at 4 °C).

2.4 Equipment

1. Manual or automated patch-clamp system.
2. Cell culture incubator.
3. Benchtop centrifuge for 14 mL centrifuge tube.
4. Water bath.
5. Automated cell counter (optional).
6. Optical microscope (optional).
7. Cell culture vacuum aspiration system (optional).

3 Methods

In-house-generated CHO Flp-In™–T-Rex™ cells expressing any ion channel in a tetracycline-inducible manner [4], are referred to as CHO-FT-IC, where IC is the overexpressed gene of interest. The method below details the procedure to prepare RTR vials of CHO-FT-IC cells from cells in culture, to allow on-demand recordings of any VGIC on a patch-clamp system.

3.1 Cell Maintenance (20 Min)

1. CHO-FT-IC cells can be cultivated in 60 cm^2 culture plates in CHO complete medium. Cells should be passaged when they achieve 80–90% confluency, typically occurring twice a week. Prior to beginning the procedure, ensure trypsin, DPBS, and the complete medium are acclimated to room temperature.
2. Discard the culture medium from the plate, and then carefully add 5 mL of DPBS onto the plate. Handle the plate gently to rinse the cells. Remove the DPBS and repeat the rinse step once more for thorough cell cleansing.
3. Discard DPBS and add 2 mL of trypsin to the plate. Exercise gentle handling of the plate to ensure all cells come into contact with trypsin. Incubate the cells for 2 min at 37 °C.
4. Use a microscope to verify that cells have detached from the plate. Proceed by resuspending the detached cells in 8 mL of CHO complete medium, and then transfer this suspension to a 14 mL conical centrifuge tube.
5. Take 10 μL apart from the resuspension for cell counting (*see* **Note 1**).
6. Centrifuge the cell suspension at 180 g for 3 min.
7. Post centrifugation, carefully discard the supernatant.
8. Resuspend the cell pellet in an adequate volume of CHO complete medium to achieve a final cell concentration of 1 million cells/mL.
9. Transfer 1 mL of the cell resuspension into a fresh 60 cm^2 culture plate, and then add 9 mL of fresh CHO complete medium.
10. Prepare an adequate number of plates to meet your maintenance needs. Ensure each plate is clearly labeled with the cell line's name, ID, and passage number, alongside the date of the passage. Note that a passage refers to a trypsinization step.

3.2 RTR Vial Preparation (3 Days)

Cells are plated on day 1 (20 min), induced on day 2 (5 min), and freeze as RTR vials on day 3 (10 min).

3.2.1 Day 1: Plating of CHO-FT-IC Cells

1. Depending on the number of RTR vials required, prepare an equivalent number of 60 cm^2 plates, keeping in mind that each RTR vial is used for a single-shot experiment.
2. Plate the CHO-FT-IC cells at a density of 1 million cells on 60 cm^2 plastic culture plates following **steps 1–7** as detailed in Subheading 3.1 (*see* **Note 2**), with the only difference that cells have to be plated in CHO-free medium.
3. Incubate the cells at 37 °C for 24 h.

*3.2.2 Day 2: Tetracycline Induction (see **Note 3**)*

Ensure the CHO-free medium is at room temperature before proceeding. Remove the culture medium from the plate of interest and replace it with 10 mL of CHO-free medium containing 10 μL of tetracycline at a concentration of 1 μg/μL (*see* **Notes 4** and **5**). Incubate the cells at 37 °C for an additional 24 h.

3.2.3 Day 3: Freezing of RTR Vials

1. Prepare the required volume of RTR freezing medium (*see* **Note 6**), accounting for 2 mL per RTR vial. Make sure to have an empty CoolCell freezing container, pre-cooled to 4 °C, ready for use.
2. 24 h post-tetracycline treatment, remove the medium containing tetracycline (*see* **Note** 7), and rinse the cells twice with 5 mL DPBS as outlined in **step 1** of Subheading 3.1.
3. After removing as much DPBS as possible, add 1.8 mL of diluted trypsin (*see* subheading 2.3). Gently agitate the plate to ensure all cells are fully coated with diluted trypsin, and then incubate for no more than 1 min at 37 °C (*see* **Note 8**).
4. This step is critical. Carefully incline the plate to pool the trypsin at the bottom of the plate. Then, carefully aspirate the trypsin using a thin glass pipette, removing as much trypsin as possible without aspirating the cells (*see* **Note 9**) (*see* Fig. 1).
5. Cells remaining on the plastic are directly resuspended in 2 mL COLD RTR freezing medium using a 1000 μL micropipette and then transferred into a properly labeled cryogenic tube.
6. Label the vial with the cell line, passage number, and date of freezing, and store it at −80 °C (*see* **Note 10**).

3.3 RTR Vial Defrosting for Patch-Clamp Recordings

Ensure the CHO-free medium is pre-warmed to 37 °C before proceeding.

1. Take out the RTR vial of interest from −80 °C storage (or from the liquid nitrogen tank) and place it on dry ice.
2. To quickly defrost the cells, place the vial in a 37 °C water bath and agitate manually until thawed (*see* **Note 11**).
3. Using a 1000 μL micropipette, resuspend the cells from the vial and transfer them into a 50 mL Falcon tube containing 20–25 mL of pre-warmed (37 °C) CHO-free medium.
4. Centrifuge the cells at 180 g for 3 min.

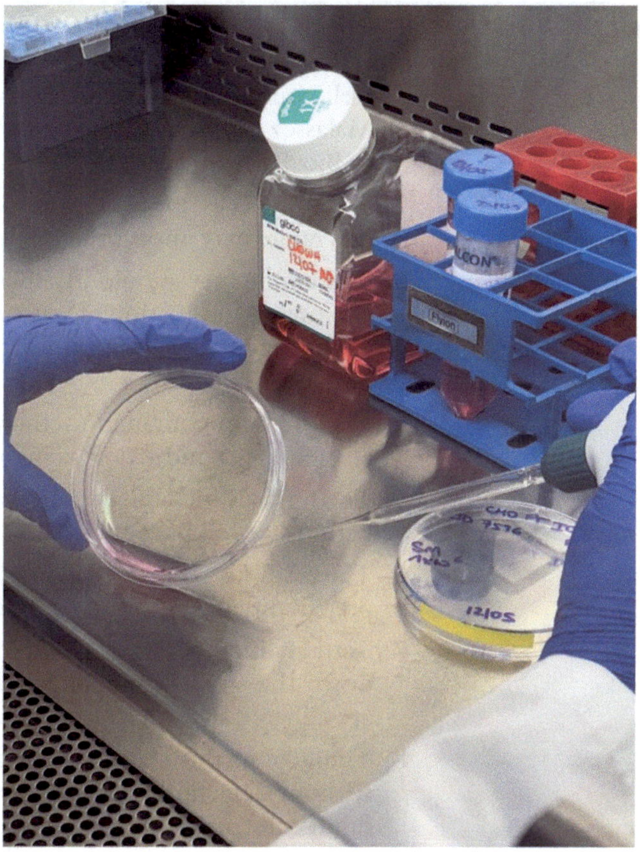

Fig. 1 Removing trypsin from plate

5. Aspirate as much of the supernatant as possible using a glass pipette, ensuring careful handling.
6. Resuspend the cell pellet in the recording medium (*see* **Note 12**).
7. For automated patch-clamp (APC) recording: If cells are being recorded with the Nanion Patchliner, resuspend them in HBSS at room temperature, adjusting the cell concentration to between 0.8 and 1 million cells/ml (*see* **Note 13**). Follow the APC robot protocol for recording.
8. For manual patch-clamp recording: Follow **steps 1–5** as mentioned above, and then resuspend the cell pellet in 4 mL of CHO-free medium. Next, four round coverslips (18 mM diameter) coated with poly-D-lysine (50 μg/mL) are placed in a 6 cm petri dish, and the resuspended cells are added to the dish. The petri dish is then incubated at 37 °C for 1 h to allow the cells to adhere to the coverslips. It is important to verify that the cells have firmly adhered before proceeding with the patching process. Follow the manual patch-clamp protocol for recording.

4 Notes

1. Cell counting may be performed either manually, utilizing a Malassez chamber, or through an automated cell counter such as Bio-Rad, adhering to the manufacturer's guidelines.
2. Typically, cells are used from passages 5–15. Passages beyond this range may impact the cell line division or general behavior.
3. Cell confluency should ideally range between 40% and 60% at the time of induction. Deviations above or below this range could potentially compromise cell survival or the effectiveness of gene induction.
4. The medium containing tetracycline must be freshly prepared at the time of induction as indicated is subheading 2.2.
5. The tetracycline concentration may require adjustment for use with other cell lines than CHO.
6. The ready-to-record (RTR) freezing medium should be freshly prepared before use and kept ***cold*** on ice.
7. Tetracycline, being an antibiotic, should not be discarded down the sink. Ensure appropriate waste disposal procedures are adhered to for tetracycline waste.
8. The goal at this stage is to ensure cells become less adherent without detaching from the plastic. To achieve this, trypsin is diluted 1:1 in DPBS and applied no longer than 1 min. Use a microscope to confirm that cells have adopted a rounded shape without detaching from the plastic.
9. This step should be executed swiftly and with precision to ensure the cells remain adhered to the plastic. It is therefore recommended to work with no more than two plates simultaneously.
10. Utilize a CoolCell freezing container for safe freezing. RTR vials remain stable for up to 1 year at −80 °C. For longer storage periods, storage in liquid nitrogen is advised.
11. Given that DMSO is toxic, removing the vial from the water bath is recommended before the cell suspension is completely melted, i.e., when a small ice cube is still visible in the vial.
12. The recording medium's composition, volume, and temperature must be tailored according to the experiment and the specific patch-clamp recording method employed.
13. Cell concentrations between 0.5 and 0.8 million cells/mL are acceptable. Variations outside of this range could impact the success rate or efficiency of recordings.

Acknowledgments

This study was supported by funding to the Blue Brain Project, a research center of the École polytechnique fédérale de Lausanne (EPFL), from the Swiss government's ETH Board of the Swiss Federal Institutes of Technology.

References

1. Sakmann B, Neher E (1984) Patch clamp techniques for studying ionic channels in excitable membranes. Annu Rev Physiol 46:455–472
2. Annecchino LA, Schultz SR (2018) Progress in automating patch clamp cellular physiology. Brain Neurosci Adv 2:2398212818776561
3. Li T, Lu G, Chiang EY et al (2017) High-throughput electrophysiological assays for voltage gated ion channels using SyncroPatch 768PE. PLoS One 12:e0180154
4. Ranjan R, Logette E, Marani M et al (2019) A kinetic map of the homomeric voltage-gated potassium channel (Kv) family. Front Cell Neurosci 13:358
5. O'Gorman S, Fox DT, Wahl GM (1991) Recombinase-mediated gene activation and site-specific integration in mammalian cells. Science 251:1351–1355
6. Yao F, Svensjö T, Winkler T et al (1998) Tetracycline repressor, tetR, rather than the tetR-mammalian cell transcription factor fusion derivatives, regulates inducible gene expression in mammalian cells. Hum Gene Ther 9:1939–1950

Chapter 6

Cell-Based Thallium-Influx Fluorescence Assay for Kv10.1 Channels

María Luisa Durán-Pastén and Enoch Luis

Abstract

The development of cell-based fluorescent assays has resulted in an incredible tool for searching new ion channels' modulators with a biophysical and clinical profile. Among all the ion channels, potassium (K^+)-permeable channels represent the most diverse and relevant for cell function, making them attractive targets for drug discovery. Some of the cell-based assays for K^+ channels take advantage of a thallium-sensitive dye whose fluorescence increased upon the binding of thallium (Tl^+), an ion able to move through K^+ channels. We optimize the FLIPR Potassium Assay Kit based on thallium influx to measure the Kv10.1 activity.

Key words Thallium influx, Cell-based fluorescence, Voltage-gated ion channels, Kv10.1, Oncochannel, Cancer, Microplate reader

1 Introduction

Ion channels are fascinating transmembrane proteins expressed in all mammal cells, thus involved in practically all cellular phenomena [1]. Several human diseases (channelopathies) originate from an anormal function or ectopic expression of ion channels, making these proteins attractive pharmacological targets [2]. Most drug discovery assays focused on searching novel ion channel modulators have been obtained from electrophysiological experiments using the patch-clamp technique, making them reliable but time-consuming [3, 4]. However, substantial efforts have been made to develop new tools to speed up ion channel drug discovery. Among them, the cell-based fluorescent assays combined with new high-throughput platforms have become powerful tools for the rapid and reliable screening of active molecules targeting ion channels [4, 5].

Potassium-permeable ion channels are the most extensive and structurally diverse subfamily of ion channels [6]. These channels influence a plethora of functions involving the setting of the resting

Simone Furini (ed.), *Potassium Channels: Methods and Protocols*, Methods in Molecular Biology, vol. 2796,
https://doi.org/10.1007/978-1-0716-3818-7_6,

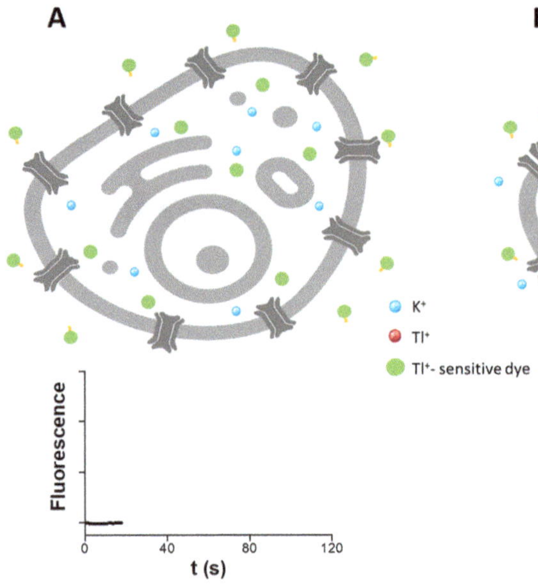

Fig. 1 Principle of the FLIPR® Potassium Assay Kit. (**a**) Scheme of the thallium-influx assay in HEK-Kv10.1 cells in the control condition. Below is a fluorescence recording before adding the stimulus. (**b**) HEK-Kv10.1 cells are stimulated with a solution containing 30 mM K_2SO_4/2 mM Tl_2SO_4. High potassium solution depolarizes the resting membrane potential, which triggers the Kv10.1 channel opening; then, the Tl^+ moves into the cells through the Kv10.1 channel. Upon inside the cell, Tl^+ binds to the Tl^+-sensitive dye, proportionally increasing the fluorescence signal. Below is the fluorescence recording after adding the stimulus solution (red arrow)

membrane potential, action potential shape, cell volume, cell cycle, etc. [6]. Among all K^+ channels, Kv10.1 is a well-recognized channel involved in carcinogenesis [7, 8]. Currently, it was the first ion channel described as an oncochannel [7, 9]. It has been reported that the Kv10.1 channel is overexpressed in around 70% of cancer samples, in contrast to a low expression in healthy tissues [8, 10]; their transmembrane and high expression profile in cancer samples make Kv10.1 an exciting target for drug discovery focus on new anti-cancer therapies.

Here, we describe the protocol implemented to study the functional activity of Kv10.1 channels through the FLIPR® Potassium Assay Kit from Molecular Devices. Briefly, this assay takes advantage of the thallium (Tl^+) permeability through open K^+ channels, and its influx into cells is measured using a thallium-sensitive fluorescent dye [4]. The dye, confined in the cytosol, increases its fluorescence intensity upon binding to Tl^+ (moving inside the cell through the Kv10.1 channels). Thus, the fluorescence intensity signal is proportional to the number of Kv10.1 channels activated by a depolarizing stimulus (*see* Fig. 1).

2 Materials

2.1 Cell Culture and Maintenance of Cells

1. HEK293 cells stably expressing the human Kv10.1 channel (*see* **Notes 1–2**).
2. Dulbecco's Modified Eagle Medium (DMEM) with GlutaMAX™ and high glucose supplemented with 10% fetal bovine serum; 1% penicillin–streptomycin antibiotic solution with a formulation containing 10,000 units of penicillin and 10 mg streptomycin/ml; and 30 μg/mL zeocin.
3. 0.25% trypsin–EDTA 1× solution.
4. Cell culture incubator at 37 °C, 5% CO_2, and 95% humidity.
5. Cell culture flasks 25 cm^2 surface treated for adherent cells.
6. Cell culture equipment: inverted microscope, biosafety cabinet, vortex, and centrifuge.
7. Serological pipettes of 5, 10, and 50 mL.
8. Water purification system.
9. 96-well, cell culture-treated, flat-bottom microplates.
10. 96-well, cell culture-treated, V-shaped-bottom microplates.

2.2 Thallium-Influx Kit Assay

1. High-throughput screening (HTS) assay. There are several HTS assays, mostly containing the following elements: dye solution, assay buffer, chloride-free assay buffer, DMSO, potassium sulfate (K_2SO_4) solution, and thallium sulfate (Tl_2SO_4) solution. In this case, FLIPR® Potassium Assay Kit from Molecular Devices® was used, which includes the following three components: component A (dye—need to be stored at −20 °C and thaw before using); component B (assay buffer for evaluation kits and explorer kits): 20 mM HEPES buffer +1× Hank's Balanced Salt Solution (HBSS), pH 7.4; and component C (store at −20 °C and thaw before using).
2. 5× chloride-free assay buffer (5×). To prepare the 1x chloride-free buffer, add 1 mL of the 5× chloride-free assay buffer in 4 mL of deionized water to obtain a 1× solution.
3. 200 mM potassium sulfate (K_2SO_4) stock solution in water.
4. 50 mM thallium sulfate (Tl_2SO_4) stock solution in water (*see* **Note 7**).
5. 3× solutions of 1× chloride-free buffer and a combination of 0, 10, 20, and 30 mM K_2SO_4 (final concentration) and 1, 2, and 3 mM of Tl_2SO_4 (final concentration) as shown in Table 1. These solutions must be prepared at 3× since when the stimulus (50 μL) is added to each well, the Tl_2SO_4 and the K_2SO_4 will be diluted 1:3.

Table 1
Optimization of Kv10.1 channel stimulant

Tl_2SO_4 (mM)	K_2SO_4 (mM)
1	0 10 20 30
2	0 10 20 30
3	0 10 20 30

6. Dimethyl sulfoxide (DMSO).
7. Pipette tips gaskets.
8. Microplate reader.

3 Methods

Prepare all the media and solutions with deionized and purified water. Store media and solutions at 4 °C.

3.1 Cell Cultures

Seed HEK-Kv10.1 cells in flat-bottom 96-well microplates at a density of 20,000 cells per well with 100 μL of supplemented DMEM, and incubate overnight at 37 °C in a 5% CO_2 incubator.

3.2 FLIPR Potassium Assay Kit

The FLIPR Potassium Assay Kit is a fluorescence assay that allows rapid measurement of potassium channel activity. This kit contains different components, which must be prepared before use, as stated below.

3.2.1 Prepare Kit Components

1. Place components A, B, and C in a water bath at 37 °C.
2. Dissolve the content of component C in 30 μL DMSO, and sonicate for 3 min.
3. Use a sterile serological pipette, and add 10 mL of component B to the vial of component A. Vortex for 1 min.
4. Add component C to component A (prepared in the previous step), and mix well in the vortex for 1–2 min until the contents are completely dissolved (protect from light).

Table 2
FlexStation 3 reader: FLIPR® Potassium Assay Kit setup parameters

Parameter	Setting
Read mode	Fluorescence
Wavelength (nm)	Excitation = 485 Emission = 525 Cutoff = 515
Plate	Costar™ clear
PTM and optics	PTM gain: Medium 6 flashes per read
Timing	3 s interval for 201 s
Compound transfer	1 Initial volume 100 μL Pipette high: 50 μL Volume 50 μL Rate 2 μL/s Time point 21 s
Compound source	Costar™ 96 Vbtm 0.3 mL
Pipette tips and layout	Number and place where the tips are placed

5. Add an equal volume (1:1) of this mix plus culture medium to each well, e.g., for 96-well plates, add 50 μL (from prepared vial A) and 50 μL of culture medium. Incubate the plate for 1 h at 37 °C with 5% CO_2.

3.3 Prepare the Stimulus Microplate (Thallium Sulfate Source Plate)

Before testing any drug or toxin, for each potassium channel, it is necessary to optimize the ideal concentrations of Tl_2SO_4 and K_2SO_4 to obtain the maximum fluorescence intensity, making it necessary to titrate both compounds (*see* **Note 6**).

1. While the cells are loading with dye, prepare a solution containing K_2SO_4 plus Tl_2SO_4 in 1× chloride-free buffer (*see* **step 2.2**).
2. Load the stimuli solution in a 96-well V-shaped-bottom microplate (*see* **Note 8**).
3. After 1 h of incubation, place the microplate with the stained cells, the microplate with the stimuli, and the tips in the microplate reader (*see* **Note 9**).
4. Set the SoftMax Pro software parameters as in Table 2. Once the data is obtained, it is necessary to make a fluorescence correction in order to subtract the basal fluorescence generated by Tl_2SO_4. For the Kv10.1 channel, the maximum

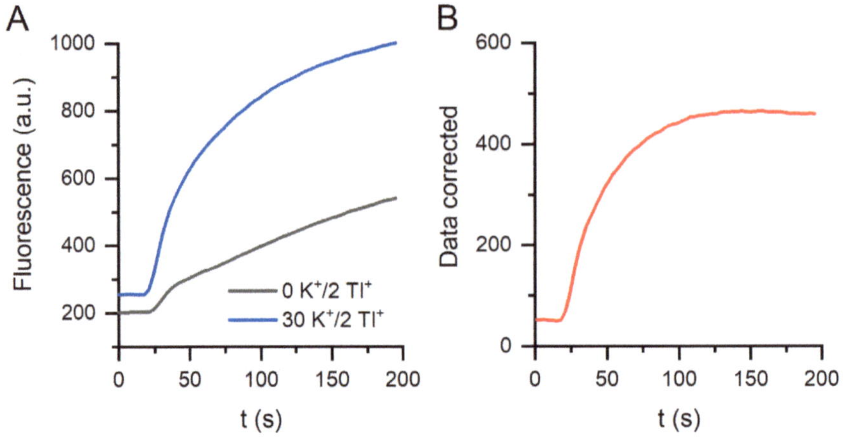

Fig. 2 FLIPR recording in HEK-Kv10.1 cells. (**a**) Typical fluorescence recordings in HEK-Kv10.1 cells induced by 30 mM K_2SO_4/2 mM Tl_2SO_4 (blue line) and by 0 mM K_2SO_4/2 mM Tl_2SO_4. The stimulus is added at 21 s. (**b**) The fluorescence signal is corrected by subtracting the basal fluorescence generated by thallium (0 mM K_2SO_4) from the maximum fluorescence obtained with 30 mM K_2SO_4

fluorescence obtained was with the combination of 2 mM Tl_2SO_4/30 mM K_2SO_4; therefore, it is necessary to subtract the basal fluorescence generated by thallium (2 mM Tl_2SO_4/ 0 mM K_2SO_4) from the maximum fluorescence obtained with 30 mM K_2SO_4 (*see* Fig. 2).

4 Notes

1. The cells must pass at least two passages after thawing for the correct channel expression. Before realizing the fluorescent screening, perform patch-clamp experiments to ensure HEK-Kv10.1 cells generate Kv10.1-like currents.
2. HEK293-WT cells have been assumed to express endogenous ion channels at extremely low levels, and their influence on exogenously expressed ion channels can be insignificant. However, these endogenous currents can sometimes represent a potential problem characterizing ion channels expressed heterologously [11]. When an endogenous ion current could represent a potential problem in this experiment, it is recommendable to use specific inhibitors of these ionic channels (e.g., TTX, TEA, 4-AP).
3. Loss of cellular adherence is problematic during cell-based assays. Sometimes the exogenously expressed ion channels can modify the adherence properties of the cells. Therefore, it is recommendable to do additional treatments to the 96-well plates to improve cell adhesion, e.g., treat microplates with laminin (1 μg/mL) for 30 min plus poly-L-lysine for another 10 min.

4. Based on our culture conditions, plastic ware, and laboratory-specific reading infrastructure, the given cell concentration has been optimized for the described cell line. Still, it should yield comparable results with other mammalian cell lines.
5. The cells in the culture must not reach >90% confluence.
6. The loading and stimulus buffer, as well as the molecule to test, should be freshly prepared before use.
7. Thallium sulfate and potassium sulfate have two equivalents of cation per mole; thus, their respective cation concentrations will be 2× once added.
8. It may be necessary to contain and dispose of thallium and all solutions prepared containing thallium sulfate as hazardous waste.
9. Probenecid, an inhibitor for the anion exchange protein, might be required for some cell lines to ensure that the dye stays inside the cell and is not pumped back out.

Acknowledgments

This work was supported by the grants SEP-CONACYT CB2017-2018-A1-S-13646.

References

1. Hille B (2001) Chapter 1 introduction. In: Ion channels of excitable membranes. Sinauer Associates, Sunderland, pp 1–22
2. Carbone E, Mori Y (2020) Ion channelopathies to bridge molecular lesions, channel function, and clinical therapies. Pflügers Arch Eur J Physiol 472:733–738
3. Verkhratsky A, Parpura V (2014) History of electrophysiology and the patch clamp. In: Methods in molecular biology, pp 1–19
4. Yu H, Li M, Wang W et al (2016) High throughput screening technologies for ion channels. Acta Pharmacol Sin 37:34–43
5. Loza-Huerta A, Milo E, Picones A et al (2021) Thallium-sensitive fluorescent assay reveals loperamide as a new inhibitor of the potassium channel Kv10.1. Pharmacol Rep 73:1744–1753
6. Taura J, Kircher DM, Gameiro-Ros I et al (2021) Comparison of K+ channel families. In: Handbook of experimental pharmacology, pp 1–49
7. Pardo LA (1999) Oncogenic potential of EAG K+ channels. EMBO J 18:5540–5547
8. Luis E, Anaya-Hernández A, León-Sánchez P et al (2022) The Kv10.1 channel: a promising target in cancer. Int J Mol Sci 23:8458
9. Prevarskaya N, Skryma R, Shuba Y (2018) Ion channels in cancer: are cancer hallmarks Oncochannelopathies? Physiol Rev 98:559–621
10. Hemmerlein B, Weseloh RM, Mello de Queiroz F et al (2006) Overexpression of Eag1 potassium channels in clinical tumours. Mol Cancer 5:41
11. Zhang J, Yuan H, Yao X et al (2022) Endogenous ion channels expressed in human embryonic kidney (HEK-293) cells. Pflugers Arch Eur J Physiol 474:665–680

Chapter 7

Use of Surface Plasmon Resonance Technique for Studies of Inter-domain Interactions in Ion Channels

Purushottam B. Tiwari, Pareesa Kamgar-Dayhoff, Prakriti Tiwari, Maria I. McKillop, and Tinatin I. Brelidze

Abstract

Ion channels are transmembrane proteins essential for cellular functions and are important drug targets. Surface plasmon resonance (SPR) is a powerful technique for investigating protein–protein and protein–small molecule ligand interactions. SPR has been underutilized for studies of ion channels, even though it could provide a wealth of information on the mechanisms of ion channel regulation and aid in ion channel drug discovery. Here we provide a detailed description of the use of SPR technology for investigating inter-domain interactions in KCNH potassium-selective and voltage-gated ion channels.

Key words SPR, Potassium channels, Ion channel gating, Ion channel function, KCNH, EAG, hERG, Kv10.1, Kv11.1

1 Introduction

Ion channels contain multiple domains with diverse functions. For instance, the pore domain provides passageways for ions to cross the membrane down their electrochemical gradient, while the voltage-sensing domain detects changes in membrane potential. Many ion channels are also sensitive to ligands and contain specialized ligand-binding domains. Inter-domain and ligand–ion channel interactions are essential for channel opening and closing (gating). Observing the changes in the inter-domain interactions in the absence and presence of ligands in unmodified full-length channels during gating in their native environment is very challenging, if not impossible, at the present state of the field. Therefore, probing the inter-domain interactions and small molecule ligand (SML)–ion channel interactions under more stripped-down conditions with a variety of complementary methods, including structural, electrophysiology, and fluorescence-based approaches, is essential to gain the understanding of ion channel gating

Simone Furini (ed.), *Potassium Channels: Methods and Protocols*, Methods in Molecular Biology, vol. 2796,
https://doi.org/10.1007/978-1-0716-3818-7_7, © The Author(s) 2024

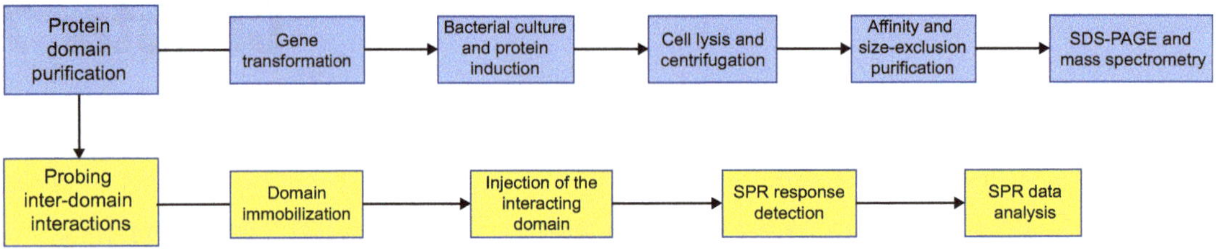

Fig. 1 Flowchart for protein domain purification and SPR experiments for probing inter-domain interactions

mechanisms. Surface plasmon resonance (SPR) is one of the techniques that offers quantitative assessment of the inter-domain interactions and direct SML binding to ion channels that could be further integrated with findings uncovered with other methods. SPR is especially well suited for studies of ion channels with soluble domains that can be easily purified in isolation. As the full-length ion channel purification becomes more mainstream, the application of SPR technique to probing inter-domain and SML–ion channel interactions could be extended to full-length ion channels as well. Here we describe the application of SPR for investigating inter-domain interactions in KCNH potassium-selective and voltage-gated ion channels (the flowchart of steps involved in this approach is shown in Fig. 1).

Similar to other potassium channels, KCNH channels are assembled from four subunits. Each subunit contains an intracellular N-terminal Per-Arnt-Sim (PAS) domain, six transmembrane segments (S1–S6), and an intracellular C-terminal cyclic nucleotide-binding homology (CNBH) domain linked to S6 via the C-linker (*see* Fig. 2a) [1–3]. The transmembrane segments are further subdivided into the voltage-sensor domains formed by S1–S4 segments and centrally located pore domain formed by S5–S6 from all four subunits. The PAS domains interact with the CNBH domains from adjacent subunits (*see* Fig. 2b), and these inter-domain interactions confer many of the functional properties of KCNH channels, including the kinetics of deactivation in KCNH2 channels [3–6]. In addition, PAS and CNBH domains can function as ligand-binding domains in KCNH channels [7–10]. The PAS and CNBH domains of KCNH channels can be purified as isolated proteins, as illustrated in Fig. 2c and described in detail in the Protein Sample Preparation section. The structure of the isolated PAS and CNBH domains and the domains in the intact full-length KCNH channels is very similar [2, 3, 11–15]. Therefore, the findings of SPR-based studies on the isolated PAS and CNBH domains have high relevance for the intact full-length channels.

The SPR technique is frequently used for investigating protein–protein and SML–protein interactions [16–18]. We have used SPR technique to probe interactions between the isolated PAS and CNBH domains of KCNH channels [15], to identify novel SMLs

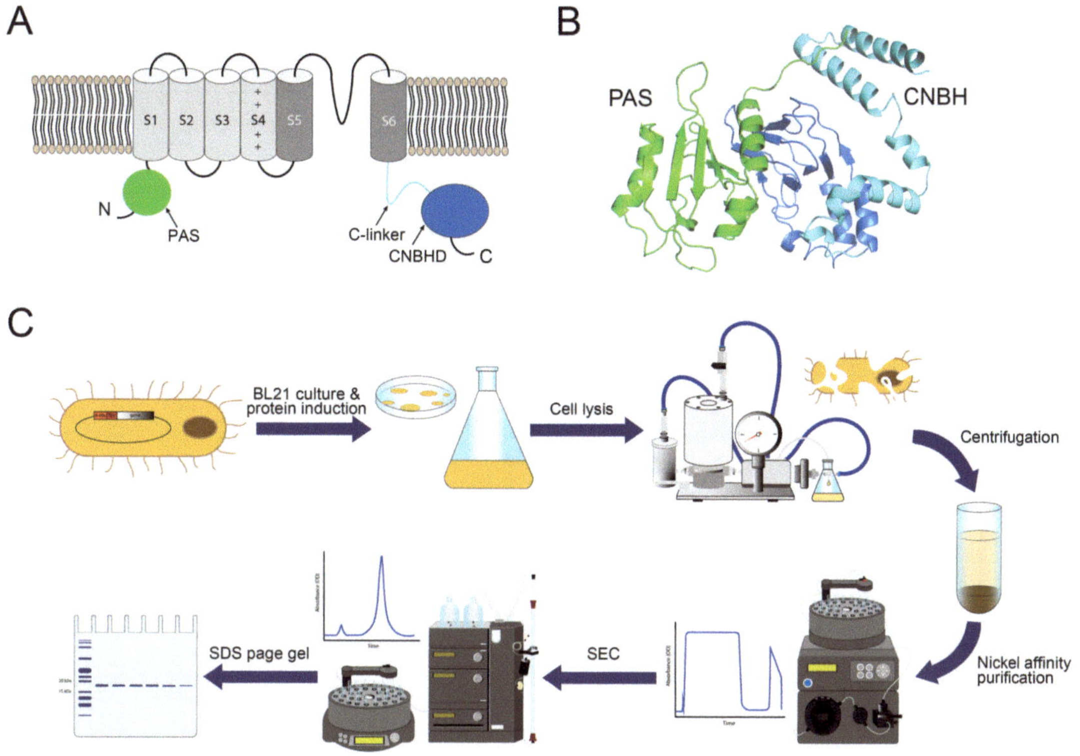

Fig. 2 KCNH channel membrane topology and schematics of PAS and CNBH domain purification. (**a**) Cartoon of a KCNH channel subunit. The PAS domain is green, the transmembrane segments S1–S4 forming the voltage-sensor domain are light gray, the transmembrane domains S5–S6 forming the pore domain are dark gray, the C-linker is cyan, and CNBH is dark blue. (**b**) The ribbon representations of the PAS and CNBH domains from the adjacent subunits of KCNH2 channels (PDB ID: 5VA2). (**c**) Graphic summary of experimental steps involved in the purification of the isolated PAS and CNBH domains of KCNH channels

that bind to these domains using chemical library screening [8], and to further characterize SML–ion channel interactions [8, 9]. SPR technique detects changes in the refractive index for the chip surface with immobilized protein upon binding of a binding partner. In this technology, a metal-coated (typically gold) glass slide is used as a sensor chip. The sensor chip can be modified and activated to immobilize proteins using various surface chemistries [19, 20]. The protein (or other biomolecules) immobilized onto the chip surface is termed as a "ligand" in the SPR technique, and the binding partner, which could be a protein, SML, DNA, or RNA, that flows over the ligand-immobilized surface in solution is called an "analyte" [21]. Since in the ion channel field "ligand" frequently refers to a small molecule binder of ion channels, to avoid the confusion, we will be referring to the immobilized binding partner as an immobilized protein instead of a ligand. Figure 3a, b shows a schematic of the optical and fluidic systems used in the SPR technology. Analyte–ligand binding changes the mass on the

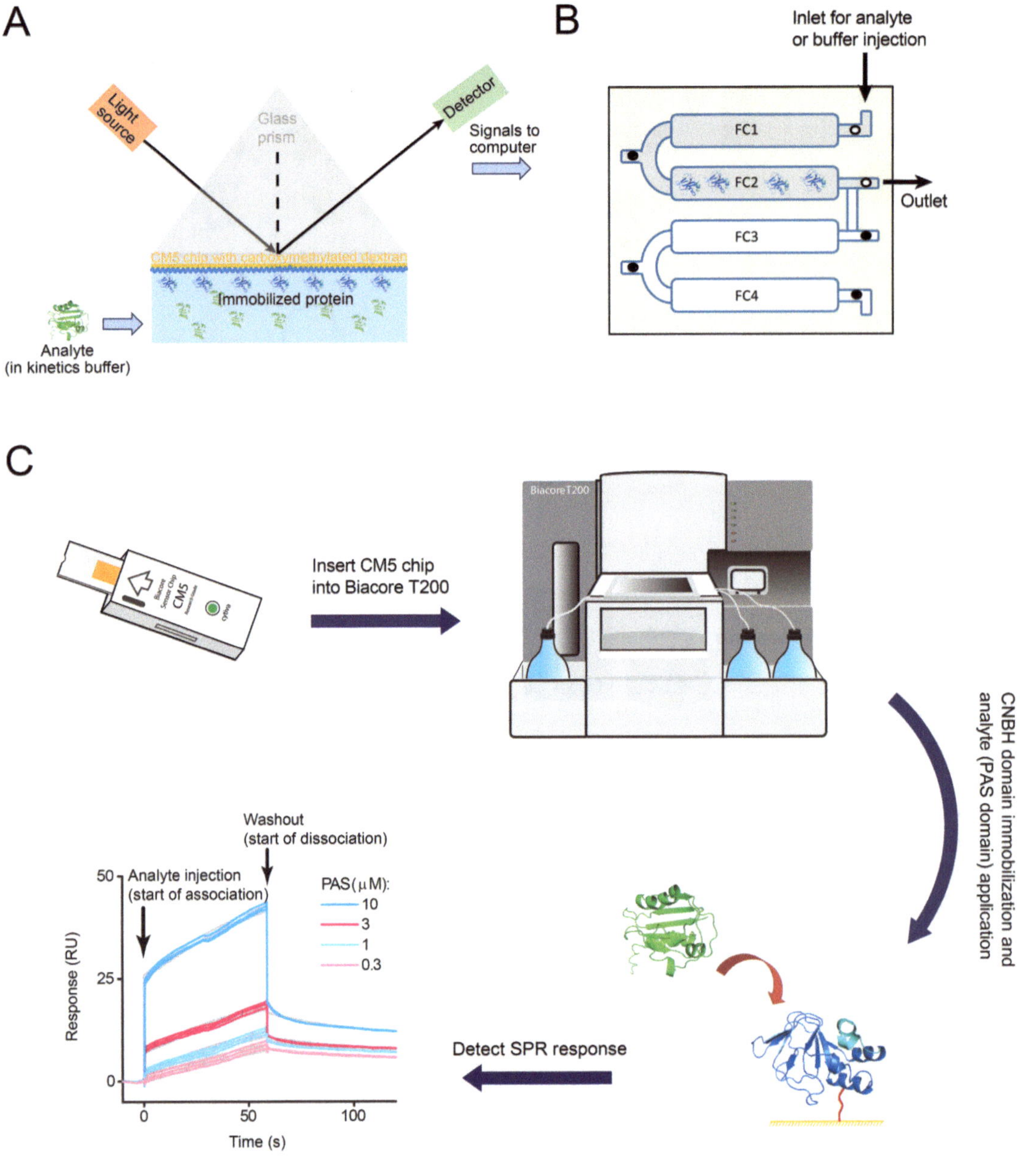

Fig. 3 Schematics of the SPR technique and experiment. (**a**) Schematic of the SPR technique, indicating the incident and reflected light off the sensor chip surface with CNBH domains immobilized on the chip surface and PAS domains applied as free analytes. (**b**) FC arrangement and fluidics of the CM5 sensor chip with the CNBH domain immobilized on FC2 and no protein applied on FC1 used as a reference FC. White circles indicate open valves and black circles indicate closed valves. (**c**) Graphic summary of experimental steps involved in the SPR signal detection for the PAS and CNBH domain binding. The gray lines in the SPR sensorgrams represent fits of the SPR data with the two-state reaction binding model using the Biaevaluation software with the K_d value of 1.4 μM

surface and the refractive index in the vicinity of the metal surface with the immobilized ligand [22–25]. The change in the refractive index is detected as SPR response, with plots of SPR response over time called SPR sensorgrams [26]. The SPR technique detects real-time interactions between two binding partners [27–29]. This technique is especially well-suited for the studies of inter-domain interactions in ion channels because no additional modification of the interacting domains is needed for the use with SPR. In addition, relatively low amounts of the protein sample are sufficient for inter-domain binding detection and determining the affinity of the binding using SPR [30]. Below we provide a detailed description of the SPR-based experiments to probe interactions between the PAS and CNBH domains of KCNH2 channels, also known as ether-á-go-go-related gene (hERG) and Kv11.1 channels.

2 Materials

2.1 Protein Sample Preparation

1. cDNA encoding the PAS domain of KCNH2 channels (GI number 7531135, amino acids 2–134) subcloned into pETM11 bacterial expression vector (*see* **Note 1**).
2. cDNA encoding the C-linker/CNBH domain of KCNH2 channels, referred as CNBH domain for simplicity (GI number 7531135, amino acids 734–867), subcloned into pETM11 bacterial expression vector.
3. BL21 (DE3) cells.
4. Buffer 1: 150 mM KCl, 1 mM tris(2-carboxyethyl)phosphine (TCEP), 10% glycerol, and 30 mM HEPES, pH 7.5 (*see* **Note 2**).
5. Isopropyl β-D-1-thiogalactopyranoside (IPTG).
6. 4-(2-Aminoethyl)benzenesulfonyl fluoride hydrochloride (AEBSF).
7. DNase I.
8. Imidazole.
9. Column for purifying proteins with immobilized metal ion affinity chromatography.
10. Size exclusion column.
11. SDS-PAGE gel (12% Bis-Tris).
12. Coomassie Brilliant Blue.

2.2 Protein Sample Dilutions

1. 96-deep-well plate.
2. Tween 20.
3. Buffer 1: 150 mM KCl, 1 mM TCEP, 10% glycerol, and 30 mM HEPES, pH 7.5.

4. Buffer 2: 150 mM KCl, 1 mM TCEP, 10% glycerol, 0.05% (v/v) Tween 20, and 30 mM HEPES, pH 7.5 (*see* **Note 3**).

2.3 Buffers for Protein Immobilization and SPR Experiments

1. Immobilization buffer: 10 mM sodium acetate, pH 5.5.
2. Immobilization running buffer: 150 mM NaCl, 10 mM HEPES, 0.05% (v/v) surfactant P20, pH 7.4.
3. Kinetics running buffer (the same as Buffer 2): 150 mM KCl, 1 mM TCEP, 10% glycerol, 0.05% (v/v) Tween 20, and 30 mM HEPES, pH 7.5.

2.4 Protein Immobilization

1. SPR equipment (*see* **Note 4**).
2. CM5 sensor chip.
3. Immobilization buffer (see above).
4. Immobilization running buffer (see above).
5. 7 mM plastic vial and seal with rubber cap.
6. 160 uL 1 M ethanolamine (*see* **Note 5**).
7. N-Hydroxysuccinimide (NHS).
8. 1-Ethyl-3-(3-dimethylaminopropyl)carbodiimide hydrochloride (EDC).

2.5 SPR Response Detection

1. Rack tray to handle sample solutions.
2. Glycine pH 2.0.
3. Opto-fluidic system to deliver solutions to chip surface and to monitor change in refractive index on the sensor surface (*see* **Note 6**). A schematic of opto-fluidic system and the steps for the SPR signal detection are shown in Fig. 3a–c.
4. Control software to run immobilization and kinetics (*see* **Note 7**).
5. Evaluation software for analysis (*see* **Note 7**).

3 Methods

3.1 Protein Sample Preparation

1. Grow bacterial cultures transformed with the pETM11 vector containing the PAS or CNBH domains at 37 °C.
2. At OD of 0.6–0.8, induce PAS and CNBH domain expression with 1 mM IPTG.
3. Continue growing bacterial cultures overnight at 18 °C.
4. Harvest bacteria by centrifugation at 5000 rpm for 15 min at 4 °C (*see* **Note 8**).

5. Resuspend the bacterial cell pellet in Buffer 1 supplemented with 1 mM AEBSF and 2.5 mg/mL DNase I.
6. Lyse the cells with a homogenizer (*see* **Note 9**).
7. Separate soluble protein by centrifugation at 30,000 rpm for 1 h at 4 °C.
8. Load the supernatant containing the soluble protein onto an affinity column, and use Ni^{2+}–nitrilotriacetic acid (NTA) chromatography to purify PAS and CNBH domains containing the 6-His tag.
9. Wash the affinity column with Buffer 1 and elute the protein with Buffer 1 supplemented with 500 mM imidazole.
10. Cleave the 6xHis tag with TEV protease (*see* **Note 10**).
11. Equilibrate a size exclusion column with Buffer 1, and further purify the PAS and CNBH domains in Buffer 1 using size exclusion chromatography (*see* **Notes 11 and 12**).
12. Elute the PAS and CNBH domains off the size exclusion column in 500 μL fractions. The fractions containing the domains are initially identified based on the elution profile.
13. Verify the molecular weight of the purified protein, and evaluate the purity of the protein sample with Coomassie Brilliant Blue-stained SDS-PAGE gels (*see* **Note 13**).
14. Mix the elusion fractions containing the PAS domains, and aliquot them in 150 μL volumes. Since the PAS domain will be used as an analyte, these volumes are appropriate for preparing dilutions with different concentrations of the analyte for injections over the immobilized CNBH domains.
15. Mix elusion fractions containing the CNBH domains, and aliquot them in 20–25 μL volumes. Since the CNBH domain will be immobilized on the SPR chip, smaller volumes of the sample are sufficient for immobilization.
16. Determine the concentration of the purified PAS and CNBH domains using Bradford protein assay. The protein concentration after the size exclusion purification is usually sufficient for the SPR experiments.
17. Determine the exact weight of the aliquoted PAS and CNBH domains with mass spectrometry (electrospray).
18. Store the purified PAS and CNBH domains at −80 °C in the small aliquots before use (*see* **Note 14**).

3.2 Protein Sample Dilutions

1. Thaw PAS domain aliquots and add 0.05% (v/v) Tween 20. For instance, add 4 μL of 10% Tween 20 to 800 μL of the PAS domain aliquots (*see* **Note 15**).
2. Prepare PAS domain dilutions at (in μM) 10, 3, 1, 0.3, and 0 concentrations in kinetics running buffer (Buffer 2). The volume for each concentration should be 330 μL.
3. For the serial protein dilutions, it is important to make sure that enough aliquots are thawed to have sufficient volume of the highest protein concentration (10 μM) for the subsequent dilutions (*see* **Notes 16 and 17**). An example of the dilutions for the PAS domain aliquots purified at 50 μM concentration is shown below. While the required volume for each analyte concentration is only 330 μL, the calculations below include extra volumes to accommodate the serial dilutions and also make pipetting easier by using round numbers for volume measurements.

Final PAS concentration	Volume of the aliquots	Buffer 2	Total volume
10 μM	150 μL of PAS aliquots	600 μL	750 μL
3 μM	180 μL of 10 μM PAS	420 μL	600 μL
1 μM	50 μL of 10 μM PAS	450 μL	500 μL
0.3 μM	180 μL of 1 μM PAS	420 μL	600 μL

4. Transfer Buffer 2 and the analyte dilutions into the 96-deep-well plate so that Buffer 2 will be injected over the immobilized protein first, followed by the injection of the analytes from low to high concentrations. A typical arrangement would be Buffer 2 in well A1, with PAS concentrations of 0.3 μM to 10 μM in wells A2–A5, respectively. The volume in each well should be 330 μL.

3.3 Protein Immobilization

1. Dissolve the proteins to be immobilized (CNBH domains) in the immobilization buffer in a 7 mM plastic vial, and seal with the rubber cap.
2. Place the sealed NHS–EDC, protein, and ethanolamine in the sample rack tray, and insert into the machine (*see* **Note 18**).
3. CM5 chip has four flow cells (FCs) (*see* Fig. 3b). Activate CM5 chip surface by injecting NHS–EDC solution onto two neighboring FCs, for example, FC1 and FC2, for 720 s (*see* **Note 19**).

4. Inject protein onto FC2 until desired protein immobilization level is roughly achieved, e.g., for **300** s, while leaving FC1 free of protein injections.
5. Inject ethanolamine solution into FC1 and FC2 for 720 s. This blocks any remaining active surface sites. Here FC1 works as the reference FC for the active FC (FC2) with immobilized proteins onto it.
6. Determine the immobilized protein level by subtracting the signal level just before injection of protein from the signal level after ethanolamine injection is completed. This difference is used to determine the density of the immobilized protein in response units (RU, 1 RU = 1 pg of protein per mm^2) [31]. For the SPR experiments described here, we immobilized ~7000 RU of the CNBH domain of KCNH2 channels.
7. The immobilization running buffer should be running in the background during protein immobilization.

3.4 SPR Response Detection

1. Inject analyte solutions (PAS solutions) at different concentrations at the flow rate of 50 μL/min into both FC1 and FC2 in the kinetics running buffer using the association and dissociation times of 60 s and 150 s, respectively (*see* **Notes 20 and 21**). Make sure to run kinetics buffer-only solution as a blank.
2. Run the kinetics running buffer in the background.
3. Run the method so that the software records data in the format of FC2–FC1. In this form of data collection, the signals corresponding to the reference FC (FC1) are subtracted from the response detected in the active FC (FC2).
4. Inject analyte and blank solutions in three technical replicates in one experiment (we recommend at least duplicate).
5. Inject regeneration solution (in this case glycine pH 2.0) for 15 s at the end of dissociation to remove any remaining analytes on the surface before another cycle of injection of analyte solution starts (*see* **Note 22**).
6. The sensorgrams recorded using the control software can be opened in evaluation software for data analysis. The signals for analysis should be both reference (signals from FC1) and blank (signals contributed by blank) subtracted.
7. Each experiment should be repeated at least three times on three different sensor chips. Therefore, the number of protein aliquots should be determined accordingly. Frequently, it is necessary to purify the domains several times to have sufficient protein to repeat the SPR experiments three times.

3.5 SPR Data Analysis

1. Use an evaluation software (*see* **Note 23**) for the initial data analysis and plotting. The sensorgrams can either be copied from the evaluation software window or raw data can be exported from the evaluation software and replotted using any other plotting software for the subsequent analysis, if desired (*see* **Note 24**).
2. Check if the data can be fitted using a steady-state affinity fitting. If there are very fast association and dissociation phases and the binding reached a plateau, this method can be used [22].
3. If data cannot be fitted using the steady-state binding model, then check whether 1:1 monophasic binding can be used to fit the kinetics data to determine binding parameters, such as the association rate constant (k_a), dissociation rate constant (k_d), and equilibrium dissociation constant (K_d).
4. If data cannot be fitted using 1:1 kinetics binding model, then check which biphasic model can better represent the experimental data. We recommend identifying the correct underlying biphasic binding mechanism as described in the prior publications [32, 33].
5. For the KCNH2 PAS and CNBH domain interactions considered here, the two-step conformational change model was determined as the best model to better predict the experimental data based on the prior publication [32], and this model was used to fit the data to determine the K_d value (*see* **Note 25**) [15].

3.6 Conclusions

Here we described the application of the SPR technique to investigating PAS and CNBH domain interactions in KCNH2 channels. This approach could be used to investigate the interactions between PAS and CNBH domains both in the absence and presence of SMLs. Importantly, many disease-associated mutations in KCNH channels are located at the PAS and CNBH domain interaction interface [2, 5, 34, 35], and the SPR technique could also be used to investigate the effect of the disease-associated mutations on the PAS and CNBH domain interactions, as we reported before [15].

The SPR technique requires protein immobilization on the sensor chip surface. This could affect the conformation of the immobilized protein and its interactions with binding partners. It is important to keep this limitation in mind when interpreting the results of the SPR experiments. To determine if the immobilization influences interactions detected with SPR, one could consider repeating the experiments using different immobilization chemistry, e.g., immobilization on the NTA sensor chip using Ni^{2+}–NTA and the 6-His tag coupling, using a CM5 chip and anti-His antibody capture method, using different chips such as CM4 and C1 chips, or reversing the analyte and immobilized protein (e.g., immobilizing PAS domains and using CNBH domains as analytes).

4 Notes

1. PAS and CNBH genes are cloned into the pETM11 bacterial expression vector with an N-terminal 6-His affinity tag followed by a tobacco etch virus (TEV) protease cleavage site.
2. All buffer solutions are prepared using deionized ultrapure water, filtered and stored in sterile 500 mL or 1 L filter storage bottles at 4 °C for no longer than 3–4 days.
3. To avoid using volumes that are too small, first make 10% (v/v) Tween 20 by mixing 2 mL of Tween 20 with 18 mL Buffer 1. Then add 10% (v/v) Tween 20 to Buffer 1 to make Buffer 2, e.g., add 10 mL of 10% (v/v) Tween 20 to 2 L Buffer 1. For increased accuracy, one might wish to take out 10 ml from 2 L of Buffer 1 prior to adding 10 mL of 10% (v/v) Tween 20. However, then the analyte (PAS protein) dilutions have to be prepared the same way to avoid a mismatch between the kinetics running buffer and analyte dilutions.
4. We use Biacore T200 (GE Healthcare, now Cytiva) for our experiments.
5. We use amine coupling kit from Cytiva that contains ethanolamine, NHS, and EDC.
6. The opto-fluidic system is embedded within the Biacore T200 unit that we use in our experiments.
7. A software to run the SPR instrument and to acquire and analyze the SPR data is supplied with the Biacore T200 as an accessory.
8. The bacterial pellet can be stored at −80 °C prior to the purification with the affinity chromatography.
9. We use EmulsiFlex-C5 homogenizer (Avestin) to lyse the cells.
10. Cleavage of the 6-His tag is optional since proteins will be immobilized using amine chemistry that does not require the 6-His tag for immobilization. The cleaved 6-His tag can be purified away from the PAS and CNBH domains with the size exclusion chromatography.
11. We use a Superdex 200 Increase column (Cytiva) for purifying the PAS and CNBH domains with the size exclusion chromatography (SEC).
12. To increase protein purity, always use SEC as the last step for the PAS and CNBH domain purification, even if the 6-His tag cleavage is not required.
13. The protein samples need to be pure when inspected on the Coomassie gel as impurities may interfere with the inter-domain interactions.
14. Having 10% glycerol in the solution increases PAS and CNBH domain stability during the transition between freezing for the

storage at −80 °C and thawing for the SPR experiments conducted at 25 °C.

15. Keep protein solutions on ice while making the dilutions.
16. It is important to determine the smallest volume of aliquots needed for the experiment, but also it is best to choose round volumes that are easy to handle, e.g., avoid using 200.4 μL and instead find out how to make dilutions for 200 μL.
17. We recommend using the same solution for the kinetics running buffer and analyte dilutions to avoid any possible mismatch that could be registered as a false SPR signal.
18. It is important to follow manufacturer's instructions on how to dissolve NHS and EDC. Make aliquots of each and freeze until use. Thaw at 4 °C.
19. Since the CM5 chip has four FCs, even after reserving one flow cell as a control, three different proteins can be immobilized on the same sensor chip, and binding interactions of these three different proteins with analytes could be detected in one experiment.
20. Biacore control software offers to write methods to inject analyte and regeneration solutions automatically. The association time (contact time) and dissociation time can be easily changed by the user. Here association time refers to the duration within which the injected analyte binds gradually to the immobilized protein on the chip surface, and dissociation time refers to the duration within which analyte injection is stopped (at the beginning of dissociation) and buffer only breaks the analyte–ligand complex.
21. While here we are using only one analyte, binding of a few more (typically up to five) different analytes applied over the range of concentrations can be tested for the same immobilized domains in the same experiment, as long as the immobilized protein is active on the surface. With time, the level of protein immobilization decreases. Therefore, examining more than five different analytes in the same experiment may not be feasible.
22. Appropriate regeneration solution is selected such that it breaks analyte–ligand complex without affecting the activity of proteins immobilized onto the chip surface. We recommend to start testing with weaker solutions such as 1 M NaCl.
23. We used Biaevaluation software version 1.0.
24. We use Origin software for replotting the SPR data; however, other commonly used software such as GraphPad Prism or Igor Pro can also be used.
25. All fitting models described here plus a few other biphasic models are included in the Biaevaluation software for data analysis.

Acknowledgments

This work was supported by the National Cancer Institute grant R01CA252969 (T.I.B.) and National Institute of General Medical Sciences grant R01GM124020 (T.I.B.). All SPR experiments were conducted at the Biacore Molecular Interaction Shared Resource (BMISR) facility at Georgetown University Medical Center. The BMISR is supported by the National Institutes of Health grant P30CA51008.

References

1. Ganetzky B, Robertson GA, Wilson GF, Trudeau MC, Titus SA (1999) Ann N Y Acad Sci 868:356–369
2. Wang W, MacKinnon R (2017) Cell 169:422–430
3. Whicher JR, MacKinnon R (2016) Science 353:664–669
4. Gustina AS, Trudeau MC (2009) Proc Natl Acad Sci USA 106:13082–13087
5. Haitin Y, Carlson AE, Zagotta WN (2013) Nature 501:444–448
6. Morais Cabral JH, Lee A, Cohen SL, Chait BT, Li M, MacKinnon R (1998) Cell 95:649–655
7. Burton MJ, Cresser-Brown J, Thomas M, Portolano N, Basran J, Freeman SL, Kwon H, Bottrill AR, Llansola-Portoles MJ, Pascal AA, Jukes-Jones R, Chernova T, Schmid R, Davies NW, Storey NM, Dorlet P, Moody PCE, Mitcheson JS, Raven EL (2020) J Biol Chem 295:13277–13286
8. Wang ZJ, Tiwari PB, Uren A, Brelidze TI (2019) BMC. Pharmacol Toxicol 20:42
9. Wang ZJ, Soohoo SM, Tiwari PB, Piszczek G, Brelidze TI (2020) J Biol Chem 295:4114–4123
10. Carlson AE, Brelidze TI, Zagotta WN (2013) J Gen Physiol 141:347–358
11. Brelidze TI, Carlson AE, Sankaran B, Zagotta WN (2012) Nature 481:530–533
12. Brelidze TI, Gianulis EC, DiMaio F, Trudeau MC, Zagotta WN (2013) Proc Natl Acad Sci USA 110:11648–11653
13. Marques-Carvalho MJ, Sahoo N, Muskett FW, Vieira-Pires RS, Gabant G, Cadene M, Schonherr R, Morais-Cabral JH (2012) J Mol Biol 423:34–46
14. Craven KB, Zagotta WN (2004) J Gen Physiol 124:663–677
15. Soohoo SM, Tiwari PB, Suzuki YJ, Brelidze TI (2022) J Biol Chem 298:101433
16. Campbell CT, Kim G (2007) Biomaterials 28:2380–2392
17. Nguyen HH, Park J, Kang S, Kim M (2015) Sensors (Basel) 15:10481–10510
18. Tiwari PB, Chapagain PP, Seddek A, Annamalai T, Uren A, Tse-Dinh YC (2020) ChemMedChem 15:623–631
19. (2023) Biacore sensor surface handbook. https://cdn.cytivalifesciences.com/api/public/content/digi-16475-pdf
20. Tiwari PB, Bencheqroun C, Lemus M, Shaw T, Kouassi-Brou M, Alaoui A, Uren A (2021) BMC Mol Cell Biol 22:17
21. Schasfoort R (2017) Introduction to surface plasmon resonance. In: Handbook of surface plasmon resonance. Royal Society of Chemistry, Cambridge
22. Hayoz S, Tiwari PB, Piszczek G, Uren A, Brelidze TI (2017) PLoS One 12:e0185359
23. Markey F (2020) Real-time analysis of biomolecular interactions. In: Nagata K, Handa H (eds) Principles of surface Plasmon resonance. Springer, Tokyo
24. Jaikaran ET, Nilsson MR, Clark A (2004) Biochem J 377:709–716
25. Homola J, Yee SS, Myszka D (2008) Surface plasmon resonance biosensors. In: Lingler FC, Taitt CR (eds) Optical biosensors. Elsevier, Amsterdam
26. Wilson WD (2002) Science 295:2103–2105
27. Jonsson U, Fagerstam L, Ivarsson B, Johnsson B, Karlsson R, Lundh K, Lofas S, Persson B, Roos H, Ronnberg I (1991) BioTechniques 11:620–627
28. Brigham-Burke M, Edwards JR, O'Shannessy DJ (1992) Anal Biochem 205:125–131
29. Pollard-Knight D, Hawkins E, Yeung D, Pashby DP, Simpson M, McDougall A, Buckle P, Charles SA (1990) Ann Biol Clin (Paris) 48:642–646

30. Bellassai N, D'Agata R, Jungbluth V, Spoto G (2019) Front Chem 7:570
31. Marchesseau S, Mani JC, Martineau P, Roquet F, Cuq JL, Pugniere M (2002) J Dairy Sci 85:2711–2721
32. Tiwari PB, Wang X, He J, Darici Y (2015) Rev Sci Instrum 86:035001
33. Tiwari PB, Uren A, He J, Darici Y, Wang X (2015) Rev Sci Instrum 86:106107
34. Chen J, Zou A, Splawski I, Keating MT, Sanguinetti MC (1999) J Biol Chem 274:10113–10118
35. Splawski I, Shen J, Timothy KW, Lehmann MH, Priori S, Robinson JL, Moss AJ, Schwartz PJ, Towbin JA, Vincent GM, Keating MT (2000) Circulation 102:1178–1185

Chapter 8

Voltage Clamp Fluorometry: Illuminating the Dynamics of Ion Channels

Daniel Sastre and David Fedida

Abstract

Ion channels comprise one of the largest targets for drug development and treatment and have been a subject of enduring fascination since first discovered in the 1950s. Over the past decades, thousands of publications have explored the cellular biology and molecular physiology of these proteins, and many channel structures have been determined since the late 1990s. Trying to connect the dots between ion channel function and structure, voltage clamp fluorometry (VCF) emerges as a powerful tool because it allows monitoring of the conformational rearrangements underlying the different functional states of the channel. This technique represents an elegant harmonization of molecular biology, electrophysiology, and fluorescence. In the following chapter, we will provide a concise guide to performing VCF on *Xenopus laevis* oocytes using the two-electrode voltage clamp (TEVC) modality. This is the most widely used configuration on *Xenopus* oocytes for its relative simplicity and demonstrated success in a number of different ion channels utilizing a variety of attached labels.

Key words Electrophysiology, Ion channel, Voltage clamp fluorometry, Two-electrode voltage clamp, Fluorescence, *Xenopus laevis*, RNA microinjection

1 Introduction

Ion channels are the second largest target for existing drugs and represent an important pharmacological target since they are expressed in virtually every cell [1]. By controlling ion movement across the cell membrane, ion channels generate and regulate electrical signals within biological tissues.

Unraveling the mysteries of electrophysiology has long been an enduring scientific pursuit since the experiments of Luigi Galvani in the late eighteenth century. The early twentieth century saw many studies that eventually led to the development of the two-electrode voltage clamp (TEVC) technique by Hodgkin, Huxley, and Katz in the giant axon of the squid [2]. This methodology consists of the use of two intracellular electrodes: the first monitors the membrane potential (Vm) and sends it through a differential amplifier to the

Simone Furini (ed.), *Potassium Channels: Methods and Protocols*, Methods in Molecular Biology, vol. 2796,
https://doi.org/10.1007/978-1-0716-3818-7_8,

second electrode, which injects the current necessary to maintain Vm at a desired value. The injected current equals the membrane current [3]. The tight control over the membrane potential represented a huge revolution in the technological evolution of electrophysiological recordings. More advanced techniques were developed in the following decades, including the widely used single-electrode patch clamp technique. However, TEVC is still the methodology of choice for recordings in large cells like *Xenopus* oocytes, in which a single intracellular electrode is insufficient to properly control fluctuations in voltage over the large membrane surface area. The oocytes of the African clawed frog (*Xenopus laevis*) have been experimentally used in scientific research previously, mainly in the embryology and development fields [4]. Moreover, they proved to be a powerful tool for heterologous expression (*see* **Note 1**) and have been used in the study of ion channels since the 1980s [5, 6].

Over the years, more ingenious techniques and technologies have been developed to propel the frontiers of ion channel study. One such breakthrough technique is voltage clamp fluorometry (VCF), which was originally developed in the Isacoff laboratory [7]. The essence of voltage clamp fluorometry lies in the powerful combination of two methodologies: electrophysiology and fluorescence imaging. Briefly, the target channel is genetically modified to contain a single reactive cysteine and expressed in *Xenopus* oocytes. Then, the cysteine residue is specifically modified with a thiol-reactive fluorophore (*see* Fig. 1). The emitted fluorescence signal is typically captured at a fixed wavelength using either a photomultiplier tube, CCD camera, or photodiode. The changes in protein structure surrounding the probe are reported as observable changes in fluorescence intensity, reflecting variations of how the probe interacts with the solvent and nearby quenching groups [8]. At the same time, the currents of the channel are recorded with a voltage clamp, usually TEVC, but other configurations are also possible (*see* **Note 2**). By seamlessly integrating both measurements, researchers can simultaneously assess the conformational changes (i.e., fluorescence signal) underlying the functional output (i.e., ionic currents) of the channel with exquisite temporal resolution. This remarkable approach has been commonly used to explore the interplay between the voltage sensor domain of K_V channels, by fluorescently tagging the S4 segment, and the current flow through their pore. Additional conformational changes related to channel activation can be studied by placing the fluorescent probe in other accessible residues. Overall, VCF has proven to be a successful approach to study the conformational changes underlying ion channel function.

This chapter serves as a comprehensive exploration of voltage clamp fluorometry using the two-electrode voltage clamp configuration on *Xenopus laevis* oocytes. We will first discuss how to extract

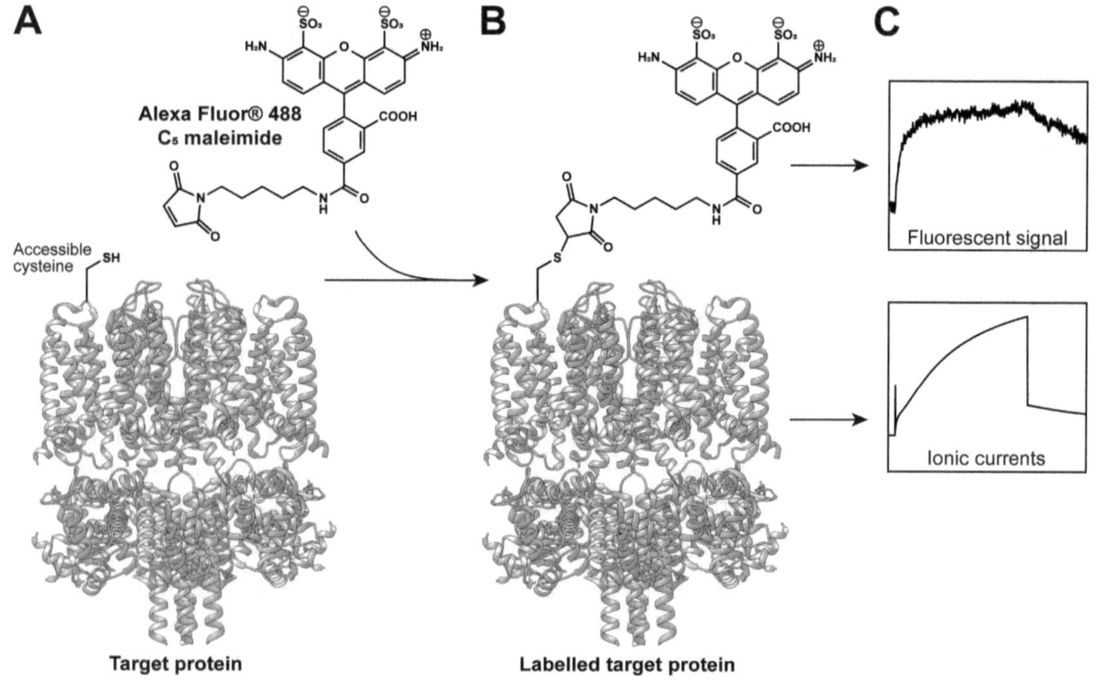

Fig. 1 Basis for the VCF technique. Upon exposure of the target protein (in the example, KCNQ1; PDB: 7TCP) to the fluorophore of interest (**a**), the maleimide group of the probe covalently binds the thiol group of an accessible cysteine residue (**b**). Following application of voltage pulses, this configuration allows the simultaneous recording of fluorescent signal, reflecting conformational rearrangements of the protein, and ionic currents, reporting channel opening (**c**)

the oocytes from adult female frogs and proceed with mRNA injection. Then, we describe how to obtain simultaneous recordings of ionic currents and fluorescence, and we will finally address data analysis.

2 Materials

All solutions are made with ultrapure water characterized by a conductivity of 0.05 μS/cm. Working solutions are stored at 4 °C and kept for use within a month.

2.1 Oocyte Isolation and Preparation

1. Anesthetic solution: 2 g/L MS-222 (tricaine methanesulfonate), 2 g/L HEPES, pH adjusted to 7.4 with NaOH.
2. Oocyte Ringer 2 (OR2) solution: 82.5 mM NaCl, 2.5 mM KCl, 1 mM $MgCl_2$, 5 mM HEPES, pH adjusted to 7.6 with NaOH, sterile filtered (0.22 μm).
3. Collagenase solution: 3 mg/mL Collagenase Type 4 in OR2 (*see* **Note 3**).

4. Incubator (e.g., MIR-153, SANYO).
5. Platform shaker (e.g., Rotamax 120, Heidolph Instruments), located inside the incubator.
6. 35 mm tissue culture dishes (e.g., Falcon® 353001, Corning) and a ridged plastic chamber.

2.2 RNA Injection and Oocyte Maintenance

1. Invitrogen™ RNase Away™ Decontamination Reagent (Thermo Fisher Scientific).
2. Stereo microscope (e.g., SMZ-168, Motic) with a fiber-optic light source (e.g., FL150 with FL150/30 dual arm floppy light guide, Meiji Techno). The total augmentation is 15–25×.
3. Micromanipulator (e.g., KITE, World Precision Instruments).
4. Nanoinjector (e.g., Manual Oocyte Microinjection Pipette, Drummond Scientific) with the right size of borosilicate capillaries (e.g., 3-000-210-G8, Drummond Scientific).
5. Pipette puller (e.g., P-97, Sutter Instrument) [9].
6. Oocyte Ringer 3 (OR3) solution: 500 mL/L Leibovitz's L-15 medium, 15 mM HEPES, 1 mM L-glutamine, 250 mg/L gentamycin, pH adjusted to 7.6 with NaOH, sterile filtered (0.22 μm).

2.3 Fluorescence Labeling and Recording

1. A completely opaque bucket filled with ice.
2. ND96: 96 mM NaCl, 3 mM KCl, 1 mM $MgCl_2$, 5 mM HEPES, 2 mM $CaCl_2$, pH adjusted to 7.4 with NaOH. If required, ND96 can be supplemented with specific ion channel blockers to reduce endogenous currents (*see* **Note 4**).
3. Depolarizing solution: 100 mM KCl, 1.5 mM $MgCl_2$, 0.5 mM $CaCl_2$, 10 mM HEPES, pH adjusted to 7.5 with KOH.
4. Invitrogen™ Alexa Fluor™ 488 C_5 maleimide (Thermo Fisher Scientific) (*see* **Note 5**): 1 mg of the dye is diluted in 138.8 μL dimethyl sulfoxide (DMSO) to a concentration of 10 mM and aliquoted in ready-to-use volumes. Aliquots are kept in the freezer at −5 to −30 °C kept away from light and moisture. An aliquot is thawed for every experiment and not frozen again.
5. Pipette solution: 3 M KCl, sterile filtered (0.22 μm).
6. Borosilicate capillaries in the right size for the pipette holders (e.g., GC120T-10, Warner Instruments).

2.4 Recording Equipment

1. The imaging and recording equipment is assembled on top of a vibration isolation table (e.g., 63–531, AMETEK TMC), inside a Faraday's cage to reduce mechanical vibrations and electrical noise. All devices must be correctly grounded to minimize interferences.

2. An appropriate perfusion chamber, like the ones supplied by Warner Instruments, and a simple perfusion system (e.g., a gravity flow system). Ensure their proper installation to avoid moisture or wetting of the microscope.
3. Two micromanipulators (e.g., MM-3, Narishige), one per each microelectrode.
4. An inverted epifluorescence microscope (e.g., Eclipse TE300 with TE-FM Epi-Fluorescence Attachment, Nikon) with an appropriate objective (e.g., Plan Apo 2× DIC, Nikon) and filter cube (e.g., XF100-2, Omega) for the experiment (*see* **Note 6**). Since VCF detects photons instead of capturing images, the microscope does not require extensive complexity besides minimizing loss through the light path. The equipment must be placed in a suitable location, away from dust, vibration, high temperature and humidity, and direct sunlight. Be mindful of dirt or dust on the lens; handle the instrument with care [10].
5. A binocular zoom stereo body (e.g., EMZ-10, Meiji Techno) mounted on the microscope with the right eyepieces (e.g., MA502, Meiji Techno) and attached to a high intensity illuminator (e.g., FL-150, Meiji Techno). The total augmentation is 15×, but anywhere between 5× and 40× will suffice.
6. Attached to the microscope is a high-pressure mercury lamp power supply (e.g., C-SHG1, Nikon) with the specified mercury lamp (e.g., HBO 103W/2N:100W, OSRAM) and lamphouse (e.g., LH-M100CB-1, Nikon) (*see* **Note 7**).
7. A shutter (e.g., Uniblitz VS35S2ZM0R1-27, Vincent Associates) installed between the mercury lamp and the sample on the optical path with an appropriate TTL-driven shutter driver (e.g., Uniblitz VCM-D1, Vincent Associates). The shutter is controlled through the computer to open a few milliseconds before and close a few milliseconds after the acquisition signal to minimize exposure of the sample to fluorescence light and photobleaching.
8. Photomultiplier and control box (i.e., Integra, Cairn Research) to measure the fluorescence intensity.
9. A voltage clamp amplifier suitable for oocyte recordings (e.g., OC-725C, Warner Instruments) and an appropriate digitizer (e.g., Digidata 1440A, Molecular Devices). Methods are written as if using an OC-725C, but other amplifiers will have similar functions [11].
10. Personal computer of choice with the proper acquisition and analysis software (e.g., Clampex 10.3 and ClampFit 10.3, Molecular Devices) [12].

3 Methods

The following section describes a protocol to achieve voltage clamp fluorometry with Alexa Fluor™ 488 C_5 maleimide using the two-electrode voltage clamp in *Xenopus laevis* oocytes (*see* Fig. 1). The molecular biology work required to generate the constructs is implied and out of the scope of this chapter (*see* **Note 8**). The following methodology has extensively been used to study I_{Ks} [13], and taken the right considerations, it can be applied to virtually any channel. It must be contemplated that the labeling conditions will vary for different target residues, ion channels of study, or fluorescent probes used (*see* **Note 5**). Hence, the protocol must be optimized for each specific experimental setup.

3.1 Basic Principles of Fluorescence

Before delving into the specifics of the methodology, it is important to highlight the essential theoretical aspects of fluorescence. The aim of this section is to provide the general knowledge to understand the key aspects of the technique and to interpret and discuss VCF results. If further details are required, we encourage the reader to engage with more specific contributions [14, 15].

Fluorescence is a form of luminescence, i.e., the emission of radiation by an excited atom or molecule following energy absorption. Fluorescence is very fast, typically occurring within 10 ns, as opposed to phosphorescence (a slower form of luminescence), which can last for a few seconds or even minutes. The molecular substrates of fluorescence are called fluorophores, i.e., their atomic composition absorbs energy of a specific excitation wavelength and re-emits it at a longer, equally specific wavelength (*see* Fig. 2a).

The basic mechanism for fluorescence is outlined in the form of a Jabłoński diagram in Fig. 2b. When exposed to a light of specific wavelengths, the probe absorbs a portion of its energy, causing the susceptible electron to transition from the ground state (S_0) to a higher energy level within its orbital (typically S_1). Absorption is considered immediate since it occurs in about 10^{-15} s, faster than many molecular motions. This state is transient and unstable, and the molecule quickly relaxes to the least energetic vibrational energy level of the excited state. After a short time (around 10 ns), the system calms down to the ground state, dissipating energy via photon emission. The transition back to the ground state commonly takes place through a high-energy level, and then the molecule quickly achieves thermal equilibrium. The vibrational relaxation in both the excited and the ground states leads to a reduced energy during emission, shifting the emission spectrum toward longer wavelengths (*see* Fig. 2a). This phenomenon is called Stokes shift, and it is determined by both the fluorophore and its interactions with the immediate environment.

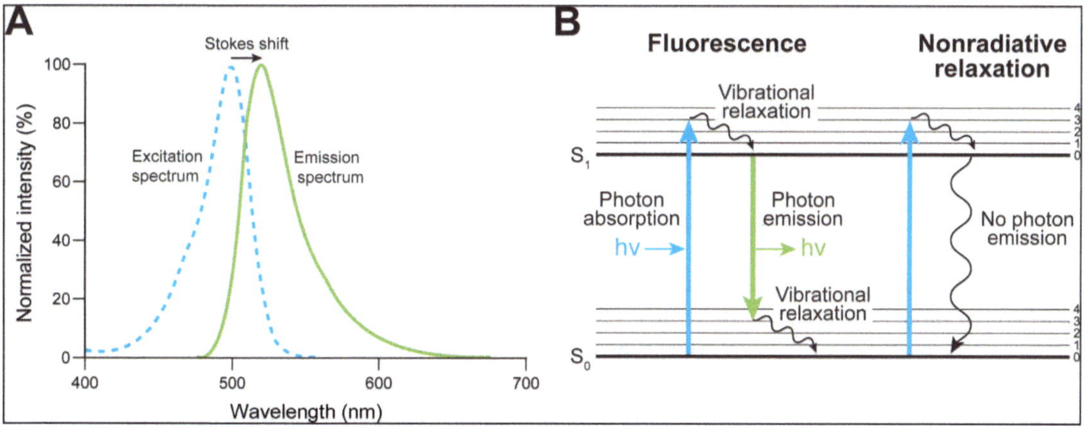

Fig. 2 Basic principles of fluorescence. (**a**) Excitation and emission spectra for Alexa Fluor™ 488. The observed change in wavelengths is called Stokes shift. (**b**) Jabłoński diagram showing the main transitions of fluorescence. For a fluorophore, both the ground state (S_0) and one excited state (S_1) are indicated. The intrinsic vibration of the molecule gives rise to a series of energy levels (named 0, 1, 2, etc.) for each state. On the left, photons of specific wavelengths are absorbed and promote the fluorophore to an excited state. Photon emission occurs during the relaxation back to the ground state. The emitted energy is reduced compared to the absorbed energy (Stokes shift) because part of it is lost during vibrational relaxation processes. On the right, the same scheme is shown without photon emission. Nonradiative relaxation can occur spontaneously or can be promoted by specific mechanisms, e.g., quenching

In a practical scenario, not all the fluorophores will come back to the ground state emitting light. The efficiency of photon emission is called the quantum yield, which is not only an intrinsic property of the fluorophore but is also affected by the solvent, pH, and temperature. The mechanisms preventing photon emission include (1) nonradiative decay, i.e., the transition back to the ground state without photon emission; (2) spin conversion of the molecules in S_1, leading to phosphorescence; (3) aggregation of fluorescent molecules; (4) photobleaching, i.e., the degradation of the fluorescent dye due to prolonged excitation; and (5) quenching. Quenching is an umbrella term to define a wide variety of processes that decrease fluorescence intensity. These underlie the changes in fluorescence in VCF, as the conformational rearrangements of the protein expose the probe to different quenchers [8]. In VCF, the most relevant quenching processes comprise (1) changes of the polar microenvironment and solvent exposure; (2) collisional quenchers in aqueous compartments, e.g., iodide or oxygen; and (3) interaction with other groups, e.g., tryptophan quenching, generation of dipole fields, and energy transfer to resonant electrons (FRET). Because the changes in the microenvironment are complex and hard to interpret, the magnitude and direction of the fluorescence changes observed in VCF are not considered proportional to those of the underlying molecular motion. Instead, the kinetics of the fluorescence recordings are more informative, as they reliably relate to the kinetics of the conformational changes they reflect.

In this section, we provided a basic theoretical background of fluorescence in the context of VCF. With this understanding, we encourage the reader to select the right fluorophore (*see* **Note 5**) and target residue (*see* **Note 8**) for their studies. In the next sections, we will dive into the specifics of the methodology, including oocyte extraction, RNA injection, and fluorescence labeling and recording.

3.2 Oocyte Isolation and Preparation

1. Sedation of female *Xenopus laevis* is achieved by immersing them for 15 min in 500 mL anesthetic solution. It is advisable to perform the procedure on ice to prolong the effects of the anesthetic. Frog unconsciousness is assessed through toe pinch and righting reflexes prior to surgery. The frog is euthanized in conformity with the animal care guidelines of the corresponding research institution, and the ovary sacks are extracted through a small abdominal incision. Alternatively, the frog can be sutured after surgery and allowed to recover instead of being sacrificed. The oocytes are kept in an OR2-filled tissue culture dish for the next steps. Optionally, it is advisable to wash the oocytes thoroughly with OR2 if they have become bloodied during the extraction. If the number of extracted oocytes goes beyond the usable amount, the exceeding oocytes might be preserved in the fridge in OR2 with antibiotic for up to a week.
2. Using fine forceps, the ovary sacks are broken down into smaller groups of 20–30 oocytes each. This step must be done carefully to prevent damage and cell death. A volume of approximately 7 mL of oocytes is washed 2–5 times with OR2 before digestion at 18 °C with 20 mL collagenase solution (*see* **Note 3**). There is significant batch-to-batch variability in the reaction (*see* **Note 9**), so oocytes must be checked under the microscope every 15–20 min until ready. The flask is gently shaken to ensure even enzymatic digestion. After the first 30–45 min, the collagenase solution is replaced for another 20 mL. The reaction is stopped when a few of the oocytes have completely shed their follicular cell layer (*see* Fig. 3a). A common duration for the whole procedure is between 45 min and 90 min since the first exposure to collagenase.
3. The oocytes are again washed with OR2 3–5 times both before and after being incubated with 20 mL OR2 at 18 °C for an hour.
4. Under the microscope, oocytes are selected for mRNA microinjection. One of the pivotal factors for success is the use of healthy, stage V–VI oocytes (*see* Fig. 3b). These can be identified by their large size (diameters of 1.0–1.2 mm) and the sharp contrast between the dark animal pole and the light (yellow or gray) vegetal pole. Stage VI oocytes will also present a clear

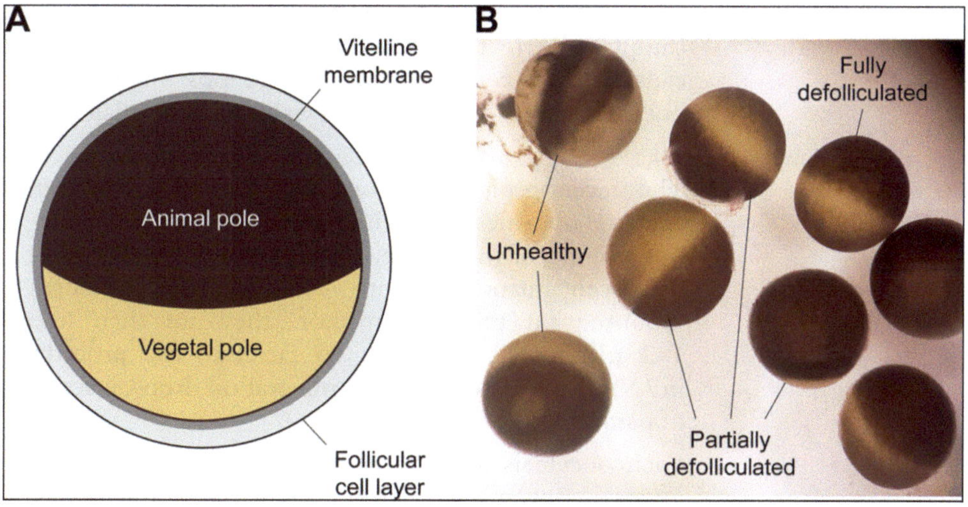

Fig. 3 Defolliculation of *Xenopus laevis* oocytes. (**a**) A diagram of a stage V–VI oocyte, indicating the animal and the vegetal poles. The vitelline membrane is not removed for VCF, but the follicular cell layer is. (**b**) Sample oocytes after treatment with collagenase IV. The unhealthy oocytes must be removed, and only the fully defolliculated ones must be selected for RNA injection. The rest of the follicular layer can be manually removed from healthy-looking partially defolliculated oocytes

band between the poles called equatorial band. Consider only the healthy-looking oocytes, i.e., spherical with no signs of damage, blemishes, asymmetries, or uneven pigmentation [6].

5. It is crucial that the selected oocytes are fully defolliculated. Any residuum of the follicular cell layer might not only obstruct fluorescent labeling but also become a source of unwanted currents because the follicular cells are electrically coupled both between each other and to the oocyte [6, 16]. If the chemical digestion had been insufficient or yielded partial results, any follicular fragments can be removed from the oocytes manually with the help of small tweezers.
6. Selected oocytes are best transferred with a fire-polished glass Pasteur pipette and stored in a 35-mm-diameter Petri dish filled with OR3 media.

3.3 RNA Injection and Oocyte Maintenance

The mRNA of interest (*see* **Note 10**) is injected in the oocytes using a nanoinjector mounted on a micromanipulator. To minimize RNA degradation, it is essential to wear gloves and utilize RNase-free pipette tips, nuclease-free water, and dedicated surface decontaminants (e.g., Invitrogen™ RNase Away™) when handling the samples.

1. Injection tips are made from borosilicate capillaries using a pipette puller. For our pipettes, we use HEAT = 595 (RAMP +25), PULL = 110, VELOCITY = 100, TIME = 150, and PRESSURE = 100 [9]. The size of the tips is adjusted by

manually breaking them under the microscope. As a rule of thumb, pipette tips should be as small as possible without causing pipette obstruction. Larger tips might damage the oocytes, compromising their viability and quality for VCF experiments.

2. The injection pipettes are airtight, backfilled with mineral oil using a non-metallic syringe needle. To ensure the proper functioning of the device, it is important to verify that no air has been introduced into the injection tip. After being mounted on the nanoinjector, they are then filled with 1–2 μL mRNA solution. To facilitate the procedure, the mRNA is placed on top of a mineral oil drop on a small piece of parafilm.
3. Each oocyte is injected with 50 nL mRNA (0.5–2 ng/nL) under a conventional stereo microscope. A micromanipulator is used to guide the nanoinjector. It is advisable to place the oocytes on a ridged plastic chamber filled with OR2, to immobilize them during the process. When lowering the injection tip, the oocyte should indent lightly upon initial contact and show a slight resistance before being pierced. Making the injection in the equatorial band between the animal and the vegetal poles optimizes RNA mobility without damaging the nucleus [6]. After injection, wait a couple of seconds before carefully withdrawing the injection tip.
4. Store the oocytes in OR3-filled sterile 35 mm culture dishes, at a concentration not higher than 20 oocytes per dish. Incubate with gentle shaking at 18 °C for 1–9 days. Currents can be recorded after 2 days, but for fluorescence experiments, it is better to wait for at least 3 days instead to maximize protein expression. For the duration of the experiment, media must be replaced daily, and oocytes must be inspected to remove the injured and dead ones. This step is important since the enzymes and chemical messengers of the unhealthy oocytes can damage the rest.

3.4 Preparation of the Two-Electrode Voltage Clamp System

1. The system uses two intracellular silver chloride electrodes made of silver wire and two bath electrodes. Although a single probe can be used for the bath, the use of a second one to establish a virtual ground is strongly encouraged, especially if the recorded currents are large (*see* **Note 11**). The silver wire of the bath probes should be tightly coiled, or, alternatively, silver–silver chloride sintered pellets can be used. To reduce junction potentials, any silver wire in touch with the pipette or bath solution should be chlorided (*see* **Note 12**).
2. Recording pipettes are made from borosilicate capillaries using a conventional pipette puller. For our pipettes, we use HEAT = 645 (RAMP +40), PULL = 0, VELOCITY = 60, TIME = 150, and PRESSURE = 500, which yields pipette tips

large enough [9]. Alternatively, the tips can be manually broken to an outer diameter of at least 3–5 μm for the voltage electrode and at least 7–9 μm for the current electrode. These diameters will ensure the respective resistances are low enough (*see* **Note 13**). Unlike other electrophysiological techniques, electrodes are not required to be fire polished or coated with Sylgard®. Pipettes are backfilled with sterile-filtered 3 M KCl using a conventional syringe. Microelectrodes should contain enough pipette solution to cover approximately 1–1.5 cm of the wire, not filling the holders.

3. The electrode holders are mounted in the micromanipulator and connected to the amplifier. The microscope should be adjusted so that the chamber is within the visual field. The solutions are loaded on the perfusion system, and both the flow and the aspiration system are checked so that the surface of the bath is stable and flow rate is appropriate for the experiment. The bath probes are submerged in the chamber or in the agar bridge wells, and the appropriate shielding and grounding of the apparatus is verified.
4. Every device of the system is turned on. The acquisition software is configured. In Clampex 10.2 [12], the main steps are (1) configuring the digitizer with the command "Configure" → "Digitizer" and changing from "Demo" to the one in use; (2) designating the file name and folder with the options on "File" → "Set data file names"; and (3) checking that the voltage protocol is correctly set up.

3.5 Fluorescence Labeling and Recording

During the whole procedure, it is crucial to be mindful of photobleaching. Some labs use blackout curtains on the Faraday's cage, while others perform the recordings in a dark room. Either way, oocytes must be handled in as low light as possible. For that purpose, different shutters in the optic path must be used (*see* Fig. 4). First, the exposure of the sample to the mercury lamp light is minimized through a computer-controlled TTL-driven shutter. Then, a filter block selects which light frequencies are used for excitation of the sample and determines the frequencies collected after sample emission (*see* **Note 6**). Finally, at least one manually activated shutter is in the path between the sample and the photomultiplier. It is advisable to close such shutter when not recording (e.g., while changing the oocyte) to preserve the photomultiplier detector from stray light.

1. A 1:1000 dilution of the dye is prepared with depolarizing solution (*see* **Note 14**). Two or three small tissue culture dishes are filled with this solution and kept on ice, in a completely opaque bucket. A few (two or three) oocytes are placed in one of these dishes and incubated with the dye for 30 min. After incubation, the oocytes are quickly rinsed in a small dish filled with ND96.

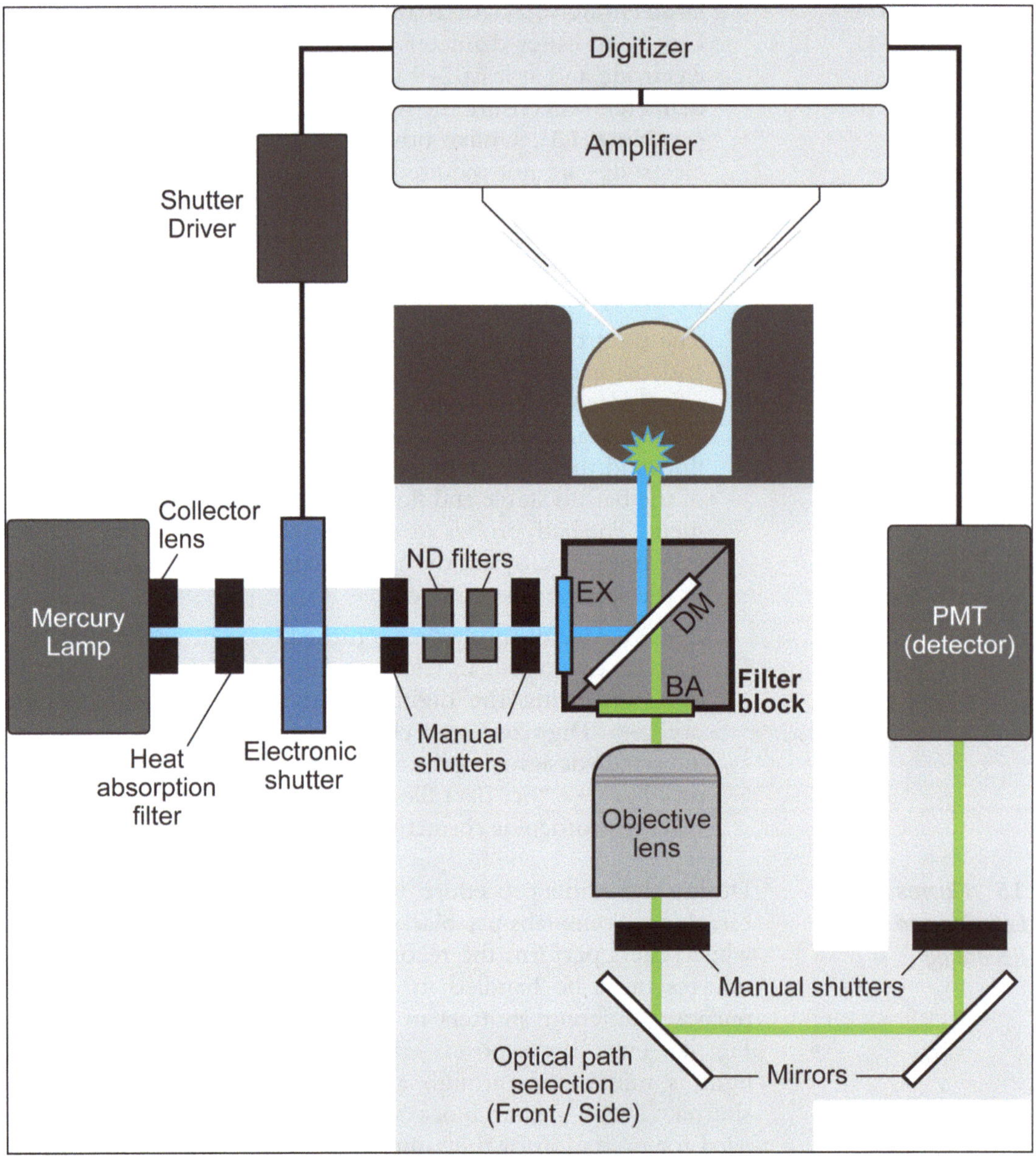

Fig. 4 Diagram of the VCF apparatus. The high intensity light produced by a mercury lamp is dimmed (collector lens, heat absorption filter, ND filters) in the path toward the filter block that will determine the excitation and emission wavelengths. In the block, the excitation filter (EX) selects the frequencies that will excite the sample. The dichroic mirror (DM) reflects those wavelengths toward the oocytes, whereas it allows the emitted light to pass the other way. Finally, the barrier filter (BA) removes unwanted frequencies from the collected light directed toward the objective length, and the signal is finally recorded by a photomultiplier (PMT). Note that within the optic path, several manual shutters, as well as a computer-driven shutter, control the flow of fluorescent light. Simultaneously to the fluorescence measurements, ionic currents are recorded with the TEVC configuration. The shutter driver, the photomultiplier, and the amplifier are controlled with a computer via the digitizer

2. Using a glass transfer pipette, one oocyte is placed in the recording chamber. It is recommended to place the oocyte with the animal pole down, because it will offer a darker background on an inverted microscope [17]. The light should be as low as possible during oocyte handling and completely turned off when not necessary.
3. Using the micromanipulator, the tips of the electrodes are lowered one at a time until touching the bath solution. Starting with the voltage electrode, the offset knob is tweaked so that the electrode potential reads 0 mV, and the same is done for the current electrode. Then the pipette resistances are tested, which should be as low as possible (around 1 MΩ or less). Higher resistances might distort either the Vm readout or the current injection and slow down the capacitive transients (*see* **Note 13**). The electrodes are carefully moved close to the oocyte but still not in contact with it. It is advisable to readjust the offset potentials if needed before impaling the cell.
4. The oocyte is carefully punctured with both electrodes. Some amplifiers offer a "BUZZ" function that facilitates disruption of the membrane by applying a voltage oscillation [11], but its use is completely optional (*see* **Note 15**). The best angle to approach the membrane is 45°. To reduce the coupling capacitance between the electrodes, they should be positioned at an angle of at least 90°. Tight closeness between electrodes should be avoided.
5. Once both electrodes have penetrated the membrane of the oocyte, the voltage meter will display the resting membrane potential. The appropriate knob in the amplifier is used to switch to voltage clamp mode, which allows control of the voltage across the oocyte membrane [12]. Most amplifiers allow a holding potential to be set, but to apply more complex protocols, the voltage should be controlled externally from a computer. It is advised to start applying test pulses to assess the quality of the recording, adjust acquisition parameters, and allow for current stabilization. The gain switch should be set as high as possible or appropriate for the current amplitude likely to be recorded (*see* **Note 13**).
6. The photomultiplier must be turned on for recording and the manual shutter opened. The appropriate filter block must be selected using the designated slider, and the light path must be changed to the "SIDE" port. During the experiment, fluorescence and current are recorded as two separate inputs (*see* Fig. 5). It is advisable to acquire the voltage signal as a third input for independent verification of the membrane voltage control. The software is configured to send command outputs not only over the voltage but also to the shutter and the PMT

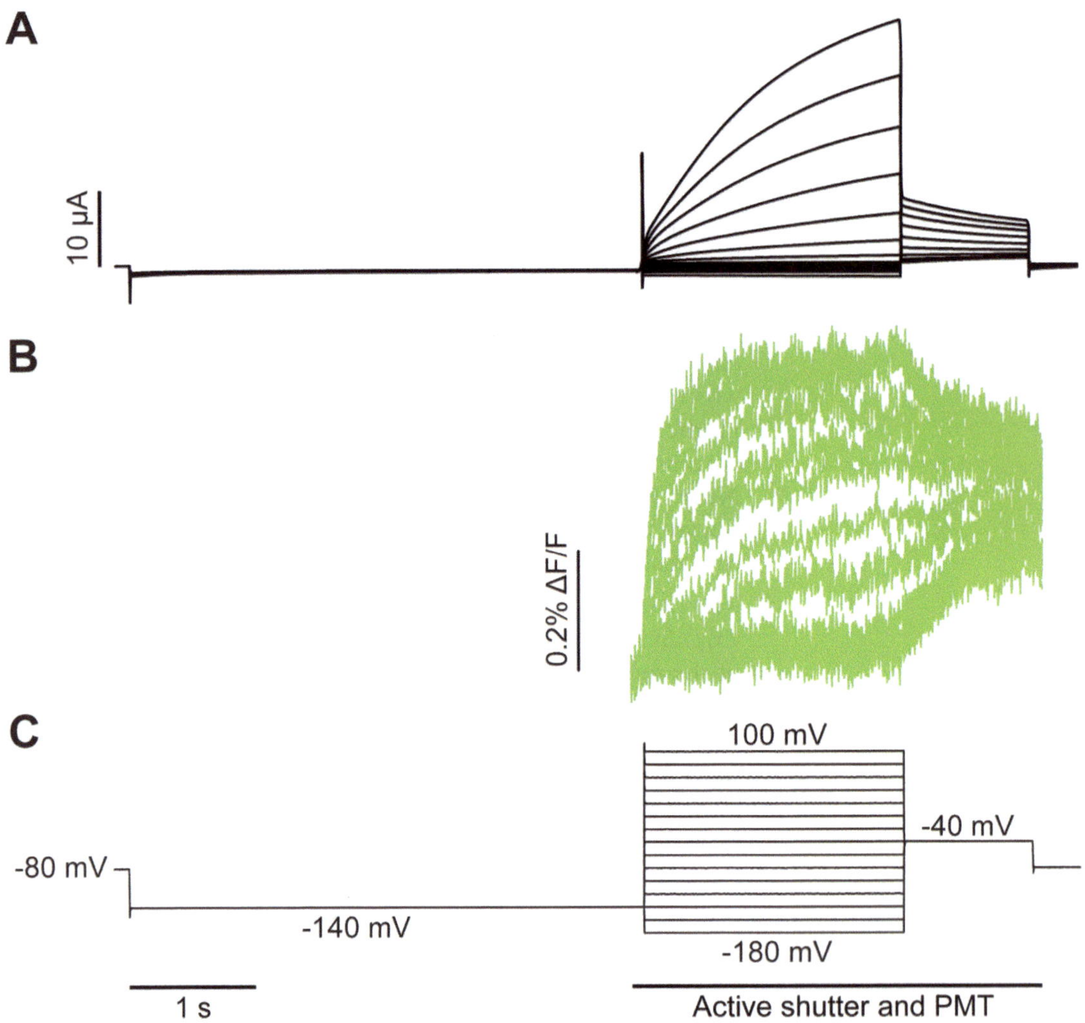

Fig. 5 Sample VCF results from the I_{Ks} channel obtained using the protocol described in this chapter. Ionic currents (**a**) and fluorescence measurements (**b**) were simultaneously recorded upon applying the indicated voltage protocol (**c**). Notice that the fluorescence traces are shorter because the PMT and the shutter open time are optimized to reduce photobleaching of the sample

gate to minimize sample exposure to fluorescent light. The same protocol is applied at least three times, and elicited fluorescent signals can be averaged to minimize noise. The number of averaging pulses depends on the signal amplitude, since it will be progressively reduced due to photobleaching. A background trace of the same duration of the protocol but kept at the holding potential is also recorded, for at least three sweeps.

7. When finished with the recordings, the cell is unclamped by turning off the gain knob before toggling the voltage clamp switch off. The order is important as not to overload the feedback amplifier. The manual shutter is turned off after the

recording. The oocyte can now be safely removed and the procedure started over with the next one. The electrodes can be reused if their resistance values have not increased significantly, which would indicate clogging with cellular debris. Hence, offset potentials should be adjusted and pipette resistance checked for each oocyte.

8. At the end of the experiment, the software should be closed first, and then the power of the devices should be turned off. The perfusion system should be emptied and washed with water.

3.6 Voltage Protocol and Data Analysis

In this section, data handling and processing is described for ClampFit 10.2 users, and some commands will be specific to the software [12]. Other analysis packages will surely offer analogous procedures.

1. Because the same protocol is applied multiple times, the recorded files are combined using the command "Analyze" → "Average files." Next, the background traces are averaged into one ("Analyze" → "Average traces").
2. In the averaged protocol, a new set of virtual, empty traces are added ("Analyze" → "Arithmetic" → "Add Traces"). In the newly created traces, the commands in "Analyze" → "Arithmetic" are used to subtract the background trace from each individual trace in the original protocol.
3. Finally, baselines are adjusted ("Analyze" → "Adjust" → "Baseline...") and traces filtered ("Analyze" → "Filter..."). A Gaussian low-pass filter around 100 Hz should be good enough.

4 Notes

1. *Xenopus laevis* oocytes have become a very widely used vector for heterologous expression in the study of ion channels. The species can be bred and maintained with relative ease and low cost, presenting a moderate death rate (approximately 2% annually) and producing mature oocytes all year round. The advantages of *Xenopus* oocytes include (1) their robustness, since they are naturally exposed and adapted to adverse environments; (2) their big size, which makes them easy to handle and manipulate; (3) the high efficiency of heterologous expression, easily achieving large amplitude currents; (4) their relatively low endogenous currents; and (5) the possibility to inject fixed ratios of mRNA. It is also worth to consider some of the most relevant drawbacks: (1) the possibility of interactions with endogenous proteins, as was the case of the injected KCNE1

that interacted with endogenous KCNQ1 [18]; (2) the presence of endogenous currents, which, despite low, in some cases is not negligible (*see* **Note 4**); and (3) the altered protein behavior due to the interspecies differences in environment and lipid composition.

2. Other techniques have further been developed, like cut-open Vaseline gap (COVG) for VCF. This method offers better control over the plasma membrane and higher temporal resolution, which makes it suitable for short transient currents like gating currents or ion channels with fast kinetics [17]. However, COVG also makes the system substantially more complex, and the benefits might not be significant for most ion channels and applications.

3. The use of OR2 during collagenase digestion and for the washes is beneficial for several reasons. On the one hand, it prevents calcium overload during digestion because it does not contain any Ca^{2+} ions. On the other hand, the OR2 washes help remove debris and enzymes from damaged cells that might affect the quality of the oocytes [6].

4. While the expression of endogenous channels in *Xenopus laevis* oocytes is usually quite low, in some cases, it is not negligible. The amplitude of their currents can also vary from batch to batch (*see* **Note 9**). The most notorious endogenous conductance is produced by calcium-activated chloride channels, which can be addressed by lowering the Ca^{2+} or Cl^{-} concentrations, replacing the pipette solution of the current electrode for 0.6 M K_2SO_4, or using blockers (e.g., 0.3 mM niflumic acid). Also relevant are cation channels, which can be blocked by extracellular divalent cation (Ca^{2+}, Mg^{2+}) concentrations above 0.1 mM, and stretch-activated channels, which can be blocked by 100 μM gadolinium. It might also be advisable to use the calcium channel blocker $LaCl_3$. It must be checked that the used blockers do not affect the activity of the channels of study [6, 19].

5. When selecting the fluorophore, it is important to consider key biophysical characteristics such as the extinction coefficient (i.e., the amount of absorbed light at a certain wavelength), frequencies for excitation and emission, quantum yield, size, hydrophobicity (i.e., water soluble and membrane impermeant to label membrane proteins), stability (i.e., low photobleaching), and pharmacology (i.e., little or no effect over the protein of study). The most common method of labeling, and the one described in this chapter, is using a thiol-reactive fluorophore targeting a cysteine residue of choice. With many commercially available options, the choice of the probe depends on the experiment. Some cysteine residues will be better labeled by

one or another, and the amplitude of the signal might also vary. Commonly used fluorescent probes include maleimide-conjugated 1-(3-carboxybenzyl)-4-(5-(4-methoxyphenyl)oxazol-2-yl) pyridinium bromide (PyMPO), tetramethylrhodamine (TMRM), and Alexa Fluor™ dyes. Make the fluorophore stocks at a high concentration to minimize the exposure of the oocyte to DMSO, which would damage the membranes and reduce the quality of the clamp.

6. The filter blocks must be selected for their best suitability to the fluorophore of use [10]. The blocks comprise two filters and a dichroic mirror. The excitation filter allows the selective passage of the light within a certain range of wavelengths to excite the sample. The range of light that can pass is called bandwidth, with higher bandwidths producing brighter images but also increasing the chances of self-fluorescence and fading. A dichroic mirror reflects the exciting light from the excitation filter toward the sample but allows the emitted light from the sample to go through. Finally, the barrier filter cleans up the fluorescent light generated by the specimen on its path to the photodetector.
7. Mercury lamps emit hazardous ultraviolet (UV) radiation, which is harmful to the eyes and skin. Direct visualization of the UV light must be avoided, and the use of neutral density filters is recommended to mitigate its brightness. It is essential to handle the lamp with care because the gas inside it can get to very high pressures. Ensuring that the device is clean, free from scratches, and not subjected to any type of additional pressure is crucial to prevent potential ruptures and gas leaks. The mercury lamp should be replaced as recommended by the manufacturer. Both the lamp and the field aperture diagram should be centered. Furthermore, flammable materials should be kept at a distance since the lamp reaches extremely high temperatures when turned on.
8. To direct the thiol-reactive fluorescent probe labeling, the target protein must be engineered to contain a single accessible cysteine residue. The probes are not membrane permeant, and thus only residues exposed to the extracellular surface can be labeled, even if exposure is limited to specific functional states of the channel (i.e., during depolarization or hyperpolarization) (*see* **Note 14**). To reduce background from the labeling, the rest of accessible cysteines must be removed, commonly by substitution to alanine. An alternative is the use of intrinsically fluorescent unnatural amino acids (fuAA), which can also be targeted to intracellular residues and do not require additional mutations [20]. The electrophysiological properties of the generated mutants must be checked to ensure their reliability to study the protein of interest.

9. It is widely known that *Xenopus laevis* oocytes present a huge variability, both batch-to-batch and seasonal. This variability impacts oocyte survival and quality, effectiveness of collagenase treatment, and efficiency of ion channel expression [6]. The reasons for this phenomenon are still poorly understood.
10. In the first heterologous expression experiments, *Xenopus* oocytes were injected total poly(A) mRNA extracts from specific tissues [5, 6], with translation of unwanted mRNAs. Nowadays, it is more convenient to use synthesized mRNA using easily available commercial kit (e.g., Invitrogen™ mMESSAGE mMACHINE™ T7 Transcription Kit, Thermo Fisher Scientific).
11. The use of a virtual ground (VG) batch clamp headstage is very common in TEVC. This configuration uses two bath electrodes, which in the OC-275C amplifier are named I_{OUT} and I_{SENSE} [11]. The latter should be placed on the side of the voltage electrode, as close to the oocyte as possible, whereas I_{OUT} should be on the same side as the current electrode and can be placed a bit further from the oocyte. In this arrangement, I_{OUT} acts as the virtual ground from the injected current, keeping the cell's exterior at the I_{SENSE} terminal at ground potential. Agar bridges are optional but are sometimes used to protect the cell membrane from the adverse effects of the silver wire or pellet of the bath electrodes.
12. Chloriding silver wires is an essential step to ensure the stable recording of high currents for long periods of time. The wire should be wiped with ethanol and water before proceeding. There are two main methods for chloridation [21]: (1) electrical chloridation, which consists of treating the electrode with a current of approximately 1 mA/cm^2 for about a minute in a solution of either 0.9% NaCl or 1 M KCl, and (2) chemical chloridation, which consists of submerging the electrode in bleach for 20–30 min or until homogeneously darkened. Fresh or well-sealed bleach yields better results. Sintered silver–silver chloride pellets do not need to be chlorided, but it is advisable to clean them occasionally. Note that recording of big currents can exhaust the AgCl coat, so it is advisable to check the electrodes regularly in case they need to be chlorided again.
13. The surface area of an oocyte, accounting for invaginations and microvilli, is around 1.5×10^7 μm^2. Hence, the large size of the oocytes produces capacitive currents that slow the voltage clamp and may mask shorter, transient currents. Following a voltage command step, the response time of the amplifier will be proportional to the resistance of the current electrode and the membrane capacity and inverse to the gain of the amplifier. Thus, a way to reduce the response time for the clamp is to use low resistance current electrodes and use the highest gain on the amplifier [6, 16].

14. In the protocol, we use a high K^+ depolarizing solution for labeling because we commonly work with a depolarization-activated ion channel (I_{Ks}) and the reactive residue increases its exposure to the dye upon increases of Vm. For other channels, and if the target residue is more exposed to the probe in the resting state or upon more negative values of Vm, it might be more convenient to use a low K^+ hyperpolarizing solution for labeling [17].
15. Due to the size of the cells, TEVC is more forgiving than patch clamp. If in need, different optional procedures will help reduce membrane damage during the recordings. First, the BUZZ function of the amplifier [11] will minimize membrane damage during cell penetration. The stability of the system can be enhanced by reducing vibration, using a mechanically isolated table or a hydraulically driven micromanipulator. Finally, it is important to consider electrical isolation by ensuring proper grounding of the equipment and using an appropriate Faraday's cage [11].

References

1. Alexander SPH, Mathie A, Peters JA et al (2021) THE concise guide to pharmacology 2021/22: ion channels. Br J Pharmacol 178
2. Hodgkin AL, Huxley AF, Katz B (1952) Measurement of current-voltage relations in the membrane of the giant axon of *Loligo*. J Physiol 116:424–448
3. Guan B, Chen X, Zhang H (2013) Two-electrode voltage clamp. In: Gamper N (ed) Ion channels: methods and protocols, Methods in molecular biology, vol 998. Springer, Berlin, pp 79–89
4. De Robertis EM, Gurdon JB (2021) A brief history of *Xenopus* in biology. Cold Spring Harb Protoc 2021:pdb.top107615
5. Sumikawa K, Houghton M, Emtage JS et al (1981) Active multi-subunit ACh receptor assembled by translation of heterologous mRNA in Xenopus oocytes. Nature 292:862–864
6. Smart TG, Krishek BJ (1995) Xenopus oocyte microinjection and ion-channel expression. In: Boulton AA, Baker GB, Walz W (eds) Patch-clamp applications and protocols, Neuromethods, vol 26. Humana Press, New Jersey, pp 259–306
7. Mannuzzu LM, Moronne MM, Isacoff EY (1996) Direct physical measure of conformational rearrangement underlying potassium channel gating. Science 271:213–216
8. Cha A, Bezanilla F (1997) Characterizing voltage-dependent conformational changes in the K channel with fluorescence. Neuron 19: 1127–1140
9. Sutter Instrument (2020) P-97 flaming/brown™ micropipette puller operation manual, Novato
10. Nikon (1997) TE-FM Epi-fluorescence attachment instructions. Minato, Tokyo
11. Warner Instruments OC-725C Oocyte Clamp Amplifier—User's Manual, Hamden
12. Molecular Devices (2016) pCLAMP™ 10 data acquisition and analysis software user guide, San José
13. Westhoff M, Eldstrom J, Murray CI et al (2019) I_{Ks} ion-channel pore conductance can result from individual voltage sensor movements. Proc Natl Acad Sci USA 116:7879–7888
14. Sauer M, Hofkens J, Enderlein J (2011) Basic principles of fluorescence spectroscopy. In: Handbook of fluorescence spectroscopy and imaging: from single molecules to ensembles. Wiley, Weinheim, pp 1–30
15. Lakowicz JR (2006) Principles of fluorescence spectroscopy. Springer, Boston
16. Molecular Devices (2021) Axon guide—electrophysiology and biophysics laboratory techniques, San José

17. Gandhi CS, Olcese R (2008) The voltage-clamp fluorometry technique. In: Lippiat JD (ed) Potassium channels. Methods in molecular biology, vol 491. Humana Press, Totowa, pp 213–231
18. Sanguinetti MC, Curran ME, Zou A et al (1996) Coassembly of KVLQT1 and minK (IsK) proteins to form cardiac IKS potassium channel. Nature 384:80–83
19. Tammaro P, Shimomura K, Proks P (2008) Xenopus oocytes as a heterologous expression system for studying ion channels with the patch-clamp technique. In: Lippiat JD (ed) Potassium channels. Methods in molecular biology, vol 491. Humana Press, a part of Springer Science + Business Media, Totowa, pp 127–139
20. Kalstrup T, Blunck R (2013) Dynamics of internal pore opening in K_V channels probed by a fluorescent unnatural amino acid. Proc Natl Acad Sci USA 110:8272–8277
21. Warner Instruments Chloriding Ag/AgCl Electrodes Disk, Pellet, or Wire, Hamden

Chapter 9

Creating Computational Models of Ion Channel Dynamics

Max E. Schoening and Jonathan R. Silva

Abstract

Markov models are widely used to represent ion channel protein configurations as different states in the model's topology. Such models allow for dynamic simulation of ion channel kinetics through the simulated application of voltage potentials across a cell membrane. In this chapter, we present a general method for creating Markov models of ion channel kinetics using computational optimization alongside a fully featured example model of a cardiac potassium channel. Our methods cover designing training protocols, iteratively testing potential model topologies for structure identification, creation of algorithms for model simulation, as well as methods for assessing the quality of fit for a finalized model.

Key words Model creation, Ion channels, Markov models, Computational optimization, Action potentials, Simulation, Modeling

1 Introduction

State-dependent continuous-time discrete-state Markov models are widely used to model voltage-dependent dynamics of ion channel dynamics in excitable cells, including neurons, cardiac myocytes, and pancreatic beta cells. Markov models reproduce the kinetic activity of ion channels at different membrane voltages and over any unit of time by representing ion channel configurations or groups of configurations as states in the model. Models are visually represented as connected bidirectional graphs, where each connection allows ion channels to transition directly between connected states.

Once fit, Markov models allow for in silico reproduction of the ion channel dynamics and can be combined with models of a cell's other ion channels for simulation of a full action potential [1, 2]. Additionally, the state-dependent nature of Markov models

Supplementary Information The online version contains supplementary material available at https://doi.org/10.1007/978-1-0716-3818-7_9.

Simone Furini (ed.), *Potassium Channels: Methods and Protocols*, Methods in Molecular Biology, vol. 2796,
https://doi.org/10.1007/978-1-0716-3818-7_9,

allows for other models to be connected for exploration of molecular interactions and their effects on ion channel activity, such as in the CiPA initiative, where models representing drug binding activity are attached to models of the rapid delayed rectifier (I_{Kr}) channel to determine the proarrhythmic potential of a variety of clinically used drugs [3]. On a macroscopic scale, single cell models can also be used to create three-dimensional virtual organs, such as a whole heart simulation, for use in more complex structural research or to create a digital twin for personalized medicine [4, 5].

In this chapter, we present a general method for creating Markov models of cardiac ion channels alongside a fully featured example model with included software (see supplementary files). Models will be constructed and identified using modern optimization techniques that have proven to be highly effective for creating Markov models and are more approachable than intuition-guided model design [6]. We hope that this method helps researchers extract more value from their own electrophysiology data while expanding accessibility.

2 Methods

2.1 Representing Ion Channel Activity with an Ensemble of Voltage-Clamp Protocols

To accurately reproduce the kinetic activity of each type of ion channel, a Markov model must be fit to training data that sufficiently describes an ion channel's activity across a variety of possible inputs [7]. In this case, the two variables of the model that can be controlled at runtime are voltage and time, allowing direct conversion of in vitro voltage-clamp protocols to Markov model training protocols.

In this section, we will present three examples of highly representative voltage-clamp protocols that are useful for Markov model fitting purposes. All training data presented is from recordings of the fast transient outward potassium current, $I_{to,\ f}$, a fast-inactivating subtype of the I_{to} ion channel that modulates the length of the cardiac action potential [8, 9].

Note that this section describes only the voltage-clamp protocols themselves, but algorithms for implementing these protocols for use in model fitting will be provided later during construction of the objective function.

2.1.1 Voltage–Conductance (GV) Curve (Fig. 1a)

The voltage–conductance (GV) curve plots the maximum conductance as a function of sequential voltage steps to different depolarized potentials, relative to the resting potential.

1. Steady-state occupancies are calculated at an initial holding voltage of −70 mV.

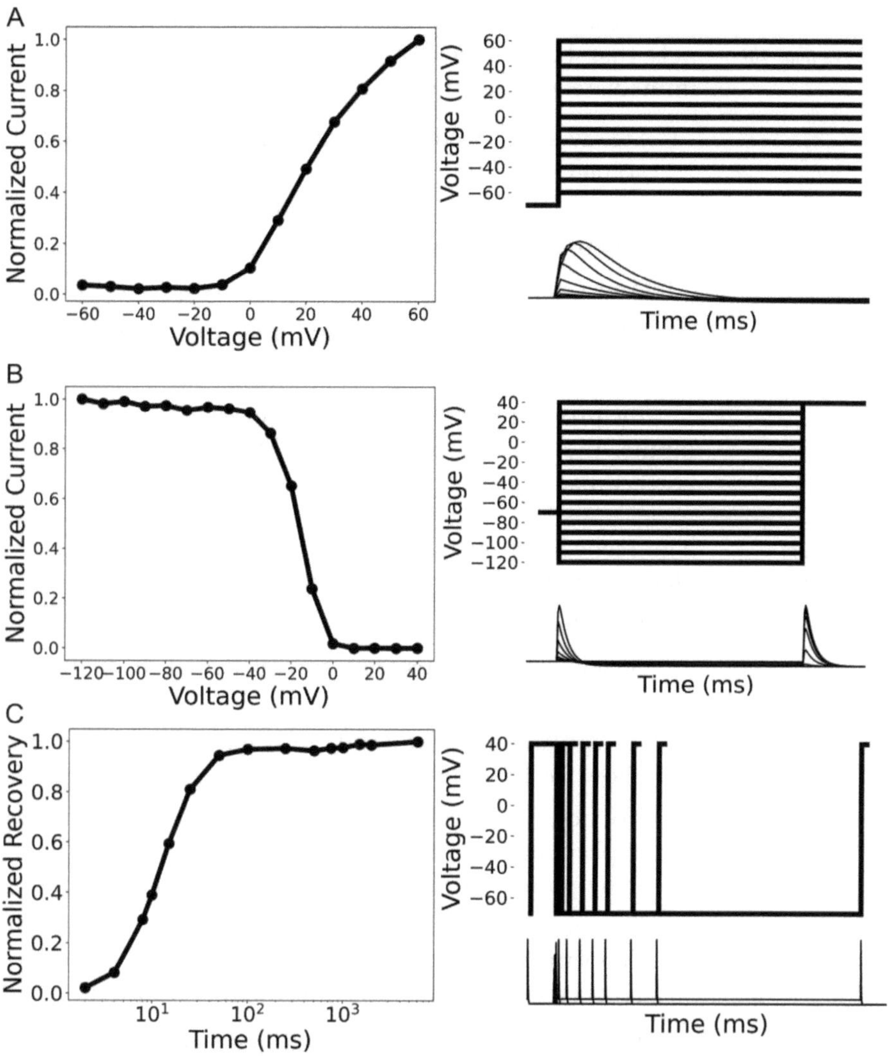

Fig. 1 Left: training datapoints. Right: voltage steps over time aligned with model current traces for each of the three sample $I_{to,\ f}$ protocols in the training ensemble. (**a**) Voltage-conductance, (**b**) steady-state inactivation, and (**c**) recovery from inactivation. Note that voltage steps from (**a**) and (**b**) apply different voltages for the same amount of time, while voltage steps in (**c**) apply the same voltages but hold at −100 mV for varying amounts of time to generate each point

2. A depolarizing pulse between −60 mV and 60 mV is applied and held for 50 ms.
3. The peak current magnitude after each depolarization is recorded.

By providing maximum conductance values across a wide range of voltages, the GV protocol trains a Markov model to adjust the amount of activation to be proportional to the voltage applied.

2.1.2 Steady-State Inactivation Curve (Fig. 1b)

The steady-state inactivation curve plots the fraction of ion channels that become inactivated and thus cannot conduct current immediately following a depolarization.

1. Steady-state occupancies are calculated at an initial holding voltage of −70 mV.
2. A pulse between −120 mv and 40 mV is applied and held for 200 ms to simulate activation at different potentials.
3. A test depolarizing pulse of 40 mV is applied for 50 ms.
4. The peak current magnitude after each test depolarization is recorded.

By providing the steady-state currents over various voltage inputs, the SSI protocol trains a Markov model to simulate the correct inactivation of ion channels after initial depolarization.

2.1.3 Recovery from Inactivation (Fig. 1c)

The recovery from inactivation protocol plots the fraction of ion channels that have fully recovered after a depolarizing pulse at a negative potential.

1. Steady-state occupancies are calculated at an initial holding voltage of −70 mV.
2. A depolarizing pulse of 40 mV is applied and held for 500 ms to allow for full activation to occur.
3. To begin recovery, a hyperpolarizing pulse of −70 mV is applied and held for 2–6000 ms, after which another depolarizing pulse of 40 mV is applied for 100 ms.
4. The peak amount of conductance that occurs after the second depolarizing pulse represents the fraction of channels that fully recovered from the initial pulse and is recorded.

By providing the proportion of recovery over various amounts of recovery time, the recovery from inactivation protocol trains a Markov model to simulate the observed refractory period in ion channels or the delay before they can fully reopen after sitting at negative potentials.

2.2 Identifying Structurally Plausible Markov Model Topologies

While initial Markov model topologies were determined with the help of human intuition and biophysical inspiration, methods for systematically generating and testing different topologies based on their complexity have enabled selection of topologies that fit best to a given ensemble of voltage protocols [6].

Although methods for determining optimal topologies based on genetic algorithms have proven to be effective for ion channel simulations [10], such techniques create additional challenges during model fitting due to the large changes in the parameter search space that occur as more states are added [6].

Therefore, our methods focus on pre-defining the number of states, sorting by topology complexity, and iteratively testing all models in order of increasing difficulty. Generation and testing of topologies can be done as follows:

1. Create a set of every possible interconnected graph with at least three but no more than N states. N represents an arbitrary upper limit to the number of states in a model, allowing for model complexity—and therefore optimization complexity—to be limited.
2. Eliminate any models that are isomorphic to prevent redundant testing, and reorder the remaining models by increasing complexity.
3. In descending order, attempt to fit each topology to the training dataset(s). Note that different ion channels may require different minimum levels of model complexity for a sufficient model fit, meaning it may be impossible for smaller models to fit more complex protocol ensembles.
4. After all models have been tested, the graphs with lower cost values correspond to better fitting models. Specifically, a model is considered fully fit when all model output datapoints fall within the standard deviation of their experimentally determined training points. However, because current methods cannot achieve truly complete fits across most models, the topology(s) with the most fully fitted output datapoints should be selected for further use.

The ensemble of voltage protocols given in the previous section corresponds to in vitro electrophysiology data from fast transient outward potassium current or $I_{to,\ f}$. Prior experimentation in Mangold et al. [6] demonstrates that this dataset fits well with the structure in Fig. 2.

2.3 Mathematically Defining a Markov Model from a Chosen Topology

A state-dependent model can be represented by its rate matrix Q, which is used to describe a system of ordinary differential equations (ODEs) representing the rate of change for each possible state.

$$\frac{d\vec{s}}{dt} = Q\vec{s} \tag{1}$$

In this section, we will manually convert the five-state Markov model topology in Fig. 2 into its corresponding r_i rate equations and Q matrix for later execution.

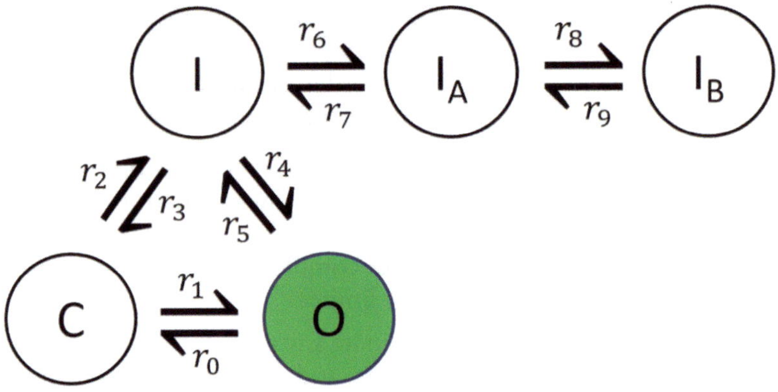

Fig. 2 A fully labeled five-state Markov model topology that is suitable for fitting to the $I_{to,\ f}$ dataset. Note that although all states have been named in the visualized model, non-open states are labeled for clarity purposes and may exhibit activity that aligns with any type of ion channel state—regardless of the state's label [10]

2.3.1 Generating the Model Rates

1. First, variables to store each individual rate must be created. In general, each rate is a function of the input voltage *V* and two optimized parameters α and β that are unique to that rate, giving a final rate function of:

$$r_x = \alpha_x * \exp(V/\beta_x) \tag{2}$$

2. In models that contain cyclical connections, microscopic reversibility must be enforced [11]. For the sample topology being used, microscopic reversibility states:

$$r_0 * r_2 * r_4 = r_1 * r_3 * r_5 \tag{3}$$

Solving this equation for any rate constrains that rate in the cycle, which enforces microscopic reversibility in the system:

$$r_0 = \frac{r_1 * r_3 * r_5}{r_2 * r_4} \tag{4}$$

3. To reduce model stiffness and improve the likelihood of a successful fit, the input voltage should also be adjusted [12]:

$$f(V) = sig\left(\frac{V - A_v}{B_v}\right) \tag{5}$$

where *sig* represents the sigmoid function and A_v, B_v are optimized parameters that are the same for all rates.

The original input voltage *V* is then directly substituted with the adjusted input voltage $f(V)$:

$$r_x = \alpha_x * \exp(f(V)/\beta_x) \tag{6}$$

4. Finally, with the structure of the rate functions and their dependent variables defined, the rate functions themselves can be constructed. Performing this algorithmically gives:

Steps	fill_rates (V, $\overrightarrow{params}$)
1	For $i = 0, \ldots, R-1, R$ do **step 2.**
2	$r_i = \alpha_i * \exp(V/\beta_i)$
3	$r_0 = (r_1 * r_3 * r_5)/(r_2 * r_4)$

where $\overrightarrow{params}$ contains the collection of ordered parameter pairs (α_i, β_i) followed by the two voltage adjustment parameters A_v, B_v, as visualized below:

$$\overrightarrow{params} = \langle \alpha_0, \beta_0, \alpha_1, \beta_1, \ldots, \alpha_{i-1}, \beta_{i-1}, \alpha_i, \beta_i, A_v, B_v \rangle \tag{7}$$

Note that with additional cycles, corresponding rates will also need to be constrained. See Menon et al. for a generalized approach [10].

2.3.2 Creating and Populating the Q Matrix

1. Let N represent the number of states in a Markov model. Define the empty matrix Q with size NxN and vector $\vec{s}$ with size N. For the purposes of constructing the Q matrix, we will populate vector $\vec{s}$ with placeholder variables representing each possible state in the model.

$$\frac{d\vec{s}}{dt} = \begin{bmatrix} & & & & \\ & & & & \\ & & & & \\ & & & & \\ & & & & \end{bmatrix} \cdot \begin{bmatrix} O \\ C \\ I \\ I_A \\ I_B \end{bmatrix} \tag{8}$$

Note that while the placeholder variables in $\vec{s}$ may take any arbitrary order before the Q matrix is filled, this order will determine both the order of the system of equations represented by the Q matrix and the output positions of the state occupancy vector when the model is executed.

2. With vector $\vec{s}$ defined, each position i,j in Q represents an interaction between states. The row number i determines the state occupancy being calculated, and the column number j determines which state the rate is originating from.

For example, the rate constant in position (0,1) represents the rate at which channels in state "C" transition to the open state "O." Referencing a graph of the model topology shows that this rate is r_1, so $Q_{0,1} = r_1$.

$$\frac{d\vec{s}}{dt} = \begin{bmatrix} & r_1 & & & \\ & & & & \\ & & & & \\ & & & & \\ & & & & \end{bmatrix} \cdot \begin{bmatrix} O \\ C \\ I \\ I_A \\ I_B \end{bmatrix} \tag{9}$$

If there is no connection between two states in a model, the rate between these states is 0. For example, position (0,3) represents the rate from I_A to O, but these two states are not connected.

$$\frac{d\vec{s}}{dt} = \begin{bmatrix} & r_1 & 0 & & \\ & & & & \\ & & & & \\ & & & & \\ & & & & \end{bmatrix} \cdot \begin{bmatrix} O \\ C \\ I \\ I_A \\ I_B \end{bmatrix} \tag{10}$$

For all $Q_{i,\ j}$ where $i \neq j$, either of the above two cases will apply, and most of the Q matrix can be populated:

$$\frac{d\vec{s}}{dt} = \begin{bmatrix} & r_1 & r_4 & 0 & 0 \\ r_0 & & r_3 & 0 & 0 \\ r_5 & r_2 & & r_7 & 0 \\ 0 & 0 & r_6 & & r_9 \\ 0 & 0 & 0 & r_8 & \end{bmatrix} \cdot \begin{bmatrix} O \\ C \\ I \\ I_A \\ I_B \end{bmatrix} \tag{11}$$

3. Entries $Q_{i,\ j}$ where $i = j$ represents the rates at which channels exit the state represented by the row number i. For example, the open state "O," represented by the first row, has two exiting transition rates, so the overall rate of change in "O" due to these outward connections is $-(r_0 + r_5)$. Therefore, the rate for position (0,0) is also $-(r_0 + r_5)$. Note that the rates in these positions will be negative as channels are exiting their current state.

$$\frac{d\vec{s}}{dt} = \begin{bmatrix} -(r_0+r_5) & r_1 & r_4 & 0 & 0 \\ r_0 & & r_3 & 0 & 0 \\ r_5 & r_2 & & r_7 & 0 \\ 0 & 0 & r_6 & & r_9 \\ 0 & 0 & 0 & r_8 & \end{bmatrix} \cdot \begin{bmatrix} O \\ C \\ I \\ I_A \\ I_B \end{bmatrix} \tag{12}$$

Repeating for all $Q_{i,\ j}$ where $i = j$ fills the Q matrix completely:

$$\frac{d\vec{s}}{dt} = \begin{bmatrix} -(r_0+r_5) & r_1 & r_4 & 0 & 0 \\ r_0 & -(r_2+r_1) & r_3 & 0 & 0 \\ r_5 & r_2 & -(r_3+r_4+r_6) & r_7 & 0 \\ 0 & 0 & r_6 & -(r_7+r_8) & r_9 \\ 0 & 0 & 0 & r_8 & -r_9 \end{bmatrix} \cdot \begin{bmatrix} O \\ C \\ I \\ I_A \\ I_B \end{bmatrix} \tag{13}$$

2.4 Computer Simulation of a Markov Model

Simulation of a Markov model occurs as a stochastic process over time steps of a pre-determined size. While various ODE integrators may be utilized to solve the model system, use of the matrix exponential has proven to significantly increase computational efficiency [12]. In this section, we will initialize the example model's state vector and implement a function to capture the simulated channel's current.

2.4.1 Calculating Initial State Occupancies

1. Using the matrix exponential requires first calculating the model's initial state occupancy values, which are then used to calculate subsequent time steps.

   The probabilistic properties of Markov models dictate that all state occupancies will sum to one for any moment in time t. By modifying the Q matrix to include this constraint, as in the top row below and setting all of the other derivatives to 0 (on the right), we can obtain the steady-state values.

$$\begin{bmatrix} 1 & 1 & 1 & 1 & 1 \\ r_0 & -(r_2+r_1) & r_3 & 0 & 0 \\ r_5 & r_2 & -(r_3+r_4+r_6) & r_7 & 0 \\ 0 & 0 & r_6 & -(r_7+r_8) & r_9 \\ 0 & 0 & 0 & r_8 & -r_9 \end{bmatrix} * \vec{s} = \begin{bmatrix} 1 \\ 0 \\ 0 \\ 0 \\ 0 \end{bmatrix} \tag{14}$$

2. Solving for the state vector $\vec{s}$:

$$\vec{s}_i = \vec{s} = \begin{bmatrix} 1 & 1 & 1 & 1 & 1 \\ r_0 & -(r_2+r_1) & r_3 & 0 & 0 \\ r_5 & r_2 & -(r_3+r_4+r_6) & r_7 & 0 \\ 0 & 0 & r_6 & -(r_7+r_8) & r_9 \\ 0 & 0 & 0 & r_8 & -r_9 \end{bmatrix}^{-1} \cdot \begin{bmatrix} 1 \\ 0 \\ 0 \\ 0 \\ 0 \end{bmatrix} \tag{15}$$

   Provides the initial starting occupancies $\vec{s}_i$ for the sample Markov topology.

2.4.2 Simulating and Capturing Ionic Current

Beginning with the model's initial starting occupancies, a complete trace over T time steps of size dt can be simulated by repeatedly using the current state vector to calculate the next step:

$$\vec{s}_t = \exp(Q) * \vec{s}_{t-1} \tag{16}$$

1. Whenever voltage changes are necessary during simulation runtime, the rate vector $\vec{r}$ and the Q matrix must be updated and re-exponentiated. The updated Q_{V2} matrix directly replaces the old Q matrix:

$$\vec{s}_t = \exp(Q_{V2}) * \vec{s}_{t-1} \tag{17}$$

With the rate matrix replaced, the state vector can continue to be iterated as before.

2. After all time steps have been calculated, the raw current trace for the simulated model is directly represented by the occupancy of the "O" open state.

$$\vec{I} = \left\langle \vec{s}_0[O], \vec{s}_1[O], \ldots, \vec{s}_{T-1}[O], \vec{s}_T[O] \right\rangle \quad (18)$$

3. Combining the above steps algorithmically gives:

Steps	***simulate_current(*** $\overrightarrow{params}$, $\vec{V}$, $\vec{T}$***)***
1	$\vec{s}_{t-1} = \vec{s}_i$
2	For all v in $\vec{V}$ do:
3	$\vec{r} = \mathit{fill\_rates}\left(v, \overrightarrow{params}\right)$
4	$Q = \mathit{generate\_q}\left(v, \overrightarrow{params}\right)$
5	$Q_{exp} = \exp(Q)$
6	For $i = 1, 2, \ldots, T_v - 1, T_v$ do:
7	$\vec{s}_t = \exp(Q) * \vec{s}_{t-1}$
8	Append $\vec{s}_t[O]$ to $\vec{I}$
9	$\vec{s}_{t-1} = \vec{s}_t$

Note that in the above algorithm, T_v represents the number of time steps to apply the current voltage and V represents the ordered list of a protocol's requested voltages. The computational efficiency benefits of using the matrix exponential rely on only re-calculating the transition matrix Q_{exp} when needed, necessitating this multi-loop approach.

2.5 Fitting a Model to Voltage-Clamp Electrophysiology Data

After mathematically defining a Markov model, it must be trained to the ensemble of training protocols to produce physiologically accurate outputs. In this section, we will create an objective cost function to measure how well the model represents its training dataset and use the objective function to fit our example Markov model.

2.5.1 Creating the Objective Function from Voltage Protocols

The objective function must provide a value that accurately represents how well the model is fit to the training data for any given set of input parameters. For our example Markov model, the input parameter set is the vector of pairs α_i, β_i for each rate r_i, followed by the two shared voltage adjustment parameters A_v, B_v:

$$\overrightarrow{params} = \langle \alpha_0, \beta_0, \alpha_1, \beta_1, \ldots, \alpha_{i-1}, \beta_{i-1}, \alpha_i, \beta_i, A_v, B_v \rangle \quad (19)$$

If the fit of the model improves, the model's outputs should more closely represent the values in its training protocol. Using the

Sum of Squared Errors (SSE) as a cost function for each protocol ensures that an optimizer minimizing each protocol's cost will maximize its fit:

$$\sum_{P} \sum_{n=0}^{N} (p_n - y_n)^2 \tag{20}$$

where N is the number of comparison datapoints in the training protocol, p_n is the point's value, y_n is the model's output for that protocol, and P is the set of all protocols.

However, extracting the model's output differs between each protocol, so we will construct error functions for each protocol individually:

1. For the GV protocol, the voltage applied to the model changes with each new datapoint, and the model's peak current after the last voltage change is saved for comparison. Using "simulate_current" from the previous section:

Steps	*voltage_conductance*$\left(\overrightarrow{params}\right)$
1	$\vec{y}_{GV} = \langle\rangle$
2	For all v in $\vec{V}_P$ do:
3	$\vec{V} = \langle v \rangle$
4	$\vec{T} = \langle 50 \rangle$
5	$\vec{I} = \boldsymbol{simulate\_current}\left(\overrightarrow{params}, \vec{V}, \vec{T}\right)$
6	Append I_{max} to $\vec{y}_{GV}$
7	$\vec{y}_{GV} = \vec{y}_{GV} / y_{GV,max}$

2. For the Recovery from Inactivation protocol, the amount of time to hold the second voltage step changes with each new datapoint, and the model's peak current during the last voltage change is saved for comparison. Again using "simulate_current" to compute:

Steps	*recovery_from_inactivation*$\left(\overrightarrow{params}\right)$
1	$\vec{y}_{RFI} = \langle\rangle$
2	For all t in $\vec{T}_P$ do:
3	$\vec{V} = \langle 40, -70, 40 \rangle$
4	$\vec{T} = \langle 500, t, 100 \rangle$
5	$\vec{I} = \boldsymbol{simulate\_current}\left(\overrightarrow{params}, \vec{V}, \vec{T}\right)$
6	Append I_{max} to $\vec{y}_{RFI}$
7	$\vec{y}_{RFI} = \vec{y}_{RFI} / y_{RFI,max}$

3. For the Steady-State Inactivation protocol, the voltage applied to the model changes with each datapoint, and the peak conductance after the last voltage step is saved for comparison. Computing the SSI error is similar to the GV protocol, only requiring that $\vec{V}$ and $\vec{T}$ are modified:

Steps	*steady_state_inactivation*$\left(\overrightarrow{params}\right)$
1	$\vec{y}_{SSA} = \langle\rangle$
2	For all v in $\vec{V}_P$ do:
3	$\vec{V} = \langle v, 40\rangle$
4	$\vec{T} = \langle 200, 50\rangle$
5	$\vec{I} = \boldsymbol{simulate\_current}\left(\overrightarrow{params}, \vec{V}, \vec{T}\right)$
6	Append I_{max} to $\vec{y}_{SSA}$
7	$\vec{y}_{SSA} = \vec{y}_{SSA} / y_{SSA,\,max}$

4. With comparison functions defined for all protocols, the objective function itself can be written:

Steps	***objective_function(*** $\overrightarrow{params}$ ***)***
1	$E = 0$
2	For each *proto* in P
3	$\vec{y}_{proto} = \boldsymbol{protocol}\left(\overrightarrow{params}\right)$
4	$\vec{y}_{err} = \vec{p}_{proto} - \vec{y}_{proto}$
5	$E = E + \vec{y}_{err} \cdot \vec{y}_{err}$

which calculates the minimizable sum of squared errors for all training protocols using the dot product.

2.5.2 Fitting the Model Using Global Optimization

With the objective function fully defined, our example Markov model is ready for fitting. During the fitting process, a global optimization algorithm approximates the set of input parameters that minimizes the value of the objective function and thus maximizes the model's similarity to our training data.

We suggest choosing the Simulated Annealing algorithm as the global optimizer. Modern simulated annealing has proven effective at finding solutions to objective functions with large search spaces and many local minima, both of which are characteristics of Markov model cost functions [13].

1. Before providing the model to an optimizer, bounds for each type of input parameter should be specified. In our example model, there are two types parameters with differing bounds:

$$\text{Rate parameters } \alpha_i, \beta_i : [-10, 10]$$

$$\text{Voltage parameters } A_v, B_v : [-80, 60]$$

By bounding all optimized parameters, the optimization search space is limited, and difficulty of convergence is greatly reduced.

2. Once bounds have been specified, the model is fit using the chosen optimization algorithm. In our example, the function for the optimizer to minimize is $\boldsymbol{objective\_function}\left(\overrightarrow{\boldsymbol{params}}\right)$. A call to an optimization algorithm typically returns the fitted parameters and the model's objective cost as follows:

$$\overrightarrow{fitted\_params}, cost = \boldsymbol{simulated\_annealing}(\boldsymbol{objective\_function}, \boldsymbol{bounds}) \quad (21)$$

2.6 Evaluating Model Fit and Uncertainty

After obtaining new parameters representing the final fitted model, each protocol can be re-run using the new optimized parameters to produce the model's representation of the ion channel activity. In this section, we will visualize and evaluate our resulting model's outputs to determine its quality and specificity.

2.6.1 Measures of Model Quality

Although the cost output by the objective function represents the literal distance between the model's output and training data, the ultimate summation of errors from each protocol results in a direct correlation between number of training protocols and overall model cost. Therefore, cost alone cannot be used to evaluate model quality.

Instead, visualization of each protocol's training data overlaid with model outputs provides a clear method to qualitatively confirm that the model's activity correctly represents the intended ion channel.

As visualized in Fig. 3, our fitted example model closely represents all training protocols and thus has a satisfactory fit. Additionally, any model output point that lies within the experimental error of the training datapoint is considered fully fit as the true value of the datapoint will always lie somewhere within its error range.

For additional confirmation, individual current traces from the fitted model can be plotted directly from $\boldsymbol{simulate\_current}$. We can instruct the model to simulate a depolarization by recording current after a large voltage increase:

$$\vec{V}_{trace} = \langle -100, -10 \rangle \quad (22)$$

$$\vec{T}_{trace} = \langle 0, 10 \rangle \tag{23}$$

$$\vec{I}_{trace} = \boldsymbol{simulate\_current}\left(\overrightarrow{\boldsymbol{fitted\_params}}, \vec{V}_{trace}, \vec{T}_{trace} \right) \tag{24}$$

And plot the resulting current trace over time (Fig. 3d), showing that the model's traces are similar to those observed in vitro. Adding more training protocols should improve the model's raw current trace, but the activity shown in Fig. 3d is to be expected with our small example training ensemble.

2.6.2 Measures of Model Uncertainty

State-dependent Markov models can often produce the same current output from significantly different parameter sets. By optimizing the same model topology and training protocol ensemble multiple times, multiple potential configurations of the same model can be fit. Comparing these fits allows us to determine the variability in model output that is due to unknown discrepancies between the fitted model and the theoretically perfect model for the training ensemble. Both the model output and its parameter

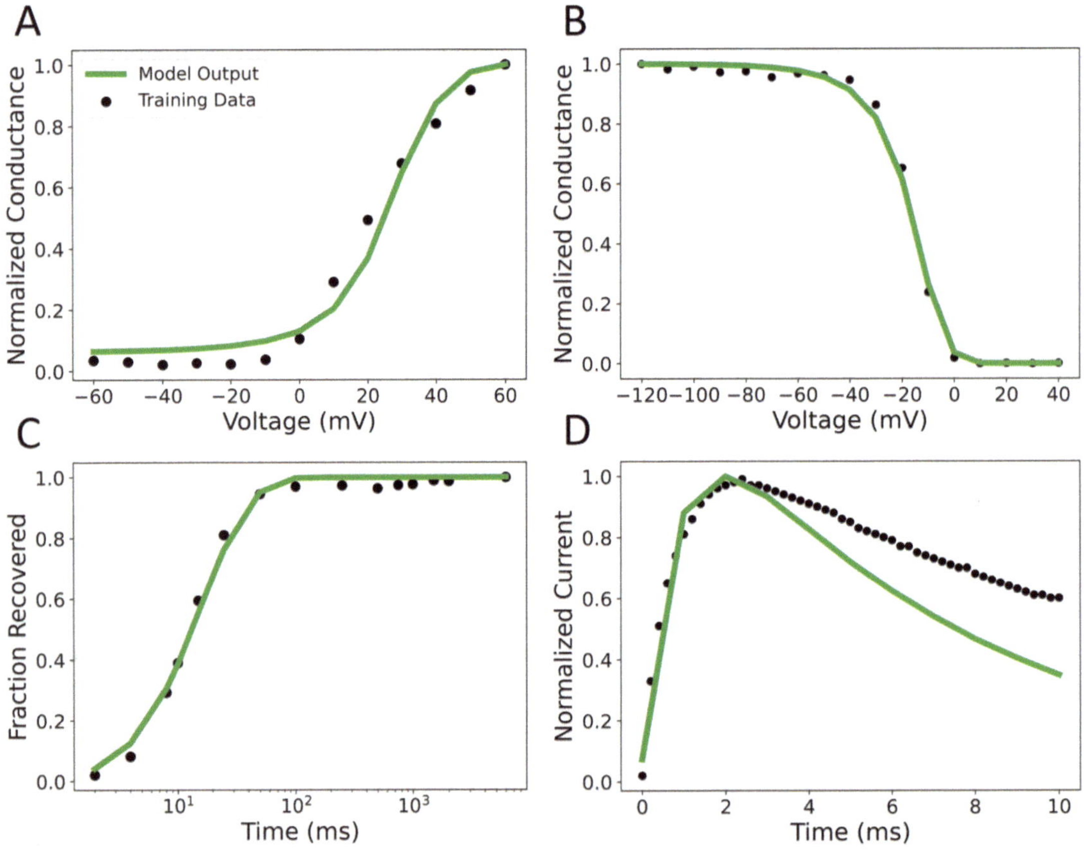

Fig. 3 Fitted model outputs alongside its training data for all training protocols. (**a**) Voltage-conductance, (**b**) recovery from inactivation, (**c**) steady-state inactivation, and (**d**) a trace of the model's reproduction of $I_{to,\ f}$

values can be compared to gain different insight into the model's ability to fit the provided data.

Comparing model outputs in Fig. 3a–c shows that although there is some uncertainty, all fits for our example model converge to a similar range of values.

By comparing model outputs, we can determine the model's overall ability to recreate the activity described in its training protocols [14].

Alternatively, comparing model parameter values in Fig. 4d shows that a wide range of different parameter values produce output that satisfies the training conditions.

Comparing model parameter values provides a measure of the discrepancy between the state-by-state activity in the model and the unknown true state activity of ion channels in a cell. Because the true state activity is theoretically perfect, this discrepancy measures the model's ability to generalize and respond to unseen protocols [14].

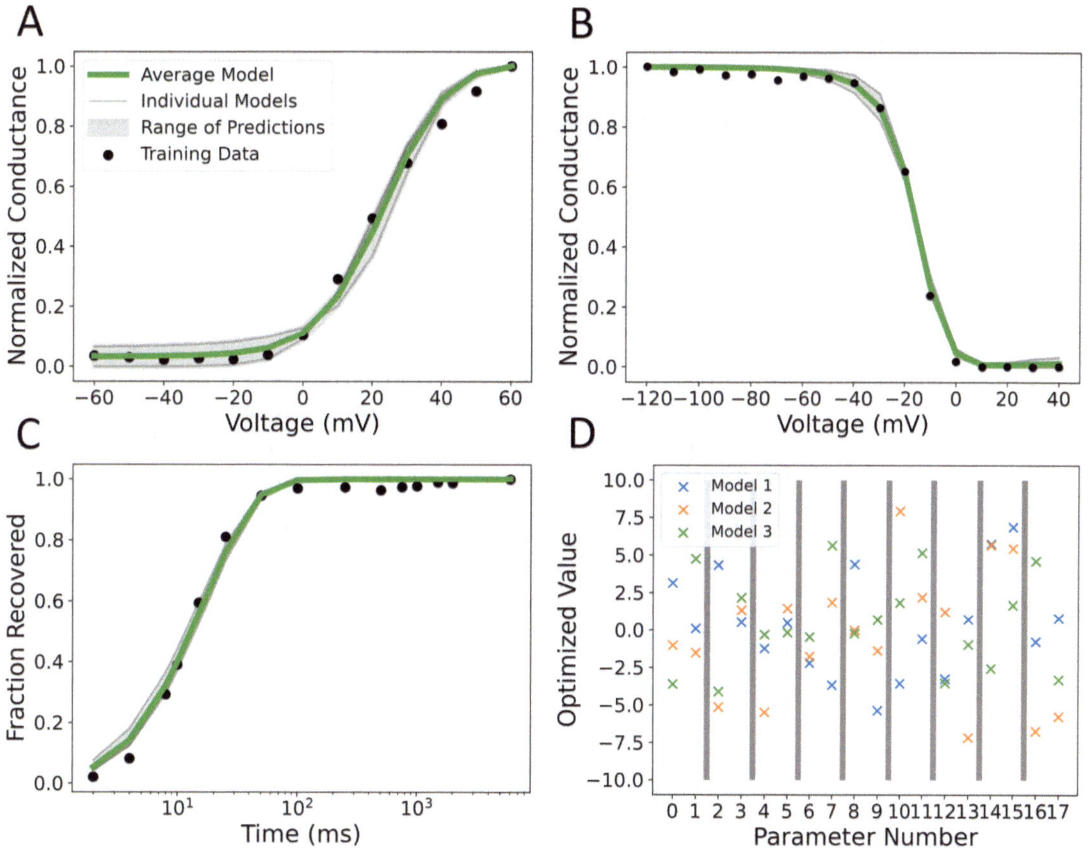

Fig. 4 Visualized uncertainty over multiple fitting attempts for all protocols. (**a**) Voltage-conductance, (**b**) recovery from inactivation, and (**c**) steady-state inactivation. Additionally, (**d**) shows the final optimized rate parameter values for each model, grouped by their respective rate equation

2.7 Additional Fitting Strategies

Our presented methodology focuses on the fundamentals required to create a basic Markov model of the $I_{to,\ f}$ cardiac ion channel, but fitting may become more difficult in alternative use cases, such as when modeling the precise state-specific activity of an ion channel or modeling different ion channels with more complex training ensembles. In this section, we will present additional strategies to improve the predictive abilities of Markov models if a better fit is needed.

2.7.1 Additional Training Protocols

The addition of training data generally leads to an improvement in the predictive ability of a model, but as discussed in Subheading 1, different training protocols apply pressure to the model's output in differing ways. Each of the two following additional training protocols will be presented in response to a potentially undesirable trait observed in a model's raw current traces.

2.7.2 Time to Peak

If a fitted model is observed as activating too quickly or slowly, the Time to Peak protocol can be added to modulate the exact time that peak current is observed during a depolarization:

1. Steady-state occupancies are calculated at a chosen initial holding voltage.
2. A depolarizing pulse across a range of desired voltages is applied.
3. The time at which peak current occurs is recorded.

By adding the Time to Peak protocol, the model's peak current can be forced to occur at a specific point in time, thereby controlling its speed of activation.

2.7.3 Time Constant of Inactivation

If a fitted model is observed as inactivating too quickly or slowly, the Time Constant of Inactivation protocol can be added to adjust the time required for the model to reach a certain level of inactivation—50% in the case of our $I_{to,\ f}$ dataset:

1. Steady-state occupancies are calculated at a chosen initial holding voltage.
2. A depolarizing pulse across a range of desired voltages is applied.
3. The time at which 50% inactivation has occurred is recorded.

By adding the Time Constant of Inactivation protocol, the model's inactivation speed is matched to the inactivation speed of the training data and can be controlled.

2.7.4 Overfitting Prevention

Overfitting of a model can occur when an optimizer achieves a very low-cost value for its training ensemble, but doesn't correctly reproduce ion channel activity at untested points—such as between training point time values and outside of the training protocol's time range.

We choose to address overfitting through the creation of a validation ensemble from the initial training dataset and the subsequent early stopping of optimization if overfitting is observed [6]. Our overfitting prevention protocol is implemented as follows:

1. Randomly select between 20% and 30% of training datapoints from each protocol in the ensemble for use in the validation dataset.
2. Replicate all protocols in the training ensemble to create a validation ensemble, and populate the training datapoints of each protocol with the chosen points in **step 1**. Remove any validation datapoints from their original training protocols.
3. Use the validation protocols to construct a validation objective function with an error equal to the sum of errors across all validation protocols plus the error returned by the original objective function. This gives a new validation error which can be used for overfitting detection.
4. Fit the model using a global optimizer as before using the original objective function but while also calculating and observing the error from the validation objective function at regular intervals.
5. If the validation error increases as training error decreases, overfitting is likely occurring, and fitting should be stopped.

Note that implementing overfitting prevention significantly reduces the amount of training data available to the optimizer, likely increasing fitting difficulty. Therefore, when using particularly small original datasets, the percentage of datapoints removed for validation may need to be reduced to achieve a sufficient fit.

3 Conclusion

In this chapter, we present a generalized method for creating and fitting custom Markov models of ion channels, alongside an explicit example fitted to the fast transient outward potassium channel. By providing a framework for the computer simulation of ion channel activity, we hope to increase the accessibility of simulation-supported research and provide additional utility to already collected electrophysiology data.

Improving the ease with which ion channel models are created can enable enhancement of current research techniques, such as the supplementation of in vitro electrophysiology data with in silico simulated recordings [3] while also enabling the use of new techniques such as digital replication of entire organs for personalized medicine [5].

References

1. Luo CH, Rudy Y (1994) A dynamic model of the cardiac ventricular action potential. I. Simulations of ionic currents and concentration changes. Circ Res 74(6): 1071–1096. https://doi.org/10.1161/01.res.74.6.1071
2. Hodgkin AL, Huxley AF (1952) A quantitative description of membrane current and its application to conduction and excitation in nerve. J Physiol 117(4):500–544. https://doi.org/10.1113/jphysiol.1952.sp004764
3. Colatsky T, Fermini B, Gintant G, Pierson JB, Sager P, Sekino Y, Strauss DG, Stockbridge N (2016) The Comprehensive in Vitro Proarrhythmia Assay (CiPA) initiative—update on progress. J Pharmacol Toxicol Methods 81: 15–20. https://doi.org/10.1016/j.vascn.2016.06.002
4. Clancy CE, Rudy Y (1999) Linking a genetic defect to its cellular phenotype in a cardiac arrhythmia. Nature 400(6744):566–569. https://doi.org/10.1038/23034
5. Trayanova NA (2011) Whole-heart modeling: applications to cardiac electrophysiology and electromechanics. Circ Res 108(1):113–128. https://doi.org/10.1161/CIRCRESAHA.110.223610
6. Mangold KE, Wang W, Johnson EK, Bhagavan D, Moreno JD, Nerbonne JM, Silva JR (2021) Identification of structures for ion channel kinetic models. PLoS Comput Biol 17(8):e1008932. https://doi.org/10.1371/journal.pcbi.1008932
7. Beattie KA, Hill AP, Bardenet R, Cui Y, Vandenberg JI, Gavaghan DJ, de Boer TP, Mirams GR (2018) Sinusoidal voltage protocols for rapid characterisation of ion channel kinetics. J Physiol 596(10):1813–1828. https://doi.org/10.1113/JP275733
8. Grant AO (2009) Cardiac ion channels. Circ Arrhythm Electrophysiol 2(2):185–194. https://doi.org/10.1161/CIRCEP.108.789081
9. Patel SP, Campbell DL (2005) Transient outward potassium current, 'Ito', phenotypes in the mammalian left ventricle: underlying molecular, cellular and biophysical mechanisms. J Physiol 569(Pt 1):7–39. https://doi.org/10.1113/jphysiol.2005.086223
10. Menon V, Spruston N, Kath WL (2009) A state-mutating genetic algorithm to design ion-channel models. Proc Natl Acad Sci USA 106(39):16829–16834. https://doi.org/10.1073/pnas.0903766106
11. Colquhoun D, Dowsland KA, Beato M, Plested AJ (2004) How to impose microscopic reversibility in complex reaction mechanisms. Biophys J 86(6):3510–3518. https://doi.org/10.1529/biophysj.103.038679
12. Teed ZR, Silva JR (2016) A computationally efficient algorithm for fitting ion channel parameters. MethodsX 3:577–588. https://doi.org/10.1016/j.mex.2016.11.001
13. Azizi N, Zolfaghari S (2004) Adaptive temperature control for simulated annealing: a comparative study. Comput Oper Res 31(14): 2439–2451. https://doi.org/10.1016/S0305-0548(03)00197-7
14. Shuttleworth JG, Lei CL, Whittaker DG et al (2023) Empirical quantification of predictive uncertainty due to model discrepancy by training with an ensemble of experimental designs: an application to ion channel kinetics [Internet]. arXiv; 2023 [Cited 6 Sept 2023]. http://arxiv.org/abs/2302.02942

Chapter 10

Human Kir2.1 Potassium Channel: Protocols for Cryo-EM Data Processing and Molecular Dynamics Simulations

Carlos A. H. Fernandes and Catherine Vénien-Bryan

Abstract

Kir channels are potassium (K^+) channels responsible for the mechanism of inward rectification, which plays a fundamental role in maintaining the resting membrane potential. There are seven Kir subfamilies, and their opening and closing mechanism is regulated by different regulatory factors. Genetically inherited defects in Kir channels are responsible for several rare human diseases, and for most of them, there are currently no effective therapeutic treatments. High-resolution structural information is not available for several members within the Kir subfamilies. Recently, our group achieved a significant breakthrough by utilizing cryo-EM single-particle analysis to elucidate the first structure of the human Kir2.1 channel. We present here the data processing protocol of the cryo-EM data of the human Kir2.1 channel, which is applicable to the structural determination of other ion channels by cryo-EM single-particle analysis. We also introduce a protocol designed to assess the structural heterogeneity within the cryo-EM data, allowing for the identification of other possible protein structure conformations present in the collected data. Moreover, we present a protocol for conducting all-atom molecular dynamics (MD) simulations for K^+ channels, which can be incorporated into various membrane models to simulate different environments. We also propose some methods for analyzing the MD simulations, with a particular emphasis on assessing the local mobility of protein residues.

Key words Kir channels, Kir2.1, Inward rectification, Cryo-electron microscopy, Molecular dynamics simulations

1 Introduction

Inward-rectifier potassium (Kir) channels are a particular group of K^+ channels that conduct K^+ ions easier in the inward direction (into the cell) than in the outward direction (out of the cell) [1]. They are responsible for the inward rectification mechanism that results from a block on the cytoplasmic side of the channels by endogenous polyamines and by Mg^{2+} ions that plug the channel pore at depolarized potentials, resulting in decreased outward currents [2]. The blockers are then removed from the pore when the K^+ ions flow into the cell at hyperpolarized potentials [3]. There are

Simone Furini (ed.), *Potassium Channels: Methods and Protocols*, Methods in Molecular Biology, vol. 2796,
https://doi.org/10.1007/978-1-0716-3818-7_10,

seven Kir subfamilies (Kir1-7), distinguishable by their varying rectification strength and their responses to cellular signals. The overall effect of rectification allows Kir channels to play a key role in eukaryotic cells by driving the resting membrane potential to E_K (K^+ equilibrium potential) when the cell is at rest, regulating pancreatic insulin secretion, contributing to renal K^+ transport, controlling muscle contraction, and regulating pacing in cardiac cells and neurons [4]. These functions are achieved by regulating the opening and closing (i.e., gating) of Kir channels through regulatory factors, such as pH (Kir1.1), the anionic lipid phosphatidylinositol 4,5-bisphosphate (PIP_2) (Kir2), PIP_2 and Gβγ subunits from heterotrimeric G-proteins (Kir3), and ATP (Kir6) [1]. For this reason, genetic alterations in Kir channels underlie many of the hereditary ion channel diseases known as channelopathies, which affect the function of multiple organ systems. A few examples of such genetically altered Kir channels include Kir1.1 (type II Bartter Syndrome) [5], Kir4.1 (seSAME syndrome) [6], Kir6.2 (neonatal diabetes) [7], and Kir2.1 channels, which are quite particular because there are four genetically inherited diseases associated with them, namely, Anderson syndrome (AS) [8], catecholaminergic polymorphic ventricular tachycardia (CPVT) [9], familiar atrial fibrillation (FAF) [10], and short QT syndrome 3 (SQT3) [11]. These diseases are chronically debilitating, and there is currently no effective therapeutic treatment, except for neonatal diabetes-related mutations in Kir6.2 channels, for which efficient treatments are available depending on the phenotype [7].

Experimentally determined structures using X-ray crystallography or cryo-electron microscopy (cryo-EM) are not available for several members of the Kir family, including Kir1, Kir4, and Kir5 subfamilies. This lack of structural information hinders a comprehensive understanding of the gating mechanism underlying these K^+ channels. Additionally, the absence of such structural data hampers the assessment of how disease-related mutations impact the structure and function of these channels. Recently, our group solved, by employing cryo-EM single-particle analysis, the first structure of the human Kir2.1 channel, encompassing both transmembrane and cytoplasmic domains (PDB ID 7ZDZ, EMDB ID EMD-14678) in the nonconductive closed state [12]. Given the major role that cryo-EM single-particle analysis technique has achieved in recent years regarding the high-resolution determination of membrane protein structures [13, 14], we present a cryo-EM data processing protocol herein, applicable to the determination of structures of other ion channels by cryo-EM single-particle analysis. This protocol is founded upon the methodology employed in the processing of the human Kir2.1 channel cryo-EM data, which contained a heavy layer of detergent (DDM) around the transmembrane domain that was treated along the processing workflow. We also present a protocol to evaluate the structural heterogeneity of cryo-EM data sets that allows the identification of other possible protein structure conformations present in the collected data set.

Proteins are very dynamical entities that present a series of motions that result in an ensemble of conformations, through which they perform their functions [15, 16]. Moreover, the membrane environment significantly influences the structure, dynamics, and function of ion channels [17–19]. Taking these assumptions into account, we present a protocol to perform all-atom molecular dynamics (MD) simulations applied to the Kir2.1 channel embedded in a membrane model. This protocol can be used for the investigation of the motion and dynamics of protein residues and extended to various ion channels and membrane proteins embedded in different membrane environments. We also propose some methods for analyzing the MD simulations. We present a protocol for evaluating the root mean square fluctuation (rmsf) of the residues, which assess the local mobility of distinct protein regions along the MD simulations. Finally, we also describe a protocol for measuring the distances of two residues (K64 and R67) to the membrane that may be relevant to Kir2.1 gating mechanism [12, 20]. This approach can be applied to assess the distances of any residue relative to the membrane from other K^+ channels.

2 Materials

1. A GPU-containing workstation with any Linux distribution with the following software installed: RELION-3.1 (or superior version) [21], crYOLO-1.7.6 (or superior version) [22], CTFFIND-4.1 [23], NAMD 3.0 [24], Chimera [25], Phenix [26], and VMD [27].

3 Methods

3.1 Cryo-EM Data Processing

3.1.1 Beam-Induced Motion Correction and CTF Estimation

1. In your workstation with any Linux distribution, create a folder called *Micrographs*, and put all the collected movies inside. Then, open the graphical user interface (GUI) of the RELION software by tipping in the Linux terminal the command *relion*. RELION will be the software employed to process the majority of cryo-EM data processing steps, excluding particle picking (performed by crYOLO; *see* Subheading 3.1.2) and resolution estimation and B-factor map sharpening (performed by Phenix; *see* **steps** 7, 9, and **10** at Subheading 3.1.5). The first step in RELION is to execute the importation of the recorded set of movies into the RELION pipeline. For that, select the *Import* tool, and in the *Movies/mics* tab, select *Yes* for *Import raw/-movies/micrographs*. In the *Raw Input files* field, browse for the

path of the folder containing the movies collected (e.g., /*Micrographs/*.tiff*). Select *Yes* for *Are these multi-frame movies?* and provide the *Pixel size (Angstrom)*, *Voltage (kV)*, and *Spherical aberration (nm)*. These parameters are obtained during the data collection. *Amplitude contrast* should be *0.1*. In the case of the human Kir2.1 channel structure, the movies were collected in electron-counting mode on a Titan Krios G3 microscope operated at *300* kV equipped with a Falcon 3 direct electron detector available at Pasteur Institute (Paris, France) using a pixel size of *0.86* Å with a spherical aberration of *2.7* nm. Browse for the MTF file of the detector. *See* **Note 1** for the MTF file of the Falcon 3 detector in electron-counting mode at 300 Kv. For the MTF file of other detectors, please see https://github.com/3dem/relion/tree/ver3.1/data.

2. Do the beam-induced motion correction of the movies collected using the RELION implementation of the MotionCor2 [28]. For that, select the *Motion Correction* tool and in the *I/O* tab, and browse for the *.star file provided by the *Import* previous job (*movies.star*). For the *First frame corrected sum*, insert 3 to remove the first two frames of the movies that normally contains higher motion than remaining frames. In *Dose per frame (e/A2)*, insert the value provided by data collection (in the case of our data set, 1.7139). For *Save sum of power spectra*, select *Yes*. For all other options, use as default. In the *Motion* tab, select *5* and *5*, respectively, for the *Number of patches X, Y* parameter, and use the other options as default. Finally, in the *Running* tab, set the number of MPI processes and threads to do this step. Since this job is not GPU-accelerated, you can use four MPI processes and six threads for a local machine; however you must take into account the amount of RAM memory of your workstation to define these parameters. The setting of the number of MPI processes and threads is dependent on the power of each workstation. The memory usage is roughly proportional to the number of MPI processes, but not the number of total threads. For a detailed discussion on how to set the number of MPI processes and threads in RELION, please see https://github.com/3dem/relion/issues/784 and https://www3.mrc-lmb.cam.ac.uk/relion/index.php/Benchmarks_%26_computer_hardware#Single_machine_.28workstation.29.

3. Select the beam-induced corrected micrographs with a global motion estimation lower than 20 Å. For that, select the *Subset selection* tool, and in the *I/O* tab, browse for the *.star file provided by the previous motion correction job (*corrected_micrographs.star*) in the *select from micrographs.star* field. Leave the other fields in blank. In the *Subset selection* tab, set *Yes* for *Select based on metadata values* and *No* for *Select on image*

statistics. In the *Metadata label for subset selection* field, type *rlnAccumMotionTotal*, and set *20* for *Maximum metadata value*. Set *No* in all available options in the *Class Options* and *Duplicate* tabs.

4. Do the contrast transfer function (CTF) parameters estimation of the motion-corrected micrographs filtered by the global motion estimation. Select the *CTF estimation* tool, and in the *I/O* tab, browse for the *.star file provided by the previous subset selection job (*micrographs.star)*, and use the other parameters as default. In the *CTFFIND-4.1*, select *Yes* for *Use CTFFIND-4.1?*, and browse the CTFFIND-4.1 executable in your machine (e.g., */usr/local/src/ctffind-4.1.3-linux64/bin/ctffind*). In *Use micrograph without dose-weighting?* select *Yes*. In the *Gctf* tab, select *No* in the *Use Gctf instead?* parameter. In the *Running* tab, set the number of MPI processes to execute this task. As this job is not GPU-accelerated, requires less memory than motion-correction job, and does not use threading, you can use one MPI process per CPU available in your workstation. Do the command *nproc* in a Linux terminal to consult the number of CPUs of your machine.
5. Select the corrected micrographs that have a CTF maximum resolution better than 6.0 Å resolution. For that, select the *Subset selection* tool, and in the *I/O* tab, browse for the *.star file provided by the previous CTF estimation job (*micrographs_ctf.star*) in the *select from micrographs.star* field. Leave the other fields in blank. In the *Subset selection* tab, set *Yes* for *Select based on metadata values* and *No* for *Select on image statistics*. In the *Metadata label for subset selection* field, type *rlnCtfMaxResolution*, and set *6* for *Maximum metadata value*. Set *No* in all available options in the *Class Options* and *Duplicate* tabs.

3.1.2 Picking the Particles in the Micrographs

1. For particle picking, use crYOLO 1.8.0 software, which performs an automated picking of particles using a deep learning neural network after training a specific picking model from the collected data set [22]. Create a folder called *cryolo* inside the folder created by RELION after MotionCorr job (which should be */MotionCorr/job002*). Subsequently, in the folder *cryolo*, create other four folders: *full_data*, *train_annotation*, *train_images*, and *boxfiles*. Copy the micrographs that were selected based on the global motion estimation and CTF-Maximum Resolution to the *full_data* folder. To visualize which micrographs were selected, open the *micrographs.star* file provided by the last subset selection job (**step 5** at Subheading 3.1.1) in any text editor, and check the field *rlnMicrographName* to isolate the name of the micrographs,

and copy them to the *full_data* before executing crYOLO; *see* **Note 2**.

2. Do the command *cryolo_boxmanager.py* to open the graphical interface of the box manager of crYOLO. Then, go to *File > Open > SPA > Folder*, then select the folder *full_data*, and click in *Choose*. In *Visualization* tab, set the box size as *200* pixels. The appropriate box size should be set at 1.5 to 2 times of the size of the largest axis of the particle. For example, the maximum longitudinal length of the Kir2 channels is approximately 120 Å. Given the pixel size of *0.86 Å*, which was defined during the data collection, the largest axis of the particle spans around 139 pixels. Consequently, a box size of *200* pixels is a suitable choice for picking this particle. For a more detailed discussion regarding the selection of the box size choice, please see https://jiang.bio.purdue.edu/how-big-should-my-particle-box-be/ and https://blake.bcm.edu/emanwiki/EMAN2/BoxSize. To enhance the visualization of the particles, consider increasing the image contrast by setting the *Low pass filter cut-off* parameter as *0.2* (or higher value, if necessary) in the *Filtering* tab for each image.
3. Manually select a subset of particles to serve as training data for the automated particle picking model. Pick around 2000–4000 particles from a variety of images taken at different defocus ranges. Ensure that the images are not selected in sequence, i.e., choose images from different positions of the image list. Be mindful to select particles from different orientations to mitigate preferential orientational problems. Save the particle selection by *File > Save*, and choose the *train_annotation* folder. To help visualize the shape of the particles and all their possible orientations on the micrographs, you can generate theoretical low-pass projections of a homologous protein structure (*see* **Note 3**).
4. Copy the micrographs used to create the training data to the *train_images* folder. In the Linux terminal, inside the *train_annotation* folder, do the command *ls *star>list.txt*. Then, replace the "*star*" to "*tiff*" in the list.txt file in any text editor, and copy this file to the *full_data* folder. Subsequently, go to the *full_data* folder, and do the following command in the Linux terminal *cat list.txt | while read line; do cp "$line" ../train_images/; done.*
5. Train the automated picking model based on the set of particles that you selected. For that, copy the *config.json file* (*see* **Note 4**) to the *cryolo* folder. Then, inside this folder, do the following command in the Linux terminal *cryolo_train.py -c config.json -w 3 -g 0*. After finishing this first task, do the command *cryolo_train.py -c config.json -w 0 -g 0 -e 10.*

6. Perform the automated particle picking in all images by performing the command *cryolo_predict.py -c config.json -w model.h5 -i full_data/ -g 0 -o boxfiles/ -t 0.3.*
7. Evaluate the automated picking results by opening the *cryolo_-boxmanager.py.* Open all the micrographs again (*File > Open > SPA > Folder*, select the folder *full_data*), and then, *File > Import box files.* You can visualize all the particles picked automatically by crYOLO. If it is needed to pick more particles, redo the automated picking command by decreasing the value of the parameter *-t* (0.25, 0.2). On the other hand, if it was picked a great amount of particles in the carbon foil or objects that are not compatible with the particles, increase the *-t* parameter value (0.35, 0.4).
8. Copy the *.box files available in the */cryolo/boxfiles/EMAN* to the *Micrographs* folder created by RELION after MotionCorr job (which should be */MotionCorr/job002/Micrographs).* In the */cryolo/boxfiles/EMAN* folder, you can do the following command in the Linux terminal *cp *.box /MotionCorr/job002/Micrographs/.*

3.1.3 2D Classification and Obtention of a De Novo 3D Initial Model

1. Import the particles picked by crYOLO in the RELION pipeline. Select the *Import* tool, and in the *Movies/mics* tab, select *No* for *Import raw/movies/micrographs.* In the tab *Others*, select *Yes* for *Import other node types.* For *Input file*, insert the path of the box files provided that you copied to the *Micrographs* folder created by RELION after MotionCorr job (*/MotionCorr/job002/Micrographs/*.box*). For *node type* parameter, select *Particle Coordinates* (**.box, *_pick.star*).
2. Do the particle extraction of the micrographs. Select the *Extraction* tool, and in the *I/O* tab, browse for the *micrographs.star* file resulted from the CTF-maximum resolution subset selection job (**step 5** at Subheading 3.1.1). In *Input coordinates*, insert the *coord_suffix.box* file provided by the Importation job from the previous step. In the *Extract* tab, inform the particle box size (*200 pixels*). In order to speed up computational time for the 2D Classification and the initial classification of 2D class averages containing suitable particles, down-scale the particles four times. For that, in the *Rescale particles*, select *Yes* and for *Rescale particles*, insert *50* pixels. For the other options, use as default. In the *Running* tab, set the number of MPI processes to execute this task. As this job is not GPU-accelerated and does not use threading, you can use one MPI process per CPU available in your workstation.
3. Perform the 2D Classification. Select the *2D Classification* tool, and in the *I/O* tab, insert the file *particles.star* resulted from the *Extraction* job from the previous step. In the

Optimization tab, the parameters that should be set are as follows: *Number of classes, 100*; *Regularization Parameter T, 2*; *Number of interactions, 50*; *Mask diameter (A), 200*; for the other parameters, use as default. In the *Compute* tab, select *Yes* in the Use GPU acceleration. If left blank, the job itself will try to allocate all available GPU resources. Otherwise, you can inform a list of which GPU will be used for this job. For example, to use GPUs 0 and 1, type *0:1*. In the *Running* tab, select the number of MPI processes and threads to execute this task. You can use one MPI process per GPU plus one and eight to ten threads per MPI process. For example, in a computer containing 2 GPUs, you may set 3 MPI processes (to use 2 GPUs) and 24 threads (8 threads per each MPI process). If you have a GPU memory sufficiently large, you can set more than one MPI process per GPU. Check the GPU utilization by doing the command *nvidia-smi* in a Linux terminal. If the GPU memory utilization remains constantly lower than 60–70%, you can try running two MPI processes per GPU. For the parameters in the other tabs, use as default.

4. Select the best 2D class averages. For that, select the *Subset selection* tool, and in the *I/O* tab, browse for the *model.star* file of the last iteration provided by the 2D Classification job (*run_it50_model.star*). In the *Class Options* tab, set *Yes* for *Re-center the class averages* and *No* for *Regroup the Particles*. In the *Subsets* tab, set *No* for all parameters. After clicking in *Run*, it will open a new window, where you can select *Sort images on rlnClassDistribution* and *Reverse sort*, to show the 2D class averages ranked according to the number of the particles, starting from the one that contains the largest number of particles. After clicking in *Display*, it will open a new window that shows all the 2D class averages. Select the best ones that assemble different orientations of the particles. After the selection, do a right click on the selection, and choose the option *Save selected classes* to create a *particles.star* file containing all the particles presented in the selected 2D class averages. To see the number of the particles containing the selection, after right clicking the selection, choose the option *Show metadata this class*. *See* **Note 5** (Fig. 1) for examples of good/selected and bad/discarded 2D class averages of the human Kir2.1 channel. For the human Kir2.1 channel data set, it was performed only one round of 2D Classification. However, for some data sets, it is useful to perform further particle cleaning at 2D level performing additional 2D classifications with selection of the good 2D class averages. For that, select the *2D classification* tool, and in *I/O* tab browse for the *particles.star* file provided by the previous 2D class averages subset selection. For the second and third rounds of 2D Classification, you can

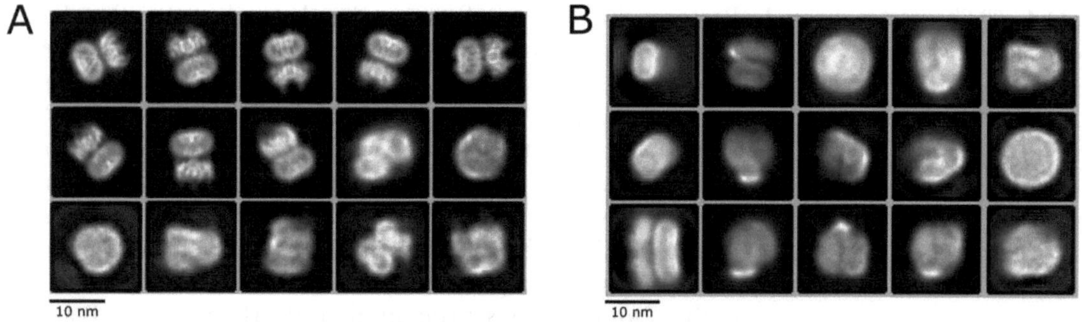

Fig. 1 Representative 2D class averages obtained from human Kir2.1 channel in the presence of the detergent DDM obtained from a cryo-EM data set obtained in a Titan Krios G3 microscope operated at 300 kV equipped with a Falcon 3 direct electron detector (BioQuantum K3 Imaging Filter) available at Pasteur Institute (Paris, France). (**a**) Good quality 2D class averages; (**b**) Bad quality 2D class averages

decrease the number of the classes required (e.g., *50*) and the number of iterations (e.g., *25*).

5. Do a new particle extraction of the micrographs without downscaling before creating a de novo 3D Initial Model. For that, select the *Particle extraction* tool, and in the *I/O* tab, browse for the *micrographs.star* file resulted from the *Subset selection* job after the last selection of the suitable 2D class averages. Leave the *Input coordinates* in blank, and set *Yes* for *re-extract refined particles.* In the field *Refined particles star file*, browse for the *particles.star* file resulted from the *Subset selection* job after the last selection of the suitable 2D class averages. In the *Extraction* tab, set the *Particle box size (pix)* to 200 (the real value of the box size for the particles), and select *No* in the *Rescale particles* field. Leave all remaining parameters as default. In the *Running* tab, set the number of MPI processors to execute this task as you did for the previous extraction job.
6. Create a de novo 3D Initial Model from the particles contained in the selected 2D class averages. For that, select *3D Initial Model* tool, and in the *I/0* tab, browse for the *particles.star* file provided by the previous extraction job. In the *Optimisation* tab, the parameters that should be set are *Mask diameter (A) 200* and *Symmetry C4*; for the other parameters, use as default. Despite it is not usual to apply symmetry in the construction of the 3D Initial Model, we obtained much better results in the quality of the model by applying symmetry C4 at this step of data processing. In the *SGD* tab, set the parameters: *Number of initial interactions, 50; Number of in-between interactions, 200; Number of final interactions, 50; Initial resolution (A), 35; and Final resolution (A), 15*; for the other parameters, use as default. In the *Compute* tab, set *Yes* for GPU acceleration, and in the *Running* tab, set the number of MPI and threads as you did for the 2D Classification.

3.1.4 3D Classification and 3D Refinement

1. Perform the 3D Classification. For that, select the *3D Classification* tool, and in the *I/0* tab, browse for the *particles.star* file provided by last *Extraction*. In the *Reference map* field, browse for the map provided by *3D Initial Model* job (*run_it300_-class001.mrc*). In the *Reference* tab, set the parameters *Initial low-pass filter (A) 40* and *Symmetry C1* (even the reference map having C4 symmetry). In the *Optimisation* tab, set *Number of classes 4* and *Mask diameter 180*. Use the other parameters as default. In the *Compute* tab, set *Yes* for GPU acceleration, and in the *Running* tab, set the number of MPI and threads as you did for the *2D Classification* and *3D Initial Model* jobs. For the options in the other tabs, use as default.
2. Open the resulting maps (*run_it025_class00?.mrc*, being *?* the number of each class) in Chimera software, and choose the best quality class by visual inspection. You should open the resulting maps (*.mrc files*) of each class in Chimera to visualize the quality of the map of each 3D class to identify the best 3D one for further processing. *See* **Note 6** (Fig. 2) to observe the good/ selected and the bad/rejected 3D classes for human Kir2.1 channel.
3. Select the subset of particles contained in the selected best 3D class in RELION. For that, select the *Subset selection* tool, and in the *I/0* tab, browse for the *model.star* file of the last iteration provided by the 3D Classification job (*run_it50_model.star*), and do the same procedure as described before for selection of 2D class averages (**step 4** at Subheading 3.1.3). By right-clicking on each class and selecting in *Show metadata this class*, it is possible to see the number of the class to identify the class that you selected after the visual evaluation in Chimera. Similar as performed in the selection of 2D Average Classes, do a right click on the selected class, and choose the option *Save selected classes*. Similar as 2D classification, you can perform several rounds of 3D Classification in order to clean particles and obtain a homogeneous set of particles able to provide a high-resolution map. To perform another round of 3D Classification, select the *3D Classification* tool, and in the *I/O tab*, browse for the *particles.star* file resulted from the previous 3D class subset selection job. In the *reference map* field, browse for the map (*run_it025_class00?.mrc* file) corresponding to the 3D class that you previously selected. For the other parameters, you set the same ones that you used in the previous rounds of 3D Classification. For the human Kir2.1 channel data set, it was performed three rounds of 3D Classification, the first two using C1 symmetry and the last one using C4 symmetry. *See* **Note 6** to observe the good/ selected and the bad/rejected 3D classes obtained in each round of 3D Classification.

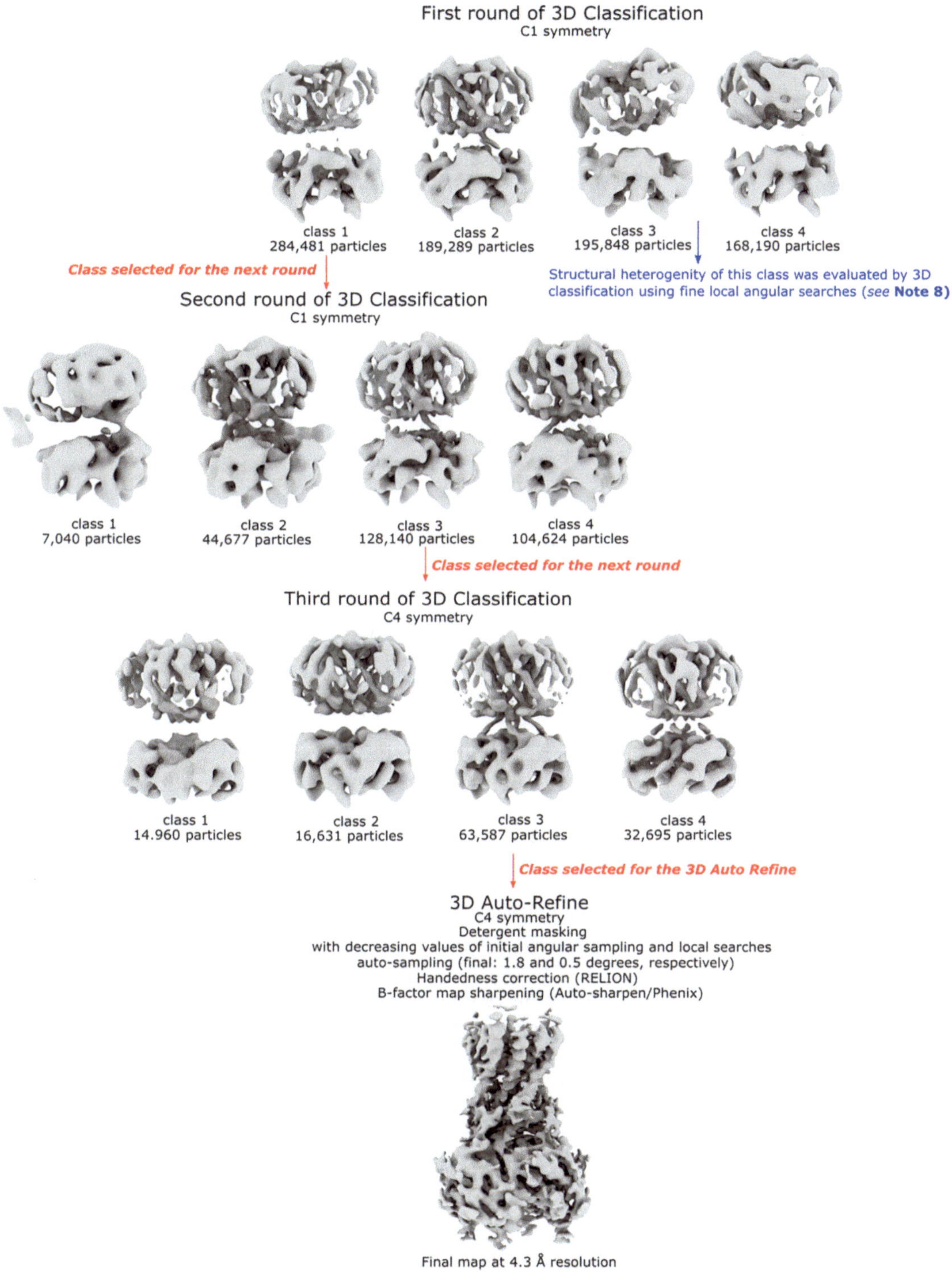

Fig. 2 Maps of all 3D classes obtained in the three rounds of the 3D Classification performed for the determination of the cryo-EM structure of the human Kir2.1 channel. Red arrows indicate the classes that were selected for a subsequent round of 3D Classification or 3D Auto Refine. Blue arrow indicates the class that was subject to a round of 3D Classification using fine angular searchers to assess its structural heterogeneity (*see* **Note 8**). The number of particles contained in each 3D Class is indicated

4. Do the 3D Auto Refine. For that, select the *3D Auto Refine* tool, and in the *I/O* tab, browse for the *particles.star* file resulted from the last 3D class subset selection job in the *Input images STAR file* field (**step 3**). In the *Reference Map* field, browse for the map corresponding to the 3D class that you previously selected. In the *Reference Map* tab, set *25* for initial low-pass filter and *symmetry C4*. In the *Optimisation tab*, set *180* in the *Mask Diameter (A)*. In the *Compute* tab, set *Yes* for GPU acceleration, and in the *Running* tab, set the number of MPI and threads to execute this task. The 3D Auto Refine job requires an odd number of MPI processes, with a minimum number of 3. The last iteration of 3D Auto Refine increases significantly the memory usage, and then you may run as many threads as you have cores in your workstation. For the parameters in the other tabs, use as default.

3.1.5 Masking the Detergent Layer

1. In the case of human Kir2.1 channel cryo-EM data set, it was noticed the presence of a heavy layer of detergent (DDM) around the TMD that impaired a proper visualization and structural analysis of this region (*see* Figs. 1 and 2 in **Notes 5** and **6** to visualize the DDM layer). Thus, this detergent layer was masked followed by a masked 3D Auto Refine. However, if this is not necessary in your case, go directly to **step 6** for further 3D Auto Refine decreasing the values of *initial angular sampling* and *local searches from auto-sampling* parameters.
2. First, create a volume of the protein without the detergent layer in Chimera, which will be used for the creation of a mask in RELION. There are two methods for that:

   Method 1: Open the *run_class001.mrc* file generated by the 3D Auto Refine Job (**step 4** of Subheading 3.1.4) in Chimera. In the *Volume Viewer* panel, set *step 1*, and select a threshold level where all structural features are clearly visible (it will be used 0.0040, e.g., in the next steps). Avoid using an excessively high threshold level that might obscure structural features or a too low one that introduces significant noise and dust. Then, in the *Volume Viewer* panel, navigate to *Tools* > *Segger (Segment Map)*, and click on *Segment*. Using *Ctrl* + *Shift* and the *left mouse* button, select the regions corresponding to the detergent layer. Ensure all areas containing the detergent layer are selected. If the areas are large enough to encompass both protein and detergent layer portions, it is possible to select such an area, and click on *Ungroup* to divide in smaller areas where it is possible to distinguish between protein and detergent fragments. Once all areas are selected, go to the *Segger* panel, and click on *Group* to group all areas containing the detergent layer. Then, go to *Regions* > *Subtract from* *the Selected Map*. This will result in a new volume in the *Volume Viewer* panel, containing only the protein without the detergent layer (*volume_imasked #2*). Save this file as a . mrc file (for instance, *mask_chimera.mrc*).

Method 2: Open the *run_class001.mrc* file generated by the 3D Auto Refine Job (**step 4** of Subheading 3.1.4) in Chimera. In the *Volume Viewer* panel, set *step 1*, and select a threshold level where all structural features are clearly visible (it will be used 0.0040, e.g., in the next steps). Avoid using an excessively high threshold level that might obscure structural features or a too low one that introduces significant noise and dust. Then, navigate to *Favorites* > *Command line*, and do the following commands in the command line: *vop threshold #0 minimum 0.0040 set 0* and, subsequently, *vop threshold #1 maximum 0.0039 setMaximum 1*. This will generate a new volume (*number* #2) in the *Volume Viewer* panel. Then, in the *Volume Viewer* panel, go to *Tools* > *Volume Eraser*. By manipulating the sphere and clicking the *Erase* button on the *Volume Eraser* panel, you can gradually remove the detergent layer from volume *#2*. Once finished, do the following command in the command line *vop subtract #2 #1*. This will result in a binarized threshold mask (*volume_difference #4* on the *Volume Viewer* panel) representing the protein without detergent layer. Save this file as a .mrc file (e.g., *mask_chimera.mrc*).

3. Create a mask for the protein without detergent layer in the RELION. For that, RELION creates the mask by applying a low-pass filter in the input map and using an initial threshold for mask binarization. Thus, it is important to select these both parameters carefully. In the Linux terminal, execute the following command *relion_image_handler --i mask_chimera.mrc --o maskfromchimera_l10.mrc --angpix 0.86 --lowpass 10*. For the pixel size, use the value provide in **step 1** at Subheading 3.1.1. For the low pass, use 10 or 15. Avoid using lower values as they may create overly tight masks. Open the *maskfromchimera_l10.mrc* file from Chimera. In the *Volume Viewer* panel, set *step 1*, and select a threshold level where the entire map is visualized without the detergent. Similar as before, avoid using an excessively high threshold level that might obscure structural features, and create a tight mask or a too low one that introduces significant noise and dust. Note down this threshold level value, as it will be used later in RELION.
4. Return to the RELION GUI and select the *Mask Creation* tool. In the *I/0* tab, browse for the mask that you created in Chimera (by method 1 or 2) in the *Input 3D Map* field. In the *Mask* tab, set the *Lowpass filter map (A)* parameter with the

same value that was used to perform the threshold evaluation in Chimera in **step 3** (*10*). For the *initial binarization threshold* field, set the threshold level that wrote down after evaluation in Chimera. For *extended binary map this many pixels* and *add a soft-edge of this many pixels* parameters, set *3* for both. In the *Running* tab, set the number of MPI processes. Open the mask resulted from this job (*mask.mrc*) in Chimera to evaluate whether it was created properly. It appears too tight; it is possible to set higher values (e.g., 6) for the *extended binary map this many pixels* and *add a soft-edge of this many pixels* parameters to loosen the mask.

5. Select the *3D Auto Refine* tool, and in the *I/O* tab, browse for the *particles.star* file resulted from the last 3D class subset selection job in the *Input images STAR file* field (**step 3** of Subheading 3.1.4). In the *Reference Map* field, browse for the *run_class001.mrc* file resulted from the 3D Auto Refine Job (**step 4** of Subheading 3.1.4), and in the *Reference Mask* field, browse for the *mask.mrc* file resulted from the previous *Mask Creation* job in RELION. For the remaining parameters, use the same values as **step 4** of Subheading 3.1.4. Open the *run_class001.mrc* provided by this task, and check if the detergent masking was performed correctly. If this masked 3D Auto Refinement did not yield satisfactory results, *see* **Note 7** to remove the detergent signal from the particles using a particle subtraction operation.
6. To enhance the resolution of the final map obtained after 3D Auto Refine, consider conducting several iterative rounds of this task with decreasing the values of *initial angular sampling* and *local-searches from auto-sampling* parameters. For that, after the first round of 3D Auto Refine, select the *3D Auto Refine* tool, and in the *I/O* tab, browse for the *particles.star* file resulted from the last 3D class subset selection job in the *Input images STAR file* field (**step 3** of Subheading 3.1.4). In the *Reference Map* field, browse for the *run_class001.mrc* file resulted from the 3D Auto Refine job from the previous job, and in the *Reference Mask*, browse for the *mask.mrc* file resulted from the previous *Mask Creation* job in RELION (if it was not necessary to perform the masked 3D Auto Refine, provide the *run_class001.mrc* obtained in **step 4** of Subheading 3.1.4 as *Reference Map*, and leave the *Reference Mask* field empty). For the remaining parameters, use the same values as in **step 4** of Subheading 3.1.4, except for the following adjustments: change *Initial angular sampling* parameter to 3.7°, and set *Local searches from auto-sampling* parameter to 0.9°. After completing this round, initiate a new *3D Auto Refine* task, using the result *run_class001.mrc* resulted from this round as the new input map in the *Reference map* field, and decrease the

Initial angular sampling, and set *Local searches from auto-sampling* parameters to 1.8° and 0.5°, respectively. The final cryo-EM map of the human Kir2.1 channel was obtained after a 3D Auto Refine job using 1.8° and 0.5° for these parameters.

7. Evaluate the resolution of the map using *Mtriage* tool in Phenix. In the Phenix GUI, search for the *Mtriage* tool, and in the *Configure* tab, browse for the files of the full map (*run_class001.mrc*) and both half-maps (*run_half?_class001_unfil.mrc*) files resulted from the last round of 3D Auto Refine. Run the job, and after a few seconds, *Mtriage* will provide the resolution of the model along with the FSC curve.
8. Occasionally, a visual map inspection can indicate that the handedness of the map is incorrect, such as when the α-helices turn the wrong way around. In such cases, it is necessary to flip the handedness of the full map (*run_class001.mrc)* resulted from the last round of 3D Auto Refine by executing the following command in a Linux terminal *relion_image_handler --i run_class001.mrc --o run_class001_invert.mrc --invert_hand*. Perform the same command to correct the handedness of both half-maps (*run_half?_class001_unfil.mrc*) also.
9. Sharpen the map using *Autosharpen* tool from Phenix. In the Phenix GUI, search for the *Autosharpen map or map coefficients* tool, and in the *Configure* tab, browse for both half-maps (*run_half?_class001_unfil.mrc*) resulted from the last round of 3D Auto Refine. Run the job with other parameters as default (ensure the *Automatically determine sharpening* parameter is selected), and after some minutes, *Autosharpen* will provide a sharpened version of the map (*sharpened_map.ccp4* file).
10. Evaluate the local resolution of the map using *Local Resolution Map* tool in Phenix. In the Phenix GUI, search for this tool, and in the *Configure* tab, browse for the files of both half-maps. Set the number of processors available on the workstation. The *Resolution* parameter is optional. To see the local resolution of the map, open the full map provided by the last round of 3D Auto Refine (*run_class001.mrc)* or the sharpened map (*sharpened_map.ccp4*), and subsequently, open the **_local_resolution.ccp4 map* resulted from the *Local Resolution* job. In the *Volume Viewer* panel, navigate to *Tools > Surface Color*. In *Color Surface* field, select the full or sharpened map, in *by* field select *volume data value*, and in *volume file* field, select the local resolution map. Then, click on *Color* to see the local resolution map of the structure.

3.1.6 Evaluation of the Structural Heterogeneity of the Cryo-EM Data Set

1. Small protein conformational differences may not be adequately separated using the default angular sampling in 3D Classification. An alternative approach to separate structures with conformational differences is to employ a previously

obtained consensus map resulted from 3D Auto Refine as input for a 3D Classification using finer local angular searches [29].

2. During the cryo-EM data processing of the human Kir2.1 channel, it was noticed in the first round of 3D Classification the presence of one class that there is a lateral opening at the interface between CTD and TMD (*see* **Note 8**, Fig. 3). To assess the structural heterogeneity of this 3D class, a subsequent 3D Classification was performed using finer angular searches.
3. The first step to perform a 3D Classification using finer angular searches is to execute a particle extraction job, as described before (**step 3** at Subheading 3.1.4), from the 3D class that it is to be assessed.
4. Select the *3D Classification* tool in the RELION GUI, and in the *I/O* tab, browse for the particles.star file provided by the previous extraction job. In the *Reference Map* field, browse for the full map provided by the last round of 3D Auto Refine (*run_class001.mrc*). In the *Optimisation tab*, set *6* as the *number of classes*, and use the same parameters as previously applied (**step 1** at Subheading 3.1.4). In the *Sampling* tab, select *Yes* for the *Perform local angular searches* field, and set *2.5* in the *Local angular search range*. Set the remaining parameters in the other tabs as previously performed (**step 1** at Subheading 3.1.4). Ensure that no symmetry is applied (*C1 symmetry*).
5. Open the resulting maps (*run_it025_class00?.mrc*, being ? the number of each class) using Chimera software, and evaluate the structural differences between the classes. *See* **Note 8** (Fig. 3) for visualizing the varying degrees of the lateral opening of the human Kir2.1 channel obtained in the six different 3D Classes resulting from a 3D Classification using fine local angular searches.

3.2 Molecular Dynamics Simulations of Human Kir2.1 Channel

All the inputs needed to perform the molecular dynamics (MD) simulations were generated using the *Input Generator Tool—Membrane Builder* in CHARMM-GUI server (https://www.charmm-gui.org/?doc=input/membrane.bilayer) [30]; more details will be described below in Subheading 3.2.2. You should create an account in CHARMM-GUI to use this tool. However, there are two preparative steps that should be performed before (Subheading 3.2.1).

3.2.1 Preparative Steps

1. Do the prediction of protein positioning in lipid bilayers. For that, download the pdb file of the cryo-EM structure of the human Kir2.1 channel (PDB ID 7ZDZ). Open the file as a text, and remove the lines concerning the K^+ and Sr^+ ions (lines 11258–11260). Then, upload the *.pdb file without the ions

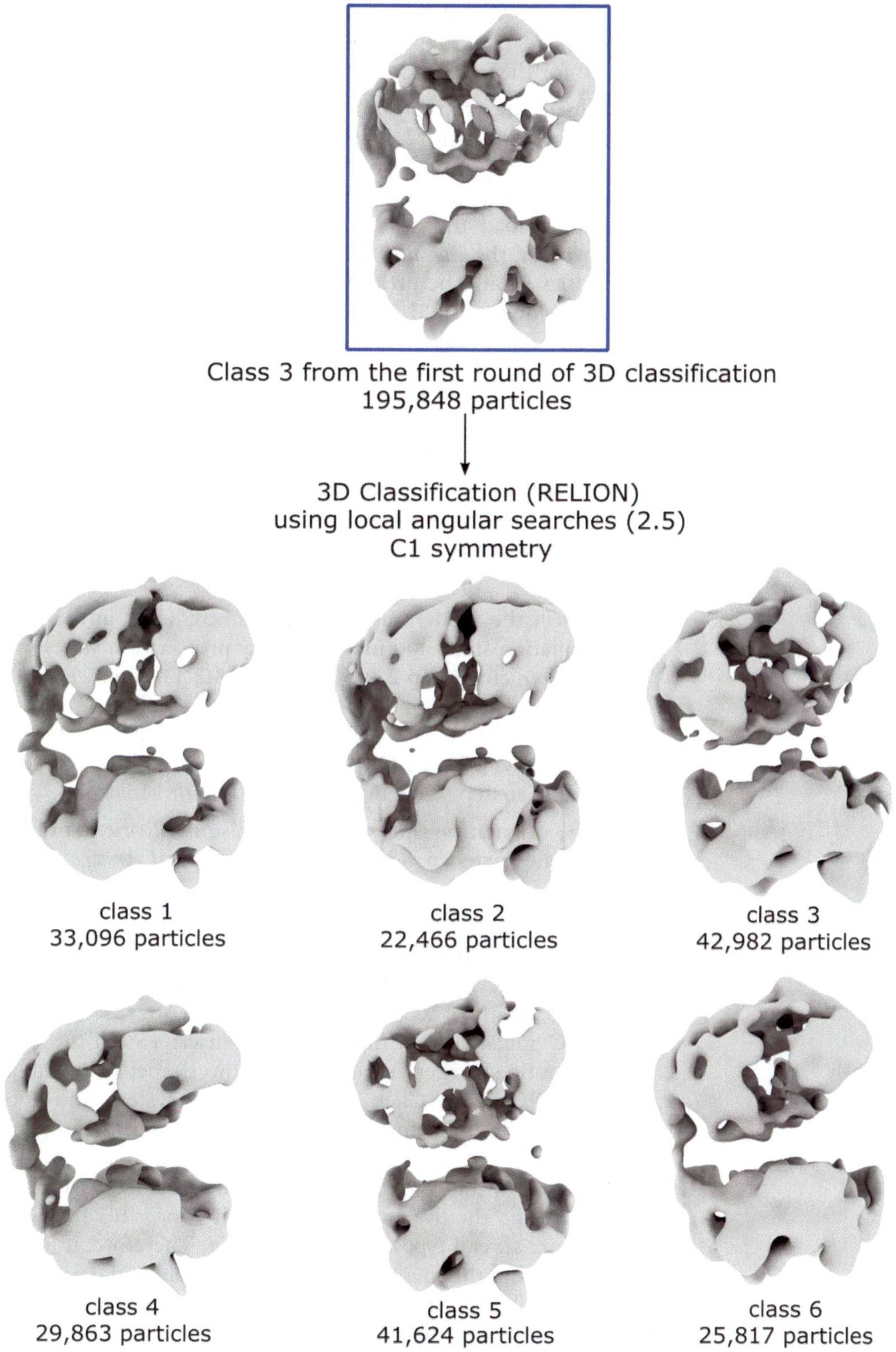

Fig. 3 3D Classification of the particles from the class 3 of the first round of 3D classification using fine angular searches performed during the cryo-EM data processing of the human Kir2.1 channel. The maps and the number of particles for the six classes obtained in this 3D classification are shown

on the PPM 2.0 web server (https://opm.phar.umich.edu/ppm_server2) [31]. Download the output pdb file; it will be used for the preparation of the inputs on CHARMM-GUI.

2. Evaluate the protonation of the side chains. The *.pdb file of the structure of Kir2.1 without the ligands needs to be converted in a *.pqr file using the APBS-PDB2PQR software suite (https://server.poissonboltzmann.org/pdb2pqr) [32]. Upload the *.pdb file, and run as default (pH 7.0). Download the *.log file*, and evaluate which Asp, His, and Glu residues should be protonated and which Cys, Tyr, Lys, and Arg residues should be deprotonated according to the predicted pKa of each residue.

3.2.2 Running MD Simulations of Kir2.1 Channel Embedded in Bilayer Membranes Formed by POPC Lipids

1. Upload the PDB file that was generated using PPM server (**step 1** at Subheading 3.2.1) in *Input Generator Tool—Membrane Builder—Bilayer Builder* in CHARMM-GUI (https://www.charmm-gui.org/?doc=input/membrane.bilayer), and go to the next step (*Select Model/Chain*).
2. Choose the four chains, and go to the next step (*Manipulate PDB*).
3. Unselect the model missing residues box. Click in the box of protonation state, and include all the protonation states previously predicted (**step 2** at Subheading 3.2.1). After, click in the box of disulfide bonds, and include the following disulfide bonds: PROA 122–PROA 154/PROB 122–PROB 154/PROC 122–PROC 154/PROD 122–PROD 154. Go to the next step (Generate PDB and Orient Molecule).
4. In the orientation options, click in Use PDB orientation, and go to the next step (*Calculate Cross-Sectional Area*).
5. Click on view structure at step2_oriented.pdb to evaluate the good orientation of the membrane in relation to the protein structure (the position of the membrane is represented by the yellow planes). At system size determination options, insert 110 at length of X and Y (initial guess). After, search below for PC lipids, click in the arrow, and insert 150 in the upperleaflet ratio (Integer) and 158 in the lowerleaflet ratio (Integer) for POPC lipids. Click in the show the system info above. Check if the following parameters are selected: heterogenous lipid, length of Z based on water thickness 22.5, and length of XY based on ratios of lipid components. Go to the next step (Determine the System Size). Here, it is possible to see that there are several types of lipids available. Therefore, it is possible to create the membrane bilayer with the lipid desired. It is also possible to create the membrane mixing different lipids and in different proportions.

6. Put a concentration of 0.15 M KCl (neutralizing option should be selected). Click in Calculate Solvent Composition. Check if the following parameters are selected: replacement method and check lipid ring, include ions and basic ion types. Go to the next step (Build Components). If the CHARMM ends abnormally, refresh the page using F5 until it works. The following two steps are informational only, so follow to the next step (*Assemble Components*).
7. In input generation options, click in the box of NAMD, and set temperature for 293 K. Check the Force field option CHARMM36m. Check if the following parameters are selected: more CHARMM minimization during input generation, generate grid information for PME FFT automatically, and NPT ensemble. All the other options should be unselected. Go to the next step (*Generate Equilibration and Dynamic Inputs*).
8. Download the download .tgz file to a computer/cluster. The equilibration inputs (step 6—composed by six substeps, 6.1, 6.2, 6.3, 6.4, 6.5, and 6.6) and production input (step 7, composed by a unique step) will not run in CHARMM-GUI web server anymore. They will run in a computer/cluster containing the software NAMD. The scripts will be inside the NAMD folder after downloading and unpacking the download .tgz file.
9. Run the equilibration inputs 6.1, 6.2, 6.3, and 6.4 using NAMD without any modifications. In the input 6.5, change the run parameter (last parameter of the file) to 500,000 (instead of 250,000 as in the original file). In the input 6.6, change the first timestep parameter to 1,135,000 (instead of 885,000 as in the original file) and the run parameter (last parameter of the file) to 500,000 (instead of 250,000 as in the original file). To run namd, do the following command in the Linux the terminal *namd3 + idlepoll + p6 + devices 0,1 name_of_input.inp > name_of_input.out &*. *+p* refers to the number of processors used for this job (in this case, choose based on what you have available on your machine). *+devices* refers to the GPUs used for this job (in this case GPUs 0 and 1; choose based on what you have available on your machine).
10. The production input 7. Change the numsteps parameter to 100,000,000 and the run parameter to 100,000,000 to achieve 200 ns of MD simulations.

3.2.3 Analysis of MD Simulations

1. The analysis of Cα root mean square deviation (rmsd) is performed in the VMD. In the VMD software, load the step5_input.psf. After, right click on the name step5_input.pdf, and load the trajectories file (step7_production.dcd). Go to *Extensions > Analysis > RMSD Visualizer Tool*, select the *Backbone*

option at *Atom Selection Modifiers*, and then click in *ALIGN*. Then, in *atom selection* write *protein and segid PROA*, click in *RMSD* and then in *Plot result*. It will be plotted in the rmsd from chain A (PROA). It will open a Multiplot window, and you can export the graph in the *File* option. For the other chains, do the same procedure, changing *PROA* to *PROB*, *PROC*, and *PROD*.

2. The analysis of Cα root mean square fluctuation (rmsf) of the residues along MD simulations is performed using an in-house tcl script (rmsf.tcl; *see* **Note 9**). This script must be in the same folder of all files produced by MD simulations. In the VMD software, do the same as before; load the step5_input.psf file, and subsequently, load the trajectories file (step7_production.dcd). Then, navigate to *Extensions* > *Tk Console*. Then, type *source rmsf.tcl*. The output (rmsf.dat) will appear in the folder, and it is numbered 1 to 1291 (there is no separation between the chains), so it is needed to separate the respective chains: Chain A, 1 to 325; Chain B, 326–646; Chain C, 647–967; Chain D, 968–1291. You can draw the graphics of Cα rmsf per chain using Xmgrace or any software of your preference.
3. The analysis of the distances between the K64 and R67 residues to the membrane (plane of the phosphate groups from the inner membrane) was performed using an in-house tcl script (distance_membrane.tcl; *see* **Note 10**). This script must be in the same folder of all files produced by MD simulations. In the Linux terminal, do the following command *vmd -dispdev text -e distance_membrane.tcl*. Now, edit the output.dat files to get the first and last (column 5) columns of these files (which contain the necessary information), as example: in a Linux terminal, do the command *awk '{print $1,$5}' distance_to_-membrane_monA_200ns_K64.dat* > *K64_A.dat*. In the output *K64_A.dat*, it will be the distance to the membrane (axis y) along the MD simulations time (axis x; in nanoseconds). You can draw the graphics using Xmgrace or any software of your preference. If you would like to evaluate the distances of other residue to the membrane, change the last lines of the script (*see* **Note 10**) to the residues that you are interested in evaluating the distances to the membrane along MD simulations.

4 Notes

1. MTF file of the Falcon 3 detector in electron-counting mode at 300 Kv (copy and save as *mtf_falcon3EC_300kV.star* file).

```
data_mtf_f3ec_300kv
loop_
```

```
_rlnResolutionInversePixel
_rlnMtfValue
0 1
0.0078125 0.9964851529
0.015625 0.9935477645
0.0234375 0.9908861019
0.03125 0.9880337235
0.0390625 0.9814730437
0.046875 0.973590513
0.0546875 0.9648087389
0.0625 0.9554497043
0.0703125 0.9455722965
0.078125 0.9349830869
0.0859375 0.9234127068
0.09375 0.9107339924
0.101562 0.8970834075
0.109375 0.882816709
0.117188 0.86832079
0.125 0.853789483
0.132812 0.8391130803
0.140625 0.8239667126
0.148438 0.8080357402
0.15625 0.7912221696
0.164062 0.7737090404
0.171875 0.755859011
0.179688 0.7380137468
0.1875 0.720315353
0.195312 0.7026623656
0.203125 0.6848208758
0.210938 0.6665986007
0.21875 0.6479621604
0.226562 0.6290412613
0.234375 0.6100404732
0.242188 0.5911270856
0.25 0.5723743125
0.257812 0.5537992
0.265625 0.5354536931
0.273438 0.5174750916
0.28125 0.5000367775
0.289062 0.483227546
0.296875 0.4669472011
0.304688 0.4508987469
0.3125 0.4346962785
0.320312 0.4180348611
0.328125 0.4008286407
0.335938 0.3832389573
0.34375 0.3655779587
0.351562 0.3481510778
```

```
0.359375 0.331139073
0.367188 0.3145842216
0.375 0.298471077
0.382812 0.2828388979
0.390625 0.2678464558
0.398438 0.2537284049
0.40625 0.2406577445
0.414062 0.2286300856
0.421875 0.2174726552
0.429688 0.2069493639
0.4375 0.1968663323
0.445312 0.1871215361
0.453125 0.1776948039
0.460938 0.1685941212
0.46875 0.164686055
0.476562 0.1607952981
0.484375 0.1565754705
0.492188 0.1517875288
```

2. To isolate the name of the micrographs and copy them to the *full_data* folder to execute crYOLO, you can perform some operations in the Linux terminal. For example, in the folder created by RELION after selection based on CTF Maximum Resolution (it should be */Select/job005/*), you can type *awk '{print $1}' micrographs.star > micrographs_name.txt*. Open the *micrographs_name.txt* file in any text editor, and remove all the extra information on the head of the file and the path to the micrograph file. Leave only the name of the micrographs written in this file. Then, copy this file in the folder *Micrographs* generated by RELION after MotionCorr (it should be */MotionCorr/job002/Micrographs/*), and, inside this folder, do the following command in the Linux terminal *cat micrographs_name.txt | while read line; do cp "$line" ../cryolo/full_data; done.*
3. To help visualize the shape of the particles and all their possible orientations on the micrographs, it is possible to generate low-pass theoretical 2D projections of the particles using a homologous structural model or a computational model obtained by AlphaFold or homology modelling. The first step is to convert the .pdb file of the structural model to an electron density map, using EMAN, for example, by doing the following command in a Linux terminal *e2pdb2mrc.py --center --res 5 model.pdb model5A.mrc*. The *--res* command refers to the resolution that you want for the output electron density map, *model.pdb* is the input homologous model, and the *model5A.mrc* is the output electron density map at 5 Å resolution. Then,

to generate the theoretical 2D projections, do the following commands in a Linux terminal, in order:

(I) `relion_image_handler --i model5A.mrc --o protein5A_lp.mrc --angpix 1 --lowpass 10`

(II) `relion_project --i protein5A_lp.mrc --o model5A_lp.proj --nr_uniform 10`

(III) `relion_display --i model5A_lp.proj.mrcs`

After the third command, ten theoretical 2D projections of the particle will be displayed in the computer screen. If you would like to see more views of particle orientation, you can set the *--nr_uniform* parameter to a higher value, *20*, for example.

4. config.json file for the model training in crYOLO (copy and save it as *config.json* file).

```
{
 "model": {
 "architecture": "YOLO",
 "input_size": 1024,
 "anchors": [
 200,
 200
 ],
 "max_box_per_image": 600,
 "num_patches": 1,
 "filter": [
 0.2,
 "filtered"
 ]
 },
 "train": {
 "train_image_folder": "train_image/",
 "train_annot_folder": "train_annotation/",
 "train_times": 10,
 "pretrained_weights": "model.h5",
 "batch_size": 4,
 "learning_rate": 0.0001,
 "nb_epoch": 50,
 "object_scale": 5.0,
 "no_object_scale": 1.0,
 "coord_scale": 1.0,
 "class_scale": 1.0,
 "log_path": "logs/",
 "saved_weights_name": "model.h5",
 "debug": true
```

```
},
"valid": {
"valid_image_folder": "",
"valid_annot_folder": "",
"valid_times": 1
}
}
```


5. Examples of good/selected and bad/discarded 2D class averages obtained during cryo-EM data processing of the human Kir2.1 channel (Fig. 1).
6. Examples of good/selected and bad/discarded 3D classes obtained during the cryo-EM data processing of the human Kir2.1 channel (Fig. 2).
7. If the masked 3D Auto Refine was not performed after the creation of a mask for the protein without the detergent layer, it is possible to perform a particle subtraction operation using RELION to remove the signal of the detergent from the particles. For that, in the RELION GUI, select the *Particle subtraction* tool. In the *I/O tab*, browse for the **_optimiser.star* file (*run_it00?_optimiser.star)* of the last interaction provided by the 3D Auto Refine (**step 4** of Subheading 3.1.4). In the *Mask of the signal to keep* field, provide the *mask.mrc* file created by RELION (**step 4** at Subheading 3.1.5), which is the mask of the protein without the detergent layer. Use the remaining parameters as default, and in the *Running* tab, set the number of MPI processes. This job will result in a *subtracted.star* file. This file will be used in the *Input images STAR file* field of the 3D Auto Refine to refine the map after particle subtraction operation. Use the same parameters as before (**step 5** at Subheading 3.1.5), including the *mask.mrc* created by RELION in the *Reference mask* field.
8. Structural heterogeneity of a particular 3D Class identified after performing a 3D Classification using fine local angular searches during the cryo-EM data processing of human Kir2.1 channel (Fig. 3).
9. Script rmsf.tcl; to analyze the Cα root mean square fluctuation (rmsf) of the residues along the molecular dynamics simulations (copy and save it as rmsf.tcl file).

```
set fp [open "rmsf.dat" w]
set sel [atomselect top "protein and name CA"]
for {set i 0} {$i < [$sel num]} {incr i} {
set rmsf [measure rmsf $sel first 1 last 1999 step 1]
puts $fp "[expr {$i+1}] [lindex $rmsf $i]"
}
close $fp
```

10. Script distance_membrane.tcl; to measure the distances of Lys64 and Arg67 to the membrane along MD simulations (copy and save it as distance_membrane.tcl file).

```
#to execute this script type vmd -dispdev text -e distance_-
membrane.tcl
proc calc_dist {sel1 sel2 out} {
#Open output file to write
set outfile [open $out w]
set numFrame [molinfo top get numframes]
#Defining references
set r1 [atomselect top $sel1]
set r2 [atomselect top $sel2]
puts $outfile "#Time up down dist distfix"
for {set f 0} {$f < $numFrame} {incr f} {
#Read each frame in trajectory
$r1 frame $f
$r1 update
$r2 frame $f
$r2 update
#Time
#set time [format "%.2f" [expr $f * 5 * 0.05]]
set time [expr $f /10.]
puts $time
#Getting Center of Mass of all PO4 (middle of the membrane)
set p1 [measure center $r1 weight mass]
set commemb [lindex $p1 2]
#Defining up and down PO4 atoms
set up []
set down []
set p [$r1 get {z}]
foreach val $p {
if { $val < $commemb } {
lappend down $val
} else {
lappend up $val
}
}
#Getting average of up and down PO4 atoms
set avup [expr ([join $up +])/[llength $up]]
set avdown [expr ([join $down +])/[llength $down]]
#protein residues
set pa [measure center $r2 weight mass]
set comprot [lindex $pa 2]
#Fixed
set comprotfix [expr $avdown - $comprot]
#Write to output
puts $outfile "$time $avup $avdown $comprot $comprotfix"
}
```

```
close $outfile
}

#set rep ${argv}
#Load topology/structure
mol new step6.6_equilibration.pdb type pdb waitfor all
#Remove the first frame
#animate delete beg 0 end 0
#Load trajectories
mol addfile step7_production.dcd type dcd waitfor all
#Call the function
#Selections: membrane protein residues and output file
calc_dist "resname POPC and name P" "segid PROA and resid 64"
"distance_to_membrane_monA_200ns_K64.dat"
calc_dist "resname POPC and name P" "segid PROA and resid 67"
"distance_to_membrane_monA_200ns_R67.dat"
calc_dist "resname POPC and name P" "segid PROB and resid 64"
"distance_to_membrane_monB_200ns_K64.dat"
calc_dist "resname POPC and name P" "segid PROB and resid 67"
"distance_to_membrane_monB_200ns_R67.dat"
calc_dist "resname POPC and name P" "segid PROC and resid 64"
"distance_to_membrane_monC_200ns_K64.dat"
calc_dist "resname POPC and name P" "segid PROC and resid 67"
"distance_to_membrane_monC_200ns_R67.dat"
calc_dist "resname POPC and name P" "segid PROD and resid 64"
"distance_to_membrane_monD_200ns_K64.dat"
calc_dist "resname POPC and name P" "segid PROD and resid 67"
"distance_to_membrane_monD_200ns_R67.dat"
quit
```

Acknowledgments

This work was supported by AFM-Téléthon #23207 for CVB and CAHF, and CAHF has received funding from the European Union's Horizon Europe Research and Innovation Program under grant agreement no. 101026386.

References

1. Hibino H, Inanobe A, Furutani K, Murakami S, Findlay I, Kurachi Y (2010) Inwardly rectifying potassium channels: their structure, function, and physiological roles. Physiol Rev 90:291–366. https://doi.org/10.1152/physrev.00021.2009
2. Doupnik CA, Davidson N, Lester HA (1995) The inward rectifier potassium channel family. Curr Opin Neurobiol 5:268–277. https://doi.org/10.1016/0959-4388(95)80038-7
3. Baronas VA, Kurata HT (2014) Inward rectifiers and their regulation by endogenous

polyamines. Front Physiol 5. https://doi.org/10.3389/fphys.2014.00325
4. Bichet D, Haass FA, Jan LY (2003) Merging functional studies with structures of inward-rectifier K+ channels. Nat Rev Neurosci 4: 957–967. https://doi.org/10.1038/nrn1244
5. Flagg TP, Tate M, Merot J, Welling PA (1999) A mutation linked with Bartter's syndrome locks Kir 1.1a (Romk1) channels in a closed state. J Gen Physiol 114:685–700. https://doi.org/10.1085/jgp.114.5.685
6. Tang X, Hang D, Sand A, Kofuji P (2010) Variable loss of Kir4.1 channel function in SeSAME syndrome mutations. Biochem Biophys Res Commun 399:537–541. https://doi.org/10.1016/j.bbrc.2010.07.105
7. Proks P, Antcliff JF, Lippiat J, Gloyn AL, Hattersley AT, Ashcroft FM (2004) Molecular basis of Kir6.2 mutations associated with neonatal diabetes or neonatal diabetes plus neurological features. Proc Natl Acad Sci USA 101: 17539–17544. https://doi.org/10.1073/pnas.0404756101
8. Plaster NM, Tawil R, Tristani-Firouzi M, Canún S, Bendahhou S, Tsunoda A, Donaldson MR, Iannaccone ST, Brunt E, Barohn R, Clark J, Deymeer F, George AL, Fish FA, Hahn A, Nitu A, Ozdemir C, Serdaroglu P, Subramony SH, Wolfe G, Fu Y-H, Ptáček LJ (2001) Mutations in Kir2.1 cause the developmental and episodic electrical phenotypes of Andersen's syndrome. Cell 105:511–519. https://doi.org/10.1016/S0092-8674(01)00342-7
9. Tester DJ, Arya P, Will M, Haglund CM, Farley AL, Makielski JC, Ackerman MJ (2006) Genotypic heterogeneity and phenotypic mimicry among unrelated patients referred for catecholaminergic polymorphic ventricular tachycardia genetic testing. Heart Rhythm 3:800–805. https://doi.org/10.1016/j.hrthm.2006.03.025
10. Xia M, Jin Q, Bendahhou S, He Y, Larroque M-M, Chen Y, Zhou Q, Yang Y, Liu Y, Liu B, Zhu Q, Zhou Y, Lin J, Liang B, Li L, Dong X, Pan Z, Wang R, Wan H, Qiu W, Xu W, Eurlings P, Barhanin J, Chen Y (2005) A Kir2.1 gain-of-function mutation underlies familial atrial fibrillation. Biochem Biophys Res Commun 332:1012–1019. https://doi.org/10.1016/j.bbrc.2005.05.054
11. Priori SG, Pandit SV, Rivolta I, Berenfeld O, Ronchetti E, Dhamoon A, Napolitano C, Anumonwo J, Di Barletta MR, Gudapakkam S, Bosi G, Stramba-Badiale M, Jalife J (2005) A novel form of short QT syndrome (SQT3) is caused by a mutation in the *KCNJ2* gene. Circ Res 96:800–807. https://doi.org/10.1161/01.RES.0000162101.76263.8c
12. Fernandes CAH, Zuniga D, Fagnen C, Kugler V, Scala R, Péhau-Arnaudet G, Wagner R, Perahia D, Bendahhou S, Vénien-Bryan C (2022) Cryo–electron microscopy unveils unique structural features of the human Kir2.1 channel. Sci Adv 8:eabq8489. https://doi.org/10.1126/sciadv.abq8489
13. Le Bon C, Michon B, Popot J-L, Zoonens M (2021) Amphipathic environments for determining the structure of membrane proteins by single-particle electron cryo-microscopy. Q Rev Biophys 54:e6. https://doi.org/10.1017/S0033583521000044
14. Vénien-Bryan C, Fernandes CAH (2023) Overview of membrane protein sample preparation for single-particle Cryo-electron microscopy analysis. Biol Life Sci IJMS 24:14785.
15. Yang L-Q, Sang P, Tao Y, Fu Y-X, Zhang K-Q, Xie Y-H, Liu S-Q (2014) Protein dynamics and motions in relation to their functions: several case studies and the underlying mechanisms. J Biomol Struct Dyn 32:372–393. https://doi.org/10.1080/07391102.2013.770372
16. Surpeta B, Sequeiros-Borja C, Brezovsky J (2020) Dynamics, a powerful component of current and future in silico approaches for protein design and engineering. IJMS 21:2713. https://doi.org/10.3390/ijms21082713
17. Kimchi O, Veatch SL, Machta BB (2018) Ion channels can be allosterically regulated by membrane domains near a de-mixing critical point. J Gen Physiol 150:1769–1777. https://doi.org/10.1085/jgp.201711900
18. Duncan AL, Reddy T, Koldsø H, Hélie J, Fowler PW, Chavent M, Sansom MSP (2017) Protein crowding and lipid complexity influence the nanoscale dynamic organization of ion channels in cell membranes. Sci Rep 7: 16647. https://doi.org/10.1038/s41598-017-16865-6
19. Zhou H-X, Cross TA (2013) Influences of membrane mimetic environments on membrane protein structures. Annu Rev Biophys 42:361–392. https://doi.org/10.1146/annurev-biophys-083012-130326
20. Cheng WWL, D'Avanzo N, Doyle DA, Nichols CG (2011) Dual-mode phospholipid regulation of human inward rectifying potassium channels. Biophys J 100:620–628. https://doi.org/10.1016/j.bpj.2010.12.3724
21. Scheres SHW (2012) RELION: implementation of a Bayesian approach to cryo-EM structure determination. J Struct Biol 180:519–

530. https://doi.org/10.1016/j.jsb.2012.09.006

22. Wagner T, Merino F, Stabrin M, Moriya T, Antoni C, Apelbaum A, Hagel P, Sitsel O, Raisch T, Prumbaum D, Quentin D, Roderer D, Tacke S, Siebolds B, Schubert E, Shaikh TR, Lill P, Gatsogiannis C, Raunser S (2019) SPHIRE-crYOLO is a fast and accurate fully automated particle picker for cryo-EM. Commun Biol 2:218. https://doi.org/10.1038/s42003-019-0437-z
23. Rohou A, Grigorieff N (2015) CTFFIND4: fast and accurate defocus estimation from electron micrographs. J Struct Biol 192:216–221. https://doi.org/10.1016/j.jsb.2015.08.008
24. Phillips JC, Braun R, Wang W, Gumbart J, Tajkhorshid E, Villa E, Chipot C, Skeel RD, Kalé L, Schulten K (2005) Scalable molecular dynamics with NAMD. J Comput Chem 26: 1781–1802. https://doi.org/10.1002/jcc.20289
25. Pettersen EF, Goddard TD, Huang CC, Couch GS, Greenblatt DM, Meng EC, Ferrin TE (2004) UCSF chimera—a visualization system for exploratory research and analysis. J Comput Chem 25:1605–1612. https://doi.org/10.1002/jcc.20084
26. Liebschner D, Afonine PV, Baker ML, Bunkóczi G, Chen VB, Croll TI, Hintze B, Hung L-W, Jain S, McCoy AJ, Moriarty NW, Oeffner RD, Poon BK, Prisant MG, Read RJ, Richardson JS, Richardson DC, Sammito MD, Sobolev OV, Stockwell DH, Terwilliger TC, Urzhumtsev AG, Videau LL, Williams CJ, Adams PD (2019) Macromolecular structure determination using X-rays, neutrons and electrons: recent developments in *Phenix*. Acta Crystallogr D Struct Biol 75:861–877. https://doi.org/10.1107/S2059798319011471
27. Humphrey W, Dalke A, Schulten K (1996) VMD: Visual molecular dynamics. J Mol Graph 14:33–38. https://doi.org/10.1016/0263-7855(96)00018-5
28. Zheng SQ, Palovcak E, Armache J-P, Verba KA, Cheng Y, Agard DA (2017) MotionCor2: anisotropic correction of beam-induced motion for improved cryo-electron microscopy. Nat Methods 14:331–332. https://doi.org/10.1038/nmeth.4193
29. Scheres SHW (2016) Processing of structurally heterogeneous cryo-EM data in RELION. In: Methods in enzymology. Elsevier, Amsterdam, pp 125–157
30. Jo S, Kim T, Iyer VG, Im W (2008) CHARMM-GUI: a web-based graphical user interface for CHARMM. J Comput Chem 29: 1859–1865. https://doi.org/10.1002/jcc.20945
31. Lomize MA, Pogozheva ID, Joo H, Mosberg HI, Lomize AL (2012) OPM database and PPM web server: resources for positioning of proteins in membranes. Nucleic Acids Res 40: D370–D376. https://doi.org/10.1093/nar/gkr703
32. Jurrus E, Engel D, Star K, Monson K, Brandi J, Felberg LE, Brookes DH, Wilson L, Chen J, Liles K, Chun M, Li P, Gohara DW, Dolinsky T, Konecny R, Koes DR, Nielsen JE, Head-Gordon T, Geng W, Krasny R, Wei G, Holst MJ, McCammon JA, Baker NA (2018) Improvements to the APBS biomolecular solvation software suite. Protein Sci 27:112–128. https://doi.org/10.1002/pro.3280

Chapter 11

Evaluation of the Role of Potassium Channels in the Proliferation, Migration, and Invasion of Blood Cells

Valeria Y. Vasileva and Vladislav I. Chubinskiy-Nadezhdin

Abstract

The potassium channels are one of the key regulators of cell membrane potential and permeability properties of blood cells. The changes in functioning of potassium channels control crucial cell processes such as proliferation, viability, migration, and invasion. The correct estimation of these processes is important for the characterization of physiological and pathophysiological cell states. Here, we present the experimental protocol for evaluation of the role of potassium ion channels in the proliferation, migration, and invasion of blood cells.

Key words Potassium channels, Blood cells, Blood cell migration, Invasion, Proliferation

1 Introduction

The potassium ion channels play multiple roles in the cells of blood origin under normal as well as in pathophysiological conditions [1–3]. Mainly potassium channels control the plasma membrane potential of the cells regulating many important cell processes and reactions. The special role of potassium channels is emphasized in the study of cancer processes. The high expression of some potassium channels is used as a marker for tumor diagnosis and prognosis. Also, many cancers have an aberrant expression of potassium channels that has an impact to proliferation, migration, and invasion of tumor cells [4]. Here, we present specific protocols for the estimation of the role of potassium ion channel modulation in proliferation, viability, migration, and invasion of blood cells. The appropriate number of the cells used for the assays is dependent on particular blood cell type and should be determined before the experiments. For example, the necessary concentrations of the human chronic myeloid leukemia cells for proliferation/migration/invasion could be found in our recent publication [5].

Simone Furini (ed.), *Potassium Channels: Methods and Protocols*, Methods in Molecular Biology, vol. 2796,
https://doi.org/10.1007/978-1-0716-3818-7_11,

2 Materials

1. Pipettes/sterile pipette tips.
2. Conical centrifuge tubes (i.e., 2 mL, 15 mL).
3. Lab centrifuge (2500–3000× *g* speed).
4. Cell counting labware (hemocytometer, automatic cell counters; *see* **Note 1**).
5. Dyes for cell viability estimation (trypan blue, propidium iodide or others, depending on cell counting labware, i.e., the possibility to register fluorescence).
6. Transwell inserts with a permeable membrane of suitable pore diameters. The pore diameter size that fits for the studied blood cell type/line should be selected from literature or experimentally.
7. Multiwell cell culture grade plates for transwell inserts. Transwell inserts that are already pre-installed to 24-well plates are commercially available.
8. Serum-free culture medium.
9. Extracellular matrix component for transwell membrane coating (for invasion assay, e.g., Corning Matrigel©, collagen or other; *see* **Note 2**).
10. Optional: specific chemoattractants for migration/invasion stimulation.

3 Methods

The protocol is suitable for isolated (primary) blood cells or blood cell cultures. Isolate the specific blood cells using the appropriate technique/grow the blood cell lines of interest using standard culturing procedures and reagents according to the cell line-specific protocols.

The blood cell cultures should be free of contamination and actively proliferating.

The potassium channel activity could be modulated using pharmacological tools or genetic manipulation (siRNA, shRNA, etc.). Before the proliferation/migration/invasion assays, the effect on potassium channel activity or expression should be confirmed.

Prepare all stock solutions using sterile solvents, and store the reagents according to the manufacturer's instructions. Prepare fresh working solutions immediately before the experiments, and vortex to ensure the full dissolution of the reagents. Carry out all procedures under sterile conditions. The protocol presented below is optimized for the study of pharmacological potassium channel modulators: if working with genetically manipulated cells, *see* **Note 3** at **step 5** below.

3.1 Cell Proliferation and Viability Assays

1. Gently (but thoroughly) pipette the cells in the culture media, and perform the counting of live/dead cells in the suspension using available lab instrumentation (*see* **Notes 1** and **4**).
2. Transfer the cell suspension to the centrifuge tube, and centrifuge the cells at 2.500–3.000× *g* for 2–3 min at room temperature.
3. Gently remove the supernatant fluid, and then add necessary volume of fresh culture medium to obtain cell concentration that is optimal for growth and proliferation.
4. Thoroughly resuspend the cells in fresh medium, and then transfer the equal volumes of the cells to the individual conical tubes.
5. Add the potassium channel modulators (or vehicle as a control), and vortex the tubes. Each experimental condition should be repeated (at least) in triplicate (*see* **Notes 3** and **5**).
6. Transfer the suspensions from each tube to cell culture plastics used for routine cell cultivation, and incubate the cells for 24 h at standard conditions.
7. After 24 h gently (but thoroughly) resuspend the cells, take the necessary volume of the cells for cell counting, and count the number of live/dead cells in each experimental condition (at least once for each of the triplicates; *see* **Notes 4** and **6**).
8. If necessary, further culture the cells followed by the repeats in cell counting/viability assays at next timepoints (*see* **Note 6**).
9. The counted live/dead cell numbers under each experimental condition allow to determine the effect of potassium channel modulation on growth and viability of the cells. If a number of experimental time points were taken, then it is possible to plot the cell growth curves (number of the live/dead cells vs time from the start of the experiment) for each experimental condition. If several concentrations of potassium channel modulators were tested, it is possible to plot the dose-response curve and to determine half-maximal effective (EC_{50}) or inhibition (IC_{50}) concentrations of the reagent. To summarize, the performed experiments allow one to quantify the effect of potassium channel modulators on the rate of cell growth and their viable status and to determine the effective concentrations of the modulators that significantly affect the proliferative properties of blood cells.

3.2 Migration and Invasion Protocol

1. Repeat **steps 1–2** from Subheading 3.1.
2. Gently remove the supernatant fluid, and then add necessary volume of fresh serum-free culture medium to obtain cell concentration that is optimal for transwell insert assay (see recommendations of the manufacturer). The number of cells

per insert varies in the range 10^5–10^6 depending on the cell type. The total amount of volume in each insert is 400 uL (*see* **step 3** and **Notes 7** and **8**).

3. Install the transwell inserts to each well of a 24-well culture plate, or use the multiwell plates with pre-installed inserts. The transwell setup consists of two chambers: upper, the transwell insert, and lower, the well of a 24-well culture plate. The chambers are divided by the permeable membrane with pores of specific sizes. For migration assay, go to **step 6.** For invasion, follow **steps 4** and **5** and omit **step 6**.
4. Add 100 uL of hydrogel to the insert, and incubate the transwell: time of incubation depends on hydrogel type and should be selected according to the manufacturer's specifications (*see* **Note 2**).
5. Wash the insert 1–2 times with empty culture medium.
6. Pre-soak the transwell membrane by adding 100 uL of serum-free culture media to the upper well of the transwell setup. Incubate for 2–3 min, and then remove the medium.
7. Add 750 uL of full (serum-containing) culture medium to the bottom of the well, and add potassium channel modulators (here and at **step 9**: if working with genetically manipulated cells, *see* **Note 3**).
8. Thoroughly resuspend the cells in serum-free medium, transfer the 400 uL volumes to the individual conical tubes, add potassium channel modulators, and vortex well (*see* **Notes 8–10**).
9. Add 400 uL of the cell suspensions to the insert.
10. Incubate for 4 h (under standard culture conditions, *see* **Note 11**).
11. Remove the transwell inserts from the wells. Gently (but thoroughly) pipette the cell suspension in the bottom of each chamber, and take the equal necessary volumes from each experimental well to perform cell counting and live/dead cell estimation (at least once for each of the triplicates, *see* **Notes 6** and **12**).
12. The counted migrated/invaded cell numbers allow to estimate the effect of potassium channel modulators (or selective genetic depletion) on the cell motility and invasive properties of blood cells of interest. If several concentrations of potassium channel modulators were tested, it is possible to plot the dose-response curve and to determine half-maximal effective (EC_{50}) or inhibition (IC_{50}) concentrations of the reagent. The variation of the time of assay (*see* **step 10**) allows to determine the rate of migration/invasion of blood cells.

4 Notes

1. Refer to the manufacturer's instructions of the labware to determine the range of cell concentrations that is required for proper cell counting. Take into account this range when performing the proliferation/migration/invasion assays. Too low or too high cell number may cause erroneous results.
2. Prepare a working solution of liquid hydrogel (Matrigel, collagen, or other) for coating the transwell inserts before the experiments according to the manufacturer's instructions.
3. If working with the cells in which a genetic depletion of potassium channel was performed, do not add channel modulators at this step.
4. If the number of counted cells is too low, then increase the concentration of the cells at the start of the experiment, or concentrate the cells (using centrifugation followed by the removal of the part of supernatant fluid) before cell counting. Important: ensure that equal volumes of the cells are used for cell counting/viability assays for each experimental condition.
5. Several potassium channel modulators could be at least partially neutralized by the serum used for cell culturing (see, e.g., [6]). Thus, the potential effects of the presence of the serum should be tested and taken into account.
6. Increase in the number of cell counting replicates for each experimental condition would result in more precise determination of the cell number/viability status.
7. The particular migration and invasion protocol aimed at using the serum gradient as a chemoattractant for blood cell migration/invasion. The use of other chemoattractants could be performed using the same protocol with modification of **step 2** (*see* Subheading 3.2 Migration and invasion protocol). Be accurate when working with transwell inserts as the transwell membrane could be easily damaged by pipette tips.
8. If another chemoattractant (than serum) is used in the migration and invasion experiment, then the fresh full culture medium should be added to the cells.
9. If another (than serum) chemoattractant is used in the migration and invasion experiment, it also should be added to the bottom well of the setup.
10. As an additional control (no or very low migration/invasion is expected) in the migration and invasion experiment, add 750 uL of serum-free medium to the bottom well (thus, in the setup there would be no serum/chemoattractant gradients).

11. The incubation time depends on the particular cell type. During that time, the cells would migrate/invade from the upper (serum- or chemoattractant-free) chamber via membrane pores to the bottom chamber.
12. If the cells have not migrated/invaded in the presence of the gradient or a significant number of the cells migrated/invaded via the transwell insert membranes even under control conditions (no serum gradient), it is possible that the membrane pore sizes are not suitable for the particular cell type (see list **item 6** in Section 2 Equipment).

Acknowledgments

The work was supported by The Russian Science Foundation (RSF) grant № 22-74-10037.

References

1. Wright JR, Mahaut-Smith MP (2021) Why do platelets express K+ channels? Platelets 32(7): 872–879. https://doi.org/10.1080/09537104.2021.1904135
2. Salsbury G, Cambridge EL, McIntyre Z, Arends MJ, Karp NA, Isherwood C, Shannon C, Hooks Y, Sanger Mouse Genetics Project, Ramirez-Solis R, Adams DJ, White JK, Speak AO (2014) Disruption of the potassium channel regulatory subunit KCNE2 causes iron-deficient anemia. Exp Hematol 42(12):1053-8.e1. https://doi.org/10.1016/j.exphem.2014.07.269
3. Vianna-Jorge R, Suarez-Kurtz G (2004) Potassium channels in T lymphocytes: therapeutic targets for autoimmune disorders? BioDrugs 18(5):329–341. https://doi.org/10.2165/00063030-200418050-00005
4. Capatina AL, Lagos D, Brackenbury WJ (2022) Targeting ion channels for cancer treatment: current progress and future challenges. Rev Physiol Biochem Pharmacol 183:1–43. https://doi.org/10.1007/112_2020_46
5. Vasileva VY, Khairullina ZM, Sudarikova AV, Chubinskiy-Nadezhdin VI (2023) Role of calcium-activated potassium channels in proliferation, migration and invasion of human chronic myeloid leukemia K562 cells. Membranes 13(6):583. https://doi.org/10.3390/membranes13060583
6. Bonito B, Sauter DR, Schwab A, Djamgoz MB, Novak I (2016) KCa3.1 (IK) modulates pancreatic cancer cell migration, invasion and proliferation: anomalous effects on TRAM-34. Pflugers Archiv Eur J Physiol 468(11–12):1865–1875. https://doi.org/10.1007/s00424-016-1891-9

Chapter 12

Non-radioactive Rb^+ Efflux Assay for Screening K_{ATP} Channel Modulators

Assmaa ElSheikh, Camden M. Driggers, and Show-Ling Shyng

Abstract

ATP-sensitive potassium (K_{ATP}) channels function as metabolic sensors that link cell membrane excitability to the cellular energy status by controlling potassium ion (K^+) flow across the cell membrane according to intracellular ATP and ADP concentrations. As such, K_{ATP} channels influence a broad spectrum of physiological processes, including insulin secretion and cardiovascular functions. K_{ATP} channels are hetero-octamers, consisting of four inward rectifier potassium channel subunits, Kir6.1 or Kir6.2, and four sulfonylurea receptors (SURs), SUR1, SUR2A, or SUR2B. Different Kir6 and SUR isoforms assemble into K_{ATP} channel subtypes with distinct tissue distributions and physiological functions. Mutations in the genes encoding K_{ATP} channel subunits underlie various human diseases. Targeted treatment for these diseases requires subtype-specific K_{ATP} channel modulators. Rubidium ions (Rb^+) also pass through K_{ATP} channels, and Rb^+ efflux assays can be used to assess K_{ATP} channel function and activity. Flame atomic absorption spectroscopy (Flame-AAS) combined with microsampling can measure Rb^+ in small volume, which provides an efficient tool to screen for compounds that alter K_{ATP} channel activity in Rb^+ efflux assays. In this chapter, we describe a detailed protocol for Rb^+ efflux assays designed to identify new K_{ATP} channel modulators with potential therapeutic utilities.

Key words ATP-sensitive potassium channel, Inward rectifier potassium channel, Sulfonylurea receptor, Kir6.2, Potassium channel opener, Flame atomic absorption spectroscopy

1 Introduction

K_{ATP} channels function as molecular transducers linking the electrical excitability of cell membranes to the intracellular energetic milieu [1–3]. They orchestrate cellular resource allocation in response to shifting metabolic demands by sensing fluctuations in intracellular ATP and ADP concentrations, with glucose increasing ATP concentrations driving closure of the K_{ATP} channel and starvation leading to high ADP/ATP ratios, which drive opening of the K_{ATP} channel [3–5]. K_{ATP} channels control cell membrane permeability of potassium ions (K^+). Under physiological conditions of high intracellular K^+ concentrations, opening of K_{ATP} channels

Simone Furini (ed.), *Potassium Channels: Methods and Protocols*, Methods in Molecular Biology, vol. 2796,
https://doi.org/10.1007/978-1-0716-3818-7_12,

allows K^+ efflux from cells and maintains membrane potential in a hyperpolarized state close to the equilibrium potential of K^+. K_{ATP} channels are found in various tissues and organs, including the heart, vascular smooth muscles, brain, pituitary gland, pancreas, and skeletal muscles [5, 6]. They are obligatory hetero-octamers composed of inward rectifier potassium subunits (Kir6) forming the K^+ conducting pore and sulfonylurea receptors (SUR) regulating channel activity. The Kir6 subunits (Kir6.1 or Kir6.2) are encoded by *KCNJ8* and *KCNJ11* genes, while SUR proteins (SUR1, SUR2A, SUR2B) are encoded by *ABCC8* and *ABCC9* genes, where SUR2A and SUR2B are two splice variants of the *ABCC9* gene product [7–11]. Tissue-specific subtypes of K_{ATP} channels assembled from specific Kir6 and SUR isoforms have unique physiological functions, and they act together to ensure energy homeostasis and metabolic health of the organism [6, 12–14].

In the heart, K_{ATP} channels predominantly consist of SUR2A and Kir6.2 subunits, and they protect the heart against metabolic stress during ischemia and hypoxia [6, 15]. Vascular K_{ATP} channels, predominantly consisting of SUR2B and Kir6.1, regulate vascular tone and blood pressure. In pancreatic β-cells, K_{ATP} channels formed by SUR1 and Kir6.2 act as a central regulator of insulin secretion to control blood glucose levels [16]. Mutations in K_{ATP} channel genes are associated with a variety of diseases, including dilated cardiomyopathy, familial atrial fibrillation, intellectual disability, Cantú syndrome, neonatal diabetes mellitus, and congenital hyperinsulinism [15, 17–21]. Current medical therapy for these diseases is limited due to cross-reactivity of existing K_{ATP} channel modulators on different K_{ATP} isoforms. Thus, there is a need to search for novel tissue-specific K_{ATP} channel modulators.

To identify novel K_{ATP} channel modulators, we need high-throughput screening methods capable of evaluating K_{ATP} channel function and activity in response to various novel agents. Traditional methods that measure potassium current through cell membranes to assess K_{ATP} channel function such as patch-clamp electrophysiological recordings suffer from low throughput [22]. Rubidium ions (Rb^+) and thallium ions (Tl^+) pass through open K_{ATP} channels [23]. Large chemical compound library screening using a Tl^+ flux assay and cells stably transfected with K_{ATP} channels has been described [24]. While this approach has led to identification of novel K_{ATP} channel modulators [25, 26], it requires substantial resources and time-consuming process of making stable cell lines.

The Rb^+ efflux assay exposes K_{ATP}-expressing cells loaded with Rb^+ to conditions that modulate K_{ATP} channel activity and measures Rb^+ concentration inside and outside the cell using flame atomic absorption spectroscopy (Flame-AAS) to investigate K_{ATP} channel function. Although Rb^+ efflux assays employing radioactive

$^{86}Rb^+$ had been widely used by the research community [27], the lack of commercially available $^{86}Rb^+$ in recent years has pushed many investigators to adopt alternative approaches. Rb^+ efflux assay based on atomic absorption spectroscopy has been previously described to monitor K_{ATP} channel activity [28–31]; however, it has not been used widely and has not been described in detail. We have refined the method to identify K_{ATP} channel inhibitors, activators, and pharmacological chaperones using cells transiently transfected with wild-type or disease mutant K_{ATP} channel DNAs and the Aurora Biomed Ion Channel Reader (ICR) Flame-AAS. This assay provides a valuable medium-throughput screening tool for individual research or clinical labs to evaluate the activity of wild-type and mutant K_{ATP} channels in intact cells in the presence or absence of K_{ATP} channel modulators.

Here, we describe step-by-step protocols for the use of non-radioactive Rb^+ efflux assays as a screening tool to identify K_{ATP} channel activators, pharmacochaperones, and inhibitors. The utilization of such assays holds promise for the discovery of new therapeutic agents capable of normalizing K_{ATP} channel activity under different pathological conditions.

1.1 Rb^+ Efflux Assay for Screening New K_{ATP} Channel Activators

K_{ATP} channels are targets of a class of compounds known as potassium channel openers (KCOs) [32, 33]. KCOs enhance the activity of K_{ATP} channels to modulate endocrine, cardiovascular, neurological, and muscular functions [34]. KCOs have therapeutic applications in a number of pathological conditions. For example, KCOs have been shown to act upon the cardiac K_{ATP} isoform to alleviate myocardial dysfunction and heart failure [32, 35]. They have demonstrated efficacy in managing hypertension and peripheral vascular disease through their influence on vascular smooth muscle K_{ATP} and are used clinically for hyperinsulinism by targeting the pancreatic K_{ATP} isoform [33]. In addition, KCOs function as bronchodilators, antiepileptic agents, therapeutics for urinary incontinence, and promoters of hair growth [36–39].

The primary molecular target site of KCOs resides within the SUR regulatory subunits [40–43]. A chemically diverse array of KCO agents exhibit variable SUR isoform selectivity [44, 45]. Recent Cryo-Electron Microscopy (Cryo-EM) structures of K_{ATP} channels bound to select KCOs have shed light on their mode of action [41, 42]. For instance, the KCO NN414 binds within the transmembrane domain of SUR1, interacting with residues from transmembrane domains 1 and 2 (TMD1 and TMD2) situated near the outer cell membrane leaflet [40]. This binding synergizes with the binding of MgATP and MgADP to the Nucleotide-Binding Domains (NBDs) of SUR1 to promote NBD dimerization and channel activation [40].

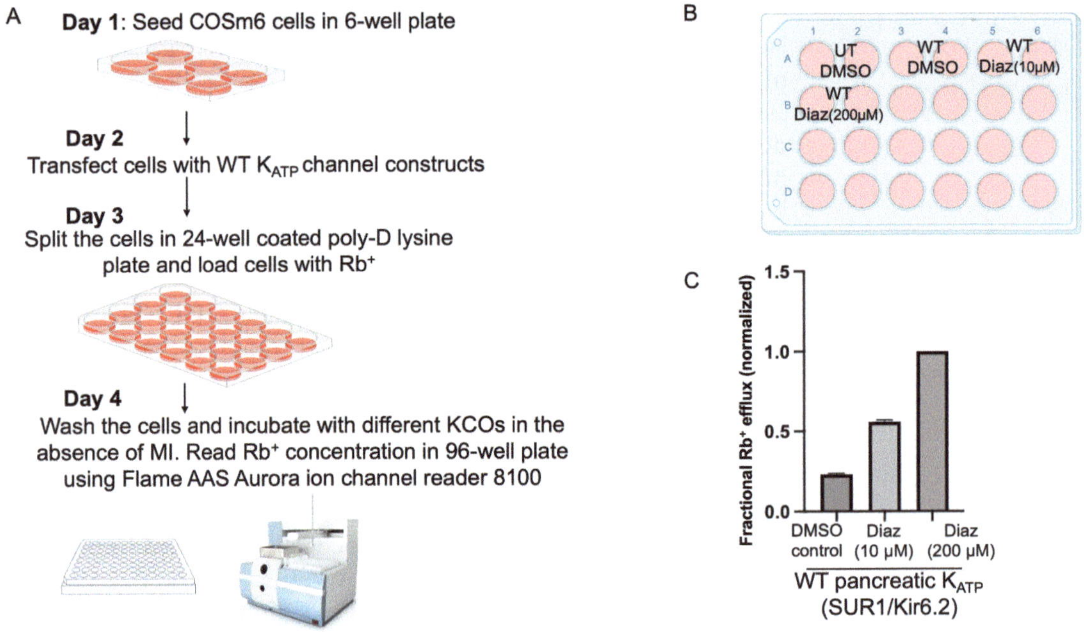

Fig. 1 Experimental protocol for the Rb^+ efflux assay as a screening tool for K_{ATP} channel activators. (**a**) The essential steps comprising the experimental workflow for employing Rb^+ efflux assay to screen novel K_{ATP} channel activators. Notably, this protocol is designed to evaluate K_{ATP} channel activation without including MI (deoxy-2-deoxy-D-glucose and oligomycin) in the efflux solutions. (**b**) An example experiment to assess the acute activation of pancreatic K_{ATP} channels by diazoxide, a well-characterized KCO. Every experimental condition is tested in duplicate to ensure data reliability. UT cells should be included in every plate to quantify background Rb^+ efflux, which will be subsequently subtracted from other experimental readings. Cells expressing WT channels (SUR1/Kir6.2) will be tested in the presence of DMSO (vehicle control) or diazoxide at 10 and 200 μM concentrations. (**c**) A bar graph presents the fractional Rb^+ efflux for cells expressing WT K_{ATP} channels under different treatment conditions, including DMSO and two concentrations of diazoxide (10 μM and 200 μM). Data have been normalized to the effect of diazoxide at 200 μM to enable direct comparison of the treatment groups. Error bars denote the standard error of the mean (SEM). Abbreviations: *WT* wild-type, *MI* metabolic inhibitors, *AAS* atomic absorption spectroscopy, *UT* untransfected COSm6 cells, *DMSO* dimethyl sulfoxide, *Diaz* diazoxide

Most KCOs, while displaying some K_{ATP} channel isoform selectivity, can significantly increase activity of all K_{ATP} isoforms, which result in undesired side effects when used for disease therapy [33, 46, 47]. Thus, there remains a great need to develop new generations of KCOs with improved specificity to target various K_{ATP} channel isoforms.

The protocol in Subheading 3.1 outlines the procedure for screening potential pancreatic K_{ATP} channel activators using a Rb^+ efflux assay, with diazoxide serving as a reference KCO (Fig. 1).

1.2 Rb^+ Efflux Assay for Screening New K_{ATP} Channel Inhibitors

Pharmacological inhibitors targeting K_{ATP} channels in pancreatic β-cells have been employed as therapeutic agents for type 2 diabetes. These drugs, such as sulphonylureas and glinides, exert their inhibitory effects on K_{ATP} channels in β-cells by interacting with the SUR1 regulatory subunits [48–50]. By reducing potassium conductance through pancreatic K_{ATP} channels, these drugs lead to membrane depolarization even in the absence of an increase in glucose. Membrane depolarization, in turn, facilitates calcium (Ca^{2+}) influx through voltage-gated Ca^{2+} channels, triggering exocytosis of insulin granules, resulting in insulin secretion [51]. Despite their efficacy, tolerability, and cost-effectiveness, these drugs are sometimes de-emphasized in clinical guidelines due to adverse effects, including uncertain cardiovascular risks from off-target interaction with SUR2A or SUR2B isoforms [52]. This highlights the need to identify new K_{ATP} channel inhibitors that carry a lower risk of such complications.

In Subheading 3.2, we describe the use of Rb^+ efflux assays to screen for new K_{ATP} channel inhibitors. Generally, the experimental procedures closely resemble those described in Subheading 3.1. The primary distinction lies in the requisite inclusion of metabolic inhibitors (MI), deoxy-glucose and oligomycin, alongside the inhibitory agents during the Rb^+ efflux assay. MI mimics glucose starvation conditions and opens K_{ATP} channels by depleting intracellular ATP stores, as deoxy-2-deoxy-D-glucose and oligomycin inhibit ATP synthesis by inhibiting glycolysis and oxidative phosphorylation, respectively. Within the framework of K_{ATP} inhibitor screening, application of test compounds along with MI allows for evaluation of the acute inhibitory effects of the tested compounds on K_{ATP} channels, while the channels are maximally activated. As an example, we will use the assay to determine the dose-dependent inhibitory effects and the half-maximal inhibitory concentration (IC_{50}) of tolbutamide, a known K_{ATP} channel inhibitor (Fig. 2).

1.3 Rb^+ Efflux Assay for Screening New K_{ATP} Channel Pharmacochaperones

Loss-of-function mutations in genes encoding pancreatic K_{ATP} channels (*ABCC8* and *KCNJ11*) represent the most prevalent genetic etiology of hyperinsulinism (K_{ATP} HI), a serious neonatal disorder marked by inordinate insulin secretion and life-threatening hypoglycemia [51, 53]. Left unmanaged, this condition can lead to severe and lifelong complications, including seizures, developmental delay, and irreversible brain damage [54, 55]. Current therapeutic approaches, such as diazoxide and somatostatin analogues, are inadequate for severe cases wherein K_{ATP} channel mutations significantly reduce channel density on β-cell surfaces by disrupting biogenesis, assembly, and trafficking (referred to as trafficking mutations) [56]. In cases unresponsive to conventional treatments, particularly those characterized by a

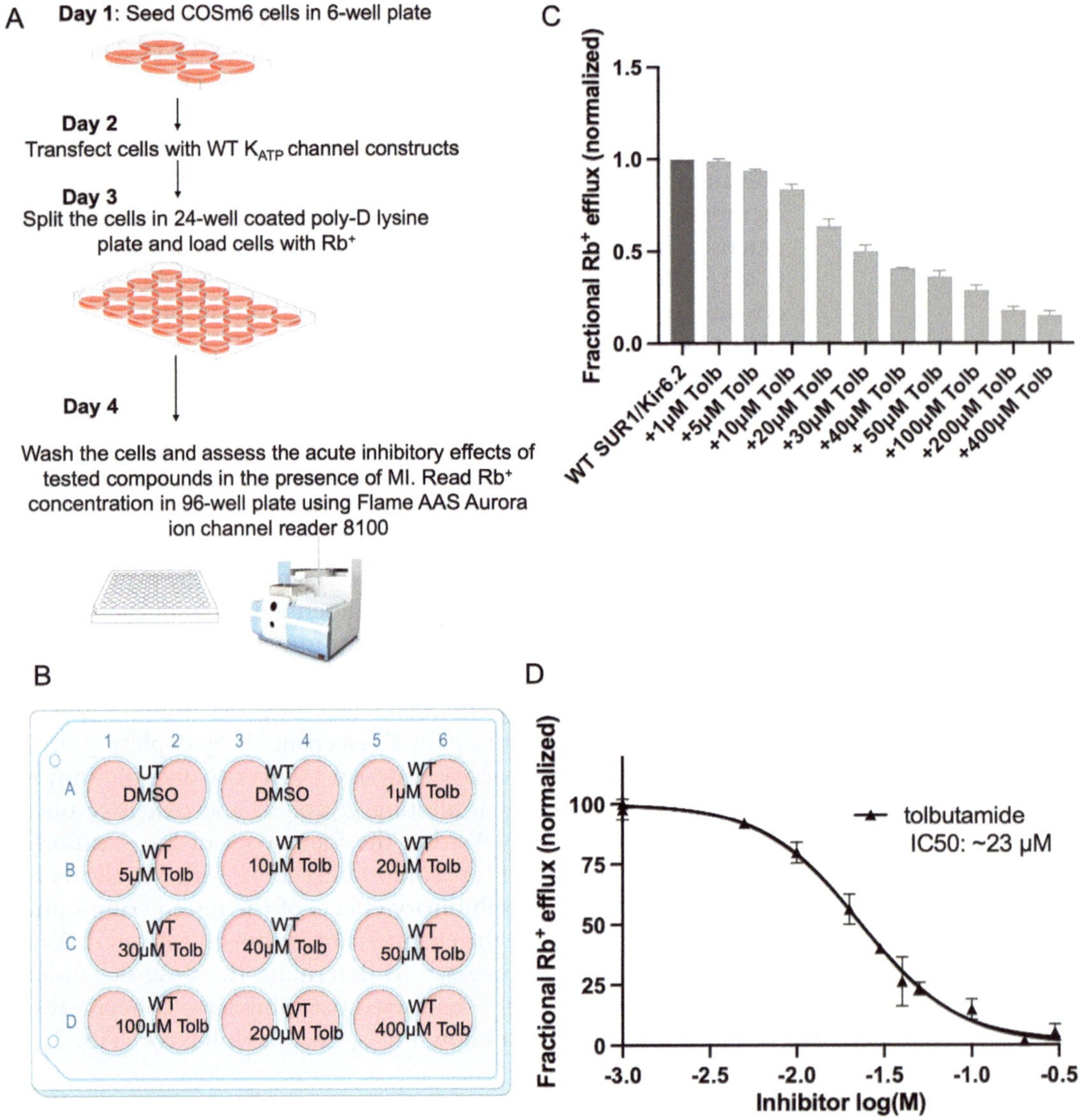

Fig. 2 Experimental protocol for the Rb^+ efflux assay as a screening tool for K_{ATP} channel inhibitors. (**a**) The experimental workflow for the use of Rb^+ efflux assay to identify novel K_{ATP} channel inhibitors. Notably, this protocol is designed to evaluate the acute inhibitory effects of different compounds in the presence of MI in the efflux solutions. (**b**) An example experiment to assess the acute inhibitory effects of tolbutamide, a known pancreatic K_{ATP} channel inhibitor. COSm6 cells expressing WT pancreatic K_{ATP} channels (SUR1/Kir6.2) will be tested in the presence of DMSO (vehicle control) or tolbutamide at 1, 5, 10, 20, 30, 40, 50, 100, 200, and 400 μM concentrations. (**c**) This bar graph presents the averaged dose response inhibitory effects of tolbutamide on cells expressing pancreatic K_{ATP}, using Rb^+ efflux assay. The fractional Rb^+ efflux from UT cells was subtracted from other experimental readings. Data have been normalized to cells expressing WT channels and treated with 0.1% DMSO. (**d**) IC_{50} of tolbutamide was analyzed utilizing a nonlinear fit through employing the Variable Slope (four parameters) model for log(inhibitor) vs. response curves in GraphPad Prism 9. The estimated IC_{50} was ~23 μM. Error bars denote the SEM. Abbreviations: *WT* wild-type, *MI* metabolic inhibitors, *AAS* atomic absorption spectroscopy, *UT* untransfected COSm6 cells, *DMSO* dimethyl sulfoxide, *Tolb* tolbutamide, *IC_{50}* half maximal inhibitory concentration

diffuse distribution pattern of the disease, partial or total pancreatectomy may be the only recourse to correct hypoglycemia and mitigate complications [55–57].

Pharmacological chaperones (PCs), small molecules capable of binding to K_{ATP} channel protein subunits to facilitate their proper biogenesis, assembly, folding, and trafficking to the cell membrane, hold promise as a therapeutic avenue for patients with K_{ATP} HI resulting from trafficking mutations [49, 58–60]. Intriguingly, certain K_{ATP} channel inhibitors, employed as treatments for type 2 diabetes mellitus, including sulfonylureas and glinides, have demonstrated PC effects, rescuing the surface expression of K_{ATP} trafficking mutants in a domain-specific manner [61–63]. The commonality among these K_{ATP} channel PCs as channel inhibitors is explained by their mechanism of action, recently elucidated through Cryo-EM structures of K_{ATP} channels bound to PCs [10, 48, 49]. These structures revealed that PCs bind to the central cavity of SUR1, closely interacting with the N-terminal domain of the Kir6.2 subunit [10, 48, 49]. While this interaction is indispensable for stabilizing the assembly of SUR1/Kir6.2 in metastable mutant channels, it concurrently restricts Kir6.2 movement needed for channel opening [64]. This high binding affinity of PCs to the rescued channel results in sustained inhibitory effects, rendering them unsuitable for treating K_{ATP} HI [51].

Tolbutamide, a low-affinity binding sulfonylurea with potential K_{ATP} channel PC properties, had its market presence discontinued due to an associated increase in the risk of cardiovascular problems [65]. Consequently, the identification of novel K_{ATP} channel PCs, characterized by less persistent inhibitory effects, is needed to help address K_{ATP}-HI cases linked to trafficking defects.

The protocol in Subheading 3.3 delineates the methodology for employing the Rb^+ efflux assay to screen for novel pancreatic K_{ATP} channel PCs. As for screening K_{ATP} channel inhibitors described above, the efflux assay will be performed with the inclusion of MI to enhance the activation of K_{ATP} channels through the depletion of intracellular ATP reserves during the efflux period. What distinguishes this protocol from the preceding one is that the tested compounds are introduced for an overnight incubation on *day 3* to allow their binding to K_{ATP} channel subunits to restore the channel surface expression. Consequently, an additional step for compound removal through washing is introduced before the start of the efflux assay on *day 4*. In Subheading 3.3, we will use the Rb^+ efflux assay to evaluate the PC effects of tolbutamide, a well-known low affinity sulfonylurea with K_{ATP} channel PC properties and less persistent inhibitory effects [59] (Fig. 3).

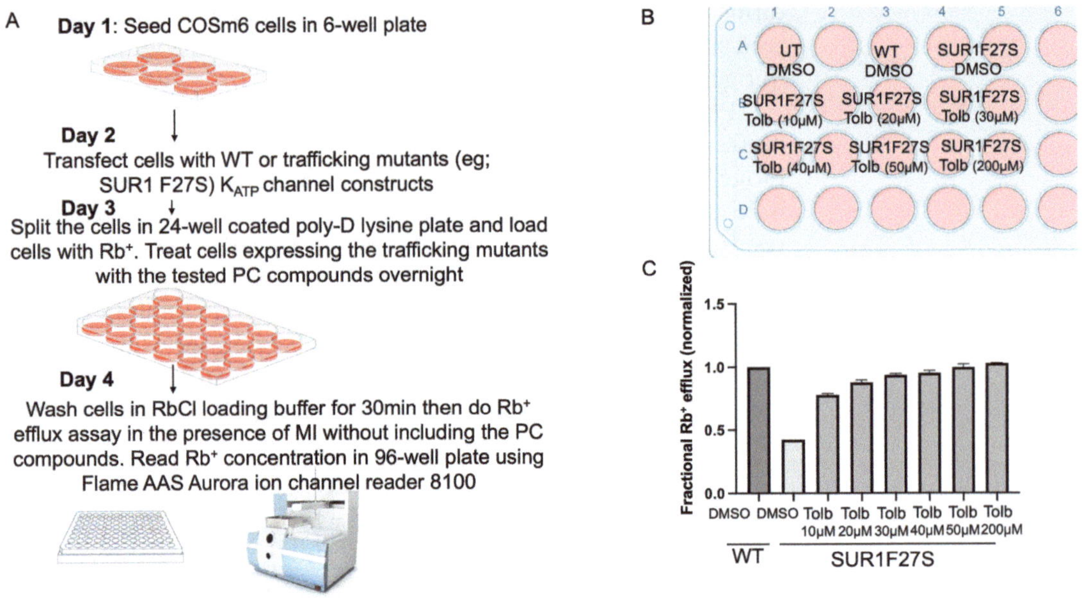

Fig. 3 Experimental protocol for the Rb^+ efflux assay as a screening tool for K_{ATP} channel pharmacochaperones (PCs). (**a**) The experimental workflow for using Rb^+ efflux assay to screen novel K_{ATP} channel PCs. Notably, in this protocol K_{ATP} PC effects are evaluated with MI present in the efflux solutions. (**b**) An example experiment to assess the PC effects of tolbutamide on a trafficking mutation SUR1-F27S. All experimental conditions are tested in duplicate, and UT cells were included to quantify background Rb^+ efflux, which will be subsequently subtracted from other experimental readings. (**c**) A bar graph showing the dose-dependent increase in the fractional Rb^+ efflux of cells expressing a trafficking mutation SUR1-F27S after being treated overnight with tolbutamide. Data have been normalized to the fractional Rb^+ efflux of cells expressing WT channels. Error bars denote the standard error of the mean (SEM). Abbreviations: *WT* wild-type, *MI* metabolic inhibitors, *AAS* atomic absorption spectroscopy, *UT*, untransfected COSm6 cells, *DMSO* dimethyl sulfoxide; *Tolb* tolbutamide

2 Materials

2.1 Cell Culture

1. COSm6 cells (*see* **Note 1**).
2. COSm6 tissue culture media: Dulbecco's Modified Eagle Medium (DMEM, high glucose), supplemented with 10% fetal bovine serum, and 100 U/mL penicillin/streptomycin.
3. Trypsin solution: 0.25% trypsin-EDTA.
4. Tissue culture vessels: 6-well, 24-well, and 96-well flat bottom tissue culture-treated polystyrene plates.

2.2 Transfection

1. Plasmids for transfection: Kir6.2 and SUR1 cDNAs in mammalian expression vectors (e.g., pcDNA vectors) (*see* **Note 2**).
2. Transfection reagents: Opti-MEM, FuGENE® 6.

2.3 Loading with RbCl

1. Plate-coating solution: Poly-D-lysine (*see* **Note 3**).
2. Phosphate-buffered saline (PBS): 137 mM NaCl, 2.7 mM KCl, 10 mM Na_2HPO_4, 1.8 mM KH_2PO_4, pH 7.4.
3. Rubidium chloride (RbCl) (*see* **Note 4**).

2.4 Rb^+ Efflux Assay

1. Ringer's solution: 5.4 mM KCl, 150 mM NaCl, 1 mM $MgCl_2$, 0.8 mM NaH_2PO_4, 2 mM $CaCl_2$, 25 mM HEPES, pH 7.4.
2. RbCl wash buffer: 5.4 mM RbCl, 150 mM NaCl, 1 mM $MgCl_2$, 0.8 mM NaH_2PO_4, 2 mM $CaCl_2$, 25 mM HEPES, pH 7.2.
3. Dimethyl sulfoxide, DMSO.
4. Metabolic inhibitors: 2-deoxy-D-glucose, oligomycin (25mg/ mL stock in DMSO).
5. Tested compounds: diazoxide or tolbutamide stock solutions in DMSO.
6. Cell lysis solution: Ringer's solution + 1% Triton™ X-100.
7. Aurora ion channel reader (ICR 8100).

3 Methods

3.1 Rb^+ Efflux Assay for Screening New K_{ATP} Channel Activators

1. Day 1: Cell seeding. Seed COSm6 cells at ~1:8–1:10 dilution in 2 mL COSm6 tissue culture media for each well of a 6-well plate. Cells are passed at ~1:8–1:10 dilution from a confluent T25 flask such that the cells will be ~70% confluent the next day ready for transfection.
2. Day 2: Transfection. Transfect COSm6 cells with human wild-type SUR1 and Kir6.2 cDNAs using Opti-MEM and FuGENE® 6. Include untransfected (UT) COSm6 cells as a negative control to establish baseline Rb^+ concentrations (*see* **Notes 5–7**).

3.1.1 Day 3: Loading with RbCl

1. Coat a 24-well plate with poly-D-lysine.
   (a) Coat the culture surface of a 24-well plate with poly-D-lysine, and incubate for 1 h (*see* **Note 3**).
   (b) Then aspirate poly-D-lysine, and rinse the culture surface with a large volume of distilled water or with PBS.
   (c) Uncover the plate and let it dry for 2 h.
2. Split COSm6 cells into the coated 24-well plate, using media containing 5.4 mM RbCl. Include UT COSm6 cells as controls.
   (a) Prepare COSm6 cell media containing 5.4 mM RbCl (*see* **Note 4**).

(b) Resuspend cells from each well of the 6-well plate in 5000 μL of media with RbCl, and distribute to 10 wells of the coated 24-well plate (approximately 450 μL per well).

(c) Include UT COSm6 cells as a control in each 24-well plate.

(d) Perform all conditions in duplicate (Fig. 1).

3.1.2 Day 4: Rb^+ Efflux Assay

1. Before starting the efflux assay, warm up the Ringer's solution, prepared without RbCl, glucose, or MI, in a water bath at 37 °C (*see* **Notes 8** and **9**).
2. Prepare control efflux solutions: Ringer's solution plus 0.1% DMSO, as a vehicle control.
3. Prepare sample efflux solutions (using diazoxide as an example): 10 μM and 200 μM diazoxide in Ringer's solution using 10 mM and 200 mM diazoxide stocks in DMSO, respectively (0.1% DMSO in all tested situations) (*see* **Notes 10** and **11**).
4. Start the Rb^+ efflux assay:
   (a) Rapidly wash each well of the 24-well plate three times with ~0.3–1 mL PBS at room temperature (*see* **Note 12**).
   (b) Add 250 μL of efflux solutions (control and samples) to each well.
   (c) Incubate the plates for 30 min at 37 °C.
   (d) After 30 min, transfer efflux solutions into each of the dedicated wells of a 96-well plate.
5. Cell lysis
   (a) Lyse cells with 1% Triton™ X-100.
   (b) Shake for 20 min at room temperature to facilitate cell detachment and lysis (*see* **Note 13**).
   (c) Transfer cell lysates into each dedicated well of a 96-well plate.
6. Read Rb^+ concentration in both efflux solution and cell lysate using Aurora ICR 8100, calculate fractional Rb^+ efflux (Rb^+ concentration in the efflux solution divided by total Rb^+ (the sum of Rb^+ in efflux solution and cell lysate)), and normalize data to the value obtained from the 200 μM diazoxide-treated positive control (*see* **Notes 14–16**).

3.2 Rb^+ Efflux Assay for Screening New K_{ATP} Channel Inhibitors

1. Day 1–3. The procedural details are described in Subheading 3.1.

3.2.1 Day 4: Rb^+ Efflux Assay

1. Prepare Ringer's solution similar to described in the preceding section without RbCl or glucose but including MI (*see* **Note 9**).
2. Prepare the sample efflux solutions (Fig. 2): Prepare efflux solutions containing tolbutamide at final concentrations of 1, 5, 10, 20, 30, 40, 50, 100, 200, and 400 μM by diluting 1000× tolbutamide stock solutions in Ringer's solution containing MI, ensuring that the final concentration of DMSO is always 0.1% (*see* **Note 17**).
3. Prepare the control efflux solutions: Use 0.1% DMSO in Ringer's solution with MI as a vehicle control.
4. Initiate the Rb^+ efflux assay to assess tolbutamide's acute effects:
   (a) Rapidly wash each well of the 24-well plate three times with ~0.3–1 mL PBS at room temperature (*see* **Note 12**).
   (b) Add 250 μL of efflux solutions (Ringer's plus MI) to each well of the 24-well plate.
   (c) Incubate the cells with efflux solutions for 30 min at 37 °C.
   (d) After 30 min, transfer efflux solutions into each of the dedicated wells of a 96-well plate.
5. Cell lysis
   (a) Lyse the COSm6 cells with 1% Triton-X 100 (250 μL per well).
   (b) Shake the cells on a shaker for 20 min at room temperature to facilitate cell detachment and lysis (*see* **Note 13**).
   (c) Transfer cell lysates into each dedicated well of a 96-well plate.
6. Data analysis
   (a) Measure the Rb^+ concentration in each well of the 96-well plate using an Aurora Ion Channel Reader 8100 (*see* **Note 15**).
   (b) Calculate the fractional Rb^+ efflux for each sample (Rb^+ concentration in the efflux solution divided by total Rb^+ (the sum of Rb^+ in efflux solution and cell lysate)).
   (c) Calculate the average of each duplicate.
   (d) Subtract the fractional Rb^+ efflux of UT COSm6 cells from all values in the same plate.

(e) Normalize values to the fractional Rb^+ efflux of cells expressing human WT in the presence of MI treated with the vehicle control.

This Rb^+ efflux assay can determine the IC50 of different inhibitory compounds by assessing dose-dependent inhibitory effects.

3.3 Rb^+ Efflux Assay for Screening New K_{ATP} Channel Pharmacochaperones

1. Day 1: Cell seeding. Described in detail in Subheading 3.1.
2. Day 2: Transfection. Transfect COSm6 cells with wild-type SUR1/Kir6.2 or SUR1-F27S/wild-type Kir6.2 cDNAs in mammalian expression vectors using Opti-MEM and FuGENE® 6 as the transfection reagent. Include UT COSm6 cells as a negative control (*see* **Note 18**).

3.3.1 Day 3: Loading with RbCl and Adding PC Treatment Overnight (ON)

1. Coat a 24-well plate with poly-D-lysine as previously mentioned in the first section.
2. Split the transfected COSm6 cells from the 6-well plate into the poly-D-lysine-coated 24-well plate (Fig. 3).
3. Resuspend cells from the 6-well plate in 5000 μL of media with RbCl, and distribute it to 10 wells of the coated 24-well plate (approximately 450 μL per well).
4. Include UT COSm6 cells as a control in each plate.
5. Perform all conditions in duplicate.
6. For cells expressing the trafficking mutation SUR1-F27S, treat them ON with different concentrations of tolbutamide (10, 20, 30, 40, 50, and 200 μM or DMSO as a vehicle control) (Fig. 3). Use tolbutamide in DMSO stock solutions at 1000× working concentration, such that the final DMSO concentration is 0.1% in all conditions (*see* **Note 19**).
7. Treat UT COSm6 cells and cells expressing the wild-type channel with DMSO (0.1%) as a vehicle control.

3.3.2 Day 4: Cell Wash and Rb^+ Efflux Assay

1. Warm up Ringer's solution (without RbCl or glucose) and add MI (*see* **Note 9**).
2. Warm up RbCl wash buffer (without MI).
3. Start the cell wash step:
   (a) Rapidly wash each well of the 24-well plate with RbCl wash buffer at room temperature.
   (b) Add 250 μL of RbCl wash buffer to each well, and incubate for 30 min at 37 °C.
   (c) After 30 min, initiate the efflux assay.

4. Start the Rb^+ efflux assay (*see* **Note 19**):
   (a) Rapidly wash each well of the 24-well plate three times with ~0.3–1 mL PBS at room temperature.
   (b) Add 250 μL of efflux solutions (Ringer's plus MI) to each well.
   (c) Incubate the cells with efflux solutions for 25 min at 37 °C (*see* **Note 20**).
   (d) After 25 min, transfer the efflux solutions into dedicated wells of a 96-well plate.
5. Cell lysis: same as in the previous sections
6. Data analysis
   (a) Measure the Rb^+ concentration in each well of the 96-well plate using an Aurora Ion Channel Reader 8100.
   (b) Calculate the fractional Rb^+ efflux for each sample.
   (c) Calculate the average of each duplicate.
   (d) Subtract the fractional Rb^+ efflux of UT COSm6 cells from all values in the same plate.
   (e) Normalize values to the fractional Rb^+ efflux of cells expressing wild-type channel (Fig. 3).

3.4 Conclusions

The Rb^+ efflux assay, which measures Rb^+ ion concentration inside and outside of cells using Flame-AAS to assess movement through K_{ATP} channels, serves as a valuable screening tool for the examination of K_{ATP} channel activity when exposed to various modulators. This assay involves exposing K_{ATP}-expressing cells, preloaded with Rb^+, to conditions that influence channel activity, allowing for the quantification of Rb^+ concentrations both inside and outside the cell. This approach facilitates the study of K_{ATP} channel function. The utilization of such assays for K_{ATP} channels expressed in mammalian systems holds significant potential for the identification of novel therapeutic agents capable of modulating K_{ATP} channel activity in diverse pathological contexts.

4 Notes

1. For our experiments, we utilize COSm6 cells, a clonal cell line derived from African green monkey COS cells for exogenous expression of wild-type or mutant K_{ATP} channels by transient transfection, although other COS cell clones or any cell lines that do not express endogenous K_{ATP} channels are also suitable for this purpose.
2. Typically, we employ mammalian expression vectors that contain the CMV promoter, like pcDNA vectors.

3. Poly-D-lysine, a chemically synthesized extracellular matrix composed of lysine polymers, is employed as a substrate for promoting cellular adhesion to tissue culture-treated surfaces. This helps cells to maintain attachment throughout the efflux process prior to cell lysis to ensure more robust and reliable efflux data.
4. We employ rubidium chloride (RbCl) (99.8%-Rb), CAS Number 7791-11-9, obtained from Strem Chemicals, Inc. RbCl is directly added to the COSm6 tissue culture media to achieve a final concentration of 5.4 mM. Media containing RbCl is filtered using a 0.22 μm PES Membrane Vacuum Filter.
5. To achieve efficient transfection of COSm6 cells seeded in 6-well plates with K_{ATP} channel constructs, follow these steps for each well (optimized based on the manufacturer's instructions for FuGENE® 6): prepare a sterile Eppendorf tube, add 50 μL of Opti-MEM, and combine with 6 μL of FuGENE® 6, mixing by gentle pipetting. Allow this mixture to stand for 5 min. In another tube, add 50 μL of Opti-MEM along with approximately 1.2 μg of SUR (SUR1 or SUR2A or SUR2B, based on your experimental design) and 1.2 μg of Kir6.1 or Kir6.2 plasmids, mix, and incubate for 5 min. If transfecting multiple wells with the same cDNA constructs, a master mix can be prepared. Combine the FuGENE® mixture with the DNA mixture, ensuring thorough mixing by gentle pipetting, and incubate at room temperature for 20–30 min inside the tissue culture hood. Subsequently, add this mixture dropwise to the cells, gently rotating the plate to ensure even distribution of the transfection mixture. Incubate at 37 °C with 5.0% CO_2 until the cells are ready for experimentation. It is essential to initially mix the plasmids in a separate tube before adding the FuGENE® to ensure the uptake of multiple plasmids required for K_{ATP} channel expression by the cell.
6. Untransfected (UT) cells are routinely included in our Rb^+ efflux experiments to determine the baseline noise, and the fractional Rb^+ efflux from these cells is subtracted from the results obtained under other experimental conditions.
7. While this protocol focuses on screening activators for pancreatic K_{ATP} channels by transfecting COSm6 cells with SUR1 and Kir6.2 isoforms, cells can be transfected with SUR2A/Kir6.2 or SUR2B/Kir6.1 constructs for screening K_{ATP} channel activators targeting cardiac and vascular isoforms, respectively.
8. When preparing Ringer's solution, glucose is not included in the Ringer's solution to prevent channel inhibition (as adding glucose will increase the intracellular ATP levels), which could diminish the effects of new activators. For KCO screening, metabolic inhibitors (MI) should not be included, as it can

activate the channel and potentially mask the effect of the tested compound. The pH of the solution is adjusted to 7.4 using potassium hydroxide, and it is advisable to filter the Ringer's solution before using for the efflux assay.

9. Ringer's solution can be prepared ahead of time and stored at 4 °C. In such cases, it should be warmed up in the water bath to 37 °C before using it in the efflux assay. In the conditions where MI is included (e.g., testing inhibitory compounds or PCs), we add 1 mM deoxy-glucose and 2.5 μg/mL oligomycin freshly to the warmed-up Ringer's solution.
10. In this protocol, diazoxide is used as an example of a pancreatic K_{ATP} channel opener, and it can also serve as a positive control in assays screening for other KCOs for the pancreatic K_{ATP} isoform. Alternatively, Pinacidil, another well-known KCO, can be utilized as a positive control in cells expressing cardiac K_{ATP} channels (SUR2A/Kir6.2), which are poorly responsive to diazoxide.
11. Ensure that the DMSO concentration is maintained at 0.1% in all conditions.
12. The washing steps should be carried out rapidly, and the PBS wash solution can be substituted with Ringer's solution.
13. Do not include any washing steps to the cells after collecting the efflux solutions and before adding your lysis buffer (1% Triton-X100), as this will dramatically affect the reading of Rb concentration in your lysates. The lysis step is not time or temperature sensitive, compared to the efflux assay.
14. For data analysis, read the Rb^+ concentration in each well of the 96-well plate using an Aurora ICR 8100. Calculate the fractional Rb^+ efflux for each sample (ratio of Rb^+ in efflux solution to total Rb^+ (Rb^+ in efflux solution plus Rb^+ in lysate)) to eliminate the potential well-to-well differences in cell densities, Rb^+ loading, and cell loss during the assay and to get more robust and direct measure of ion channel activity. Determine the average of the fractional Rb^+ efflux for each duplicate. Subtract the fractional Rb^+ efflux of UT COSm6 cells from all values in the same plate. Normalize the values to the fractional Rb^+ efflux of human wild-type K_{ATP} expressing COSm6 cells treated with 200 μM Diazoxide (Fig. 1).
15. Aurora ion channel reader (ICR 8100) uses flame atomic absorption spectroscopy (Flame-AAS), an analytical technique to measure Rb^+ concentration with a sensitivity of 0.1 mg/L and accurately within a linear range from 0.1 mg/L to ~15 mg/L. When a beam of light at 780 nm passes through a flame containing atomized Rb, the energy (light) is absorbed by Rb^+. The amount of light absorbed can be measured, and the concentration of the element in the sample can be calculated based on a calibration curve of Rb standards at known

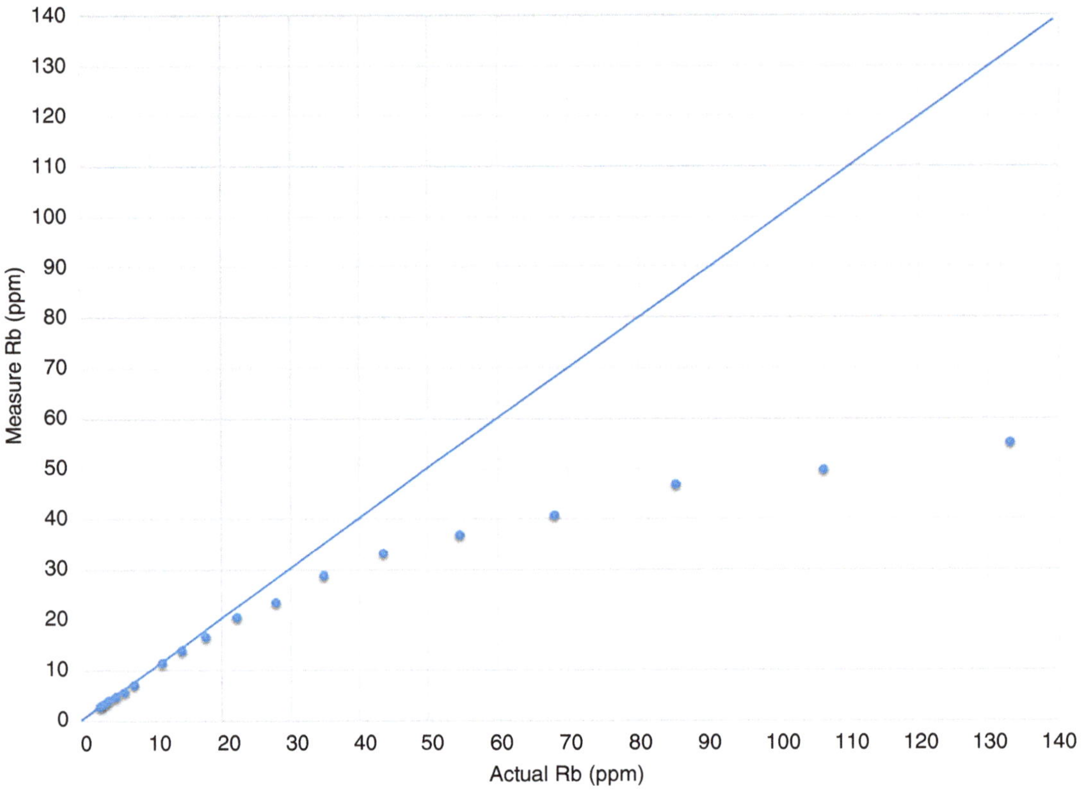

Fig. 4 Measured Rb^+ concentration compared to actual Rb^+ concentration for a set of standard solutions of known Rb^+ concentration for the Aurora ICR 8100. The blue line represents the linear relationship derived from the standard curve which was generated with standards at 0, 1, 2.2, and 5 parts per million (ppm). The measured concentration starts to deviate from the actual concentration at ppm ~>15, and as the concentration increases, the Rb^+ concentration is increasingly underestimated. Higher concentrated samples should be diluted before reading for accurate measurements, but efflux and total cell lysate samples using this method typically have Rb^+ concentration within the linear range

concentrations (Aurora Biomed, 0, 1, 2, 5 ppm). The calibration curve must be included at every flame ignition cycle for accurate measurements, and samples can be diluted (or concentrated by evaporation) as needed to make sure the concentrations of samples are within the linear portion of the calibration curve. To demonstrate the relationship between measured Rb^+ concentration and actual Rb^+ concentration, we measured Rb^+ ppm of Rb^+ solutions with known concentrations (Fig. 4).

16. For Rb^+ efflux assay using Flame-AAS, achieving twofold augmentation in the efflux of Rb^+ in response to channel activation compared to the basal efflux levels is considered satisfactory of good quality of the assay, due to the observed low standard deviations associated with Rb^+ measurements when employing Flame-AAS [66–68].

17. To assess the acute inhibitory effects of K_{ATP} channel modulators, we include the test compounds during the efflux assay rather than treating cells with the compounds overnight. Each of the test compounds will be included in Ringer's solution containing MI to assess the ability of each compound to inhibit MI-induced channel activation. Also, glibenclamide, a known K_{ATP} channel inhibitor, can be included in such experiments as a positive control.
18. In this context, we employ the SUR1-F27S trafficking mutation as an example because a prior study demonstrated that the rescued channels exhibit a normal response to Mg-ADP and diazoxide-mediated activation [63].
19. For studying the PC effects of novel compounds, we treat cells with test compounds overnight, for approximately 16–17 h. This extended duration allows these compounds to exert their corrective influence on the mutant channels. Because K_{ATP} PCs are also channel inhibitors, we do not apply these compounds during the actual efflux assay as including such test compounds in the efflux solution would prevent the activity of any channels rescued by the test compounds to the cell surface. Instead, we are interested in test compounds that would dissociate easily from the channels rescued to the cell surface to recover channel function. Therefore, it is important also to conduct a thorough wash of the PC compounds using RbCl wash buffer before starting the efflux assay.
20. Note we reduced the efflux time in this protocol (PC testing) to 25 min (instead of 30 min) because the trafficking mutant SUR1-F27S, while deficient in trafficking compared to wild-type channels, still has some basal expression in the absence of PCs such that long incubation results in high basal efflux in non-treated cells expressing the trafficking mutation that could obscure the PC effects of test compounds.

Conflict of Interest

The authors declare that the research was conducted in the absence of any commercial or financial relationships that could be construed as a potential conflict of interest.

Funding

This work was supported by the National Institutes of Health grants DK066485 and GM145784 (to Show-Ling Shyng) and a predoctoral scholarship from the Egyptian Government to Assmaa ElSheikh.

References

1. Ashcroft FM (1988) Adenosine 5′--triphosphate-sensitive potassium channels. Annu Rev Neurosci 11:97–118
2. Aguilar-Bryan L, Nichols CG, Wechsler SW, Clement JP, Boyd AE 3rd, González G et al (1995) Cloning of the beta cell high-affinity sulfonylurea receptor: a regulator of insulin secretion. Science 268(5209):423–426
3. Minami K, Miki T, Kadowaki T, Seino S (2004) Roles of ATP-sensitive K+ channels as metabolic sensors: studies of Kir6. x null mice. Diabetes 53(suppl_3):S176–S180
4. Thomas C, Tampé R (2018) Multifaceted structures and mechanisms of ABC transport systems in health and disease. Curr Opin Struct Biol 51:116–128
5. Aguilar-Bryan L, Bryan J (1999) Molecular biology of adenosine triphosphate-sensitive potassium channels. Endocr Rev 20(2): 101–135
6. Foster MN, Coetzee WA (2016) KATP channels in the cardiovascular system. Physiol Rev 96(1):177–252
7. Shyng S-L, Nichols C (1997) Octameric stoichiometry of the KATP channel complex. J Gen Physiol 110(6):655–664
8. Aguilar-Bryan L, Clement JP IV, Gonzalez G, Kunjilwar K, Babenko A, Bryan J (1998) Toward understanding the assembly and structure of KATP channels. Physiol Rev 78:227
9. Chutkow WA, Simon MC, Le Beau MM, Burant CF (1996) Cloning, tissue expression, and chromosomal localization of SUR2, the putative drug-binding subunit of cardiac, skeletal muscle, and vascular KATP channels. Diabetes 45(10):1439–1445
10. Martin GM, Yoshioka C, Rex EA, Fay JF, Xie Q, Whorton MR et al (2017) Cryo-EM structure of the ATP-sensitive potassium channel illuminates mechanisms of assembly and gating. elife 6:e24149
11. Sung MW, Yang Z, Driggers CM, Patton BL, Mostofian B, Russo JD et al (2021) Vascular KATP channel structural dynamics reveal regulatory mechanism by Mg-nucleotides. Proc Natl Acad Sci 118(44):e2109441118
12. Inagaki N, Gonoi T, Clement JP IV, Namba N, Inazawa J, Gonzalez G et al (1995) Reconstitution of I KATP: an inward rectifier subunit plus the sulfonylurea receptor. Science 270(5239):1166–1170
13. Tinker A, Aziz Q, Thomas A (2014) The role of ATP-sensitive potassium channels in cellular function and protection in the cardiovascular system. Br J Pharmacol 171(1):12–23
14. Flagg TP, Enkvetchakul D, Koster JC, Nichols CG (2010) Muscle KATP channels: recent insights to energy sensing and myoprotection. Physiol Rev 90(3):799–829
15. Shyng SL (2022) K(ATP) channel function: more than meets the eye. Function (Oxf) 3(1):zqab070
16. Standen NB, Quayle JM, Davies NW, Brayden JE, Huang Y, Nelson MT (1989) Hyperpolarizing vasodilators activate ATP-sensitive K+ channels in arterial smooth muscle. Science 245(4914):177–180
17. Nichols CG, Singh GK, Grange DK (2013) KATP channels and cardiovascular disease. Circ Res 112(7):1059–1072
18. Singh GK, McClenaghan C, Aggarwal M, Gu H, Remedi MS, Grange DK et al (2022) A Unique high-output cardiac hypertrophy phenotype arising from low systemic vascular resistance in Cantu syndrome. J Am Heart Assoc 11(24):e027363
19. Harakalova M, Van Harssel JJ, Terhal PA, Van Lieshout S, Duran K, Renkens I et al (2012) Dominant missense mutations in ABCC9 cause Cantu syndrome. Nat Genet 44(7):793–796
20. Babenko AP, Polak M, Cavé H, Busiah K, Czernichow P, Scharfmann R et al (2006) Activating mutations in the ABCC8 gene in neonatal diabetes mellitus. N Engl J Med 355(5): 456–466
21. Huopio H, Shyng S-L, Otonkoski T, Nichols C (2002) KATP channels and insulin secretion disorders. Am J Physiol Endocrinol Metab 283(2):E207–EE16
22. Hamill OP, Marty A, Neher E, Sakmann B, Sigworth FJ (1981) Improved patch-clamp techniques for high-resolution current recording from cells and cell-free membrane patches. Pflugers Arch 391(2):85–100
23. Chaudhari MI, Vanegas JM, Pratt LR, Muralidharan A, Rempe SB (2020) Hydration mimicry by membrane ion channels. Annu Rev Phys Chem 71:461–484
24. Raphemot R, Weaver CD, Denton JS (2013) High-throughput screening for small-molecule modulators of inward rectifier potassium channels. JoVE (J Visual Exp) 71:e4209
25. Raphemot R, Swale DR, Dadi PK, Jacobson DA, Cooper P, Wojtovich AP et al (2014) Direct activation of β-cell KATP channels with a novel xanthine derivative. Mol Pharmacol 85(6):858–865
26. Gao J, McClenaghan C, Matreyek KA, Grange DK, Nichols CG (2023) Rapid characterization of the functional and pharmacological

consequences of Cantú syndrome K(ATP) channel mutations in intact cells. J Pharmacol Exp Ther 386(3):298–309

27. de-Allie FA, Bolsover SR, Nowicky AV, Strong PN (1996) Characterization of Ca2+-activated 86Rb+ fluxes in rat C6 glioma cells: a system for identifying novel IKCa-channel toxins. Br J Pharmacol 117(3):479–487
28. Wang K, McIlvain B, Tseng E, Kowal D, Jow F, Shen R et al (2004) Validation of an atomic absorption rubidium ion efflux assay for KCNQ/M-channels using the ion Channel Reader 8000. Assay Drug Dev Technol 2(5): 525–534
29. Bruin JE, Erener S, Vela J, Hu X, Johnson JD, Kurata HT et al (2014) Characterization of polyhormonal insulin-producing cells derived in vitro from human embryonic stem cells. Stem Cell Res 12(1):194–208
30. Li JB, Huang X, Zhang RS, Kim RY, Yang R, Kurata HT (2013) Decomposition of slide helix contributions to ATP-dependent inhibition of Kir6.2 channels. J Biol Chem 288(32): 23038–23049
31. Driggers CM, Kuo Y-Y, Zhu P, ElSheikh A, Shyng S-L (2023) Structure of an open KATP channel reveals tandem PIP2 binding sites mediating the Kir6. 2 and SUR1 regulatory interface. bioRxiv:2023.08. 01.551546
32. Jahangir A, Shen W, Terzic A (2001) Potassium channel openers: therapeutic potential in cardiology and medicine. Expert Opin Pharmacother 2(12):1995–2010
33. Jahangir A, Terzic A (2005) K(ATP) channel therapeutics at the bedside. J Mol Cell Cardiol 39(1):99–112
34. Lawson K, Dunne MJ (2001) Peripheral channelopathies as targets for potassium channel openers. Expert Opin Investig Drugs 10(7): 1345–1359
35. Fujita A, Kurachi Y (2000) Molecular aspects of ATP-sensitive K+ channels in the cardiovascular system and K+ channel openers. Pharmacol Ther 85(1):39–53
36. Pelaia G, Gallelli L, Vatrella A, Grembiale R, Maselli R, De Sarro G et al (2002) Potential role of potassium channel openers in the treatment of asthma and chronic obstructive pulmonary disease. Life Sci 70(9):977–990
37. Wickenden AD (2002) Potassium channels as anti-epileptic drug targets. Neuropharmacology 43(7):1055–1060
38. Fey TA, Gopalakrishnan M, Strake JG, King LL, Brioni JD, Sullivan JP et al (2003) Effects of ATP-sensitive K+ channel openers and tolterodine on involuntary bladder contractions in a pig model of partial bladder outlet obstruction. Neurourol Urodyn 22(2):147–155
39. Messenger A, Rundegren J (2004) Minoxidil: mechanisms of action on hair growth. Br J Dermatol 150(2):186–194
40. Wang M, Wu JX, Ding D, Chen L (2022) Structural insights into the mechanism of pancreatic KATP channel regulation by nucleotides. Nat Commun 13(1):2770
41. Wu J-X, Ding D, Chen L (2022) The emerging structural pharmacology of ATP-sensitive potassium channels Mol Pharm:MOL-PHARM-EMC-2022-000570
42. Driggers CM, Shyng SL (2023) Mechanistic insights on KATP channel regulation from cryo-EM structures. J Gen Physiol 155(1)
43. Ding D, Wu J-X, Duan X, Ma S, Lai L, Chen L (2022) Structural identification of vasodilator binding sites on the SUR2 subunit. Nat Commun 13(1):2675
44. Coghlan MJ, Carroll WA, Gopalakrishnan M (2001) Recent developments in the biology and medicinal chemistry of potassium channel modulators: update from a decade of progress. J Med Chem 44(11):1627–1653
45. Mannhold R (2004) KATP channel openers: structure-activity relationships and therapeutic potential. Med Res Rev 24(2):213–266
46. Clement A, Christensen SL, Jansen-Olesen I, Olesen J, Guo S (2023) The ATP sensitive potassium channel (K(ATP)) is a novel target for migraine drug development. Front Mol Neurosci 16:1182515
47. Desai J, Key L, Swindall A, Gaston K, Talati AJ (2021) The danger of diazoxide in the neonatal intensive care unit. Ther Adv Drug Saf 12: 20420986211011338
48. Martin GM, Sung MW, Yang Z, Innes LM, Kandasamy B, David LL et al (2019) Mechanism of pharmacochaperoning in a mammalian KATP channel revealed by cryo-EM. elife 8: e46417
49. Martin GM, Sung MW, Shyng S-L (2020) Pharmacological chaperones of ATP-sensitive potassium channels: mechanistic insight from cryoEM structures. Mol Cell Endocrinol 502: 110667
50. Martin GM, Kandasamy B, DiMaio F, Yoshioka C, Shyng S-L (2017) Anti-diabetic drug binding site in a mammalian KATP channel revealed by Cryo-EM. elife 6:e31054
51. ElSheikh A, Shyng SL (2023) K(ATP) channel mutations in congenital hyperinsulinism: progress and challenges towards mechanism-based therapies. Front Endocrinol (Lausanne) 14: 1161117

52. Leonard CE, Hennessy S, Han X, Siscovick DS, Flory JH, Deo R (2017) Pro-and antiarrhythmic actions of sulfonylureas: mechanistic and clinical evidence. Trends Endocrinol Metab 28(8):561–586
53. Ashcroft FM (2005) ATP-sensitive potassium channelopathies: focus on insulin secretion. J Clin Invest 115(8):2047–2058
54. Arnoux J-B, de Lonlay P, Ribeiro M-J, Hussain K, Blankenstein O, Mohnike K et al (2010) Congenital hyperinsulinism. Early Hum Dev 86(5):287–294
55. Arnoux JB, Verkarre V, Saint-Martin C, Montravers F, Brassier A, Valayannopoulos V et al (2011) Congenital hyperinsulinism: current trends in diagnosis and therapy. Orphanet J Rare Dis 6:63
56. Banerjee I, Salomon-Estebanez M, Shah P, Nicholson J, Cosgrove KE, Dunne MJ (2019) Therapies and outcomes of congenital hyperinsulinism-induced hypoglycaemia. Diabet Med 36(1):9–21
57. Banerjee I, Raskin J, Arnoux JB, De Leon DD, Weinzimer SA, Hammer M et al (2022) Congenital hyperinsulinism in infancy and childhood: challenges, unmet needs and the perspective of patients and families. Orphanet J Rare Dis 17(1):61
58. Leidenheimer NJ, Ryder KG (2014) Pharmacological chaperoning: a primer on mechanism and pharmacology. Pharmacol Res 83:10–19
59. Martin GM, Rex EA, Devaraneni P, Denton JS, Boodhansingh KE, DeLeon DD et al (2016) Pharmacological correction of trafficking defects in ATP-sensitive potassium channels caused by sulfonylurea receptor 1 mutations. J Biol Chem 291(42):21971–21983
60. Yang H-Q, Echeverry FA, ElSheikh A, Gando I, Anez Arredondo S, Samper N et al (2022) Subcellular trafficking and endocytic recycling of KATP channels. Am J Phys Cell Phys 322(6):C1230–C1C47
61. Yan F, Lin CW, Weisiger E, Cartier EA, Taschenberger G, Shyng SL (2004) Sulfonylureas correct trafficking defects of ATP-sensitive potassium channels caused by mutations in the sulfonylurea receptor. J Biol Chem 279(12):11096–11105
62. Yan FF, Casey J, Shyng SL (2006) Sulfonylureas correct trafficking defects of disease-causing ATP-sensitive potassium channels by binding to the channel complex. J Biol Chem 281(44):33403–33413
63. Yan FF, Lin YW, MacMullen C, Ganguly A, Stanley CA, Shyng SL (2007) Congenital hyperinsulinism associated ABCC8 mutations that cause defective trafficking of ATP-sensitive K+ channels: identification and rescue. Diabetes 56(9):2339–2348
64. Sung MW, Driggers CM, Mostofian B, Russo JD, Patton BL, Zuckerman DM et al (2022) Ligand-mediated structural dynamics of a mammalian pancreatic K(ATP) channel. J Mol Biol 434(19):167789
65. Schwartz TB, Meinert CL (2004) The UGDP controversy: thirty-four years of contentious ambiguity laid to rest. Perspect Biol Med 47(4):564–574
66. Terstappen GC (1999) Functional analysis of native and recombinant ion channels using a high-capacity nonradioactive rubidium efflux assay. Anal Biochem 272(2):149–155
67. Zhang J-H, Chung TDY, Oldenburg KR (1999) A simple statistical parameter for use in evaluation and validation of high throughput screening assays. SLAS Discov 4(2):67–73
68. Terstappen GC (2011) Nonradioactive rubidium efflux assay technology for screening of ion channels. Label-Free Technologies for Drug Discovery:111–124

Chapter 13

The Dynamic Clamp Technique: A Robust Toolkit for Investigating Potassium Channel Function

Chiara Bartolucci and Luca Sala

Abstract

The dynamic clamp technique has emerged as a powerful tool in the field of cardiac electrophysiology, enabling researchers to investigate the intricate dynamics of ion currents in cardiac cells. Potassium channels play a critical role in the functioning of cardiac cells and the overall electrical stability of the heart. This chapter provides a comprehensive overview of the methods and applications of dynamic clamp in the study of key potassium currents in cardiac cells. A step-by-step guide is presented, detailing the experimental setup and protocols required for implementing the dynamic clamp technique in cardiac cell studies. Special attention is given to the design and construction of a dynamic clamp setup with Real Time eXperimental Interface, configurations, and the incorporation of mathematical models to mimic ion channel behavior. The chapter's core focuses on applying dynamic clamp to elucidate the properties of various potassium channels in cardiac cells. It discusses how dynamic clamp can be used to investigate channel kinetics, voltage-dependent properties, and the impact of different potassium channel subtypes on cardiac electrophysiology. The chapter will also include examples of specific dynamic clamp experiments that studied potassium currents or their applications in cardiac cells.

Key words Potassium channels, Dynamic clamp, Action potential, Cardiac electrophysiology, Pluripotent stem cells

1 Introduction

The patch clamp is the gold standard technique in cellular electrophysiology, allowing the investigation of currents and voltages in excitable cells. Two main modalities exist: voltage clamp (V-Clamp), where the operator controls the membrane potential and elicits ionic currents, and current clamp (I-Clamp), where action potentials (APs) are triggered in cells with current pulses or recorded from intrinsically active cells. Over the years, an alternative approach called Dynamic Clamp technique (DC, [1]) was also developed; the DC is a powerful configuration that allows the injection of artificial/synthetic conductances in excitable cells to influence their electrical activity in real time. DC links in vitro

Simone Furini (ed.), *Potassium Channels: Methods and Protocols*, Methods in Molecular Biology, vol. 2796,
https://doi.org/10.1007/978-1-0716-3818-7_13,

cellular electrophysiology with real-time computer simulations, providing advanced electrophysiological readouts not possible in classical configurations. Complex applications initially emerged for neurophysiology [2–5] and subsequently were translated into cardiac physiology [6–9], where the DC has been widely used to model potassium channel conductances to better study their role in cardiac repolarization and arrhythmias.

More recently, DC has also been successfully applied to human induced pluripotent stem cell-derived cardiomyocytes (hiPSC-CMs) [10–13]. To overcome the depolarized resting membrane potential (RMP) mainly derived from their relative immaturity, an in silico inward rectifier K^+ current (I_{K1}) is injected to hyperpolarize the membrane and set the RMP close to the potassium equilibrium potential.

In this chapter, we will provide the technical requirements to set up DC applications and describe relevant examples of how the DC has been used in guinea pig and human stem-cell derived cardiomyocytes to provide important advances for scientists willing to:

1. Modify the biophysical properties of a current to simulate in vitro the functional effect of genetic mutations
2. Introduce an ionic current in a species in which the current is not naturally present to investigate species-specific mechanisms of arrhythmogenesis
3. Compensate cellular immaturity by injecting currents, with particular emphasis on chamber-specific conductances

2 Materials

2.1 Cardiomyocyte Preparation

2.1.1 Guinea Pig Left Ventricular CMs

1. Solution A: 143 mM NaCl, 5.4 mM KCl, 0.5 mM $MgCl_2$, 0.25 mM NaH_2PO_4, 5 mM HEPES, 5.5 mM D-glucose, adjusted to pH 7.4 with NaOH.
2. Solution B: 143 mM NaCl, 5.4 mM KCl, 1.8 mM $MgCl_2$, 0.5 mM $MgCl_2$, 0.25 mM NaH_2PO_4, 5 mM HEPES, 5.5 mM D-glucose, adjusted to pH 7.4 with NaOH.
3. Solution C: 143 mM NaCl, 5.4 mM KCl, 0.1 mM $MgCl_2$, 0.5 mM $MgCl_2$, 0.25 mM NaH_2PO_4, 5 mM HEPES, 5.5 mM D-glucose, adjusted to pH 7.4 with NaOH. After, add Collagenase II or Liberase TH Research Grade at 0.08–0.28 Wünsch units/mL.
4. Solution D: 70 mM KOH, 50 mM glutamic acid, 40 mM KCl, 20 mM taurine, 20 mM KH_2PO_4, 3 mM $MgCl_2$, 10 mM HEPES, 10 mM D-glucose, 0.5 mM EGTA, adjusted to pH 7.4 with KOH.

5. Ventricular cardiomyocytes from adult Dunkin-Hartley guinea pigs.

2.1.2 hiPSC-CMs

1. Recombinant human vitronectin 5 μg/mL to coat the single- or multiwell plate for hiPSC culture.
2. hiPSC lines generated from patients or healthy donors.
3. 0.5 mM EDTA.
4. Small molecules to induce differentiation of hiPSCs toward hiPSC-CMs: CHIR99021 and Wnt-C59.
5. B27 supplement and B27 supplement without insulin.
6. TrypLE Select 10X digestion enzyme.
7. Human embryonic stem cell-qualified Matrigel matrix diluted into DMEM:F12 at the concentration recommended by the manufacturer.
8. RPMI 1640 medium.
9. KnockOut serum replacement.
10. 10 mm Ø glass coverslips.

2.2 Patch Clamp Solutions

1. Tyrode's solution containing 144 mM NaCl, 5.6 mM KCl, 1.8–2.5 mM $CaCl_2$, 1.2 mM $MgCl_2$, 5 mM HEPES, and 11 mM D-glucose, pH adjusted to 7.4 with NaOH. Store at +4 °C until use.
2. Intracellular solution for hiPSC-CMs: 125 mM K-aspartate, 20 mM KCl, 10 mM NaCl, 5 mM Na_2-ATP, 10 mM HEPES, adjusted to pH 7.3 with KOH.
3. Intracellular solution for AP recording in hiPSC-CMs: same as previous solution plus 0.22 mM amphotericin B dissolved in DMSO.
4. Intracellular solution for guinea pig CMs: 100 mM K-aspartate, 45 mM KCl, 0.4 mM $CaCl_2$ (calculated free Ca^{2+} concentration of 10^{-7} M), 3 mM $MgCl_2$, 5 mM HEPES-KOH, 1 mM EGTA-KOH, 0.4 mM GTP-Na^+ salt, 5 mM ATP-K^+ salt, and 5 mM creatine phosphate Na^+ salt, adjusted to pH 7.2 with KOH.

2.3 Dynamic Clamp System

1. A personal computer with a dynamic clamp software installed, e.g., Real Time eXperiment Interface (RTXI, www.rtxi.org) using a real-time Linux operating system.
2. Multipurpose Analog/Digital-Digital/Analog (A/D-D/A) device installed in the computer (e.g., PCI-6024E; National Instruments, USA) and shielded connector block (e.g., BNC-2120; National Instruments, USA). RTXI uses the open-source library of Data Acquisition (DAQ) drivers, Linux Control, and Measurement Device Interface (Comedi, www.

comedi.org), to interface with DAQ devices (supported hardware can be found on the Comedi Website). Many DAQ cards compatible with PCI, PCI express, or PXI interface are available; its choice should depend on the number of A/D-D/A channels required, the data resolution (e.g., 12 or 16 bit), the sampling resolution, speed, and whether a simultaneous/sequential sampling of multiple input channels is required. A list of DAQs compatible with RTXI is available in the user manual under Data Acquisition Cards: http://rtxi.org/docs/manual/.

2.4 Dynamic Clamp Setup

1. Standard patch clamp setup equipped with an amplifier connected to a headstage, a digidata, a personal computer, a temperature-controlled chamber, a multi-line solution heater coupled with a perfusion system, and a rapid solution changer.
2. Linux-based personal computer for RTXI Installation.
3. Digital acquisition device.

3 Methods

3.1 Cell Preparation

3.1.1 Guinea Pig Left Ventricular Cardiomyocytes

1. Ventricular myocytes from adult Dunkin-Hartley guinea pigs (body weight ~300 g) are isolated by using a Langendorff-based retrograde coronary perfusion method and enzymatic digestion with Collagenase II or Liberase TH Research Grade (Roche) at 0.08–0.28 Wünsch units/mL, and adjust the enzyme concentration according to manufacturer's instructions to obtain an optimal yield of viable CMs with intact membranes. The procedure should comply with the most updated ethical guidelines for the use of laboratory animals (http://data.europa.eu/eli/dir/2010/63/2019-06-26).
2. Gradually adapt the dissociated CMs to physiological solutions by progressively increasing the extracellular Ca^{2+} concentration to reach physiological levels (1.8–2.5 mM).
3. Rod-shaped, Ca^{2+}-tolerant ventricular myocytes are used within 12 h from dissociation. Measurements are performed only in quiescent ventricular CMs with clear-cut striations.

3.1.2 hiPSC-CMs

1. Human induced pluripotent stem cells (hiPSCs) are cultured on human recombinant vitronectin and passaged twice a week when the colonies reach 70% confluency. Cardiac differentiation is started when the colonies reach 80–95% confluency.
2. hiPSC-CMs are differentiated with a small molecule-based protocol derived from that published by Lian et al. [14], with minor modifications to efficiently fit multiple hiPSC lines as previously described [15].

3. After 14–18 days, beating hiPSC-CMs are cryopreserved to generate large stocks of immature, early stage CMs. Before the experiments, the CMs are thawed and cultured for at least additional 14 days in a maturation medium derived from Feyen et al. [16].
4. After 14 days from the thawing (~day 28–32 from the start of the differentiation), dissociate the monolayer as single cells with TrypLE 10X, and seed the cells on Matrigel-coated 10 mm Ø glass coverslips at a density of ~ 0.8–1 × 10^4 cells/ cm^2. Isolated hiPSC-CMs are then used for patch clamp experiments from 3 to 10 days after the seeding.

3.2 Dynamic Clamp Setup

A standard patch clamp setup should be used for DC recordings. Specifically, the setup must be equipped with a Digidata (e.g., Molecular Devices 1440A, 1550B, etc.), an amplifier (e.g., Axopatch 200B, Multiclamp 700B, etc.) connected to its headstage, and a personal computer equipped with a commercial or open-source patch clamp software interface (Fig. 1). In addition, specifically for the DC setup, additional hardware is required.

1. RTXI installation. RTXI installation requires a real-time Linux operating system. It can be installed through a "Live USB" or directly run from the USB, resulting in a portable real-time software system. Details on the installation can be found on the web page (http://www.rtxi.org/install).
2. Dynamic clamp hardware setup. RTXI and the amplifier communicate through a data acquisition device. Since the software for operating many patch clamp amplifier models is usually available only for Windows or Mac operating systems but not Linux, a second computer may be required to control the amplifier. For its configuration, it is required to:
   (a) Install the DAQ card into the personal computer that will run RTXI and connect the BNC connector block with the appropriate cable.
   (b) Connect the Analog Output 0 (labeled AO 0) of the connector block to the external Input of the amplifier with a BNC cable.
   (c) Connect the Analog Input 0 (labeled AI 0) to the Output of the amplifier.
3. RTXI and the patch clamp amplifier. RTXI receives the voltage signal, read through a headstage, from the amplifier. To achieve an adequate signal-to-noise ratio, RTXI should be configured to match the scale factor with that of the amplifier. This is done by manually changing the scaling factors for the input/output signals in the RTXI interface. The scale factors are described in the amplifier's manual. For example, if the output gain on the

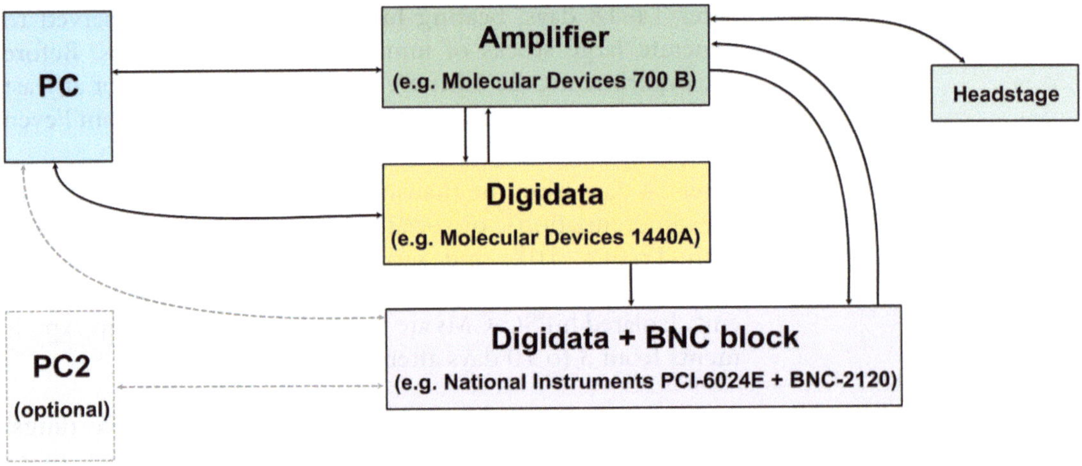

Fig. 1 Schematic representation of the hardware and connections required for performing DC experiments. Depending on the configuration and interface used, the second PC can be optionally configured

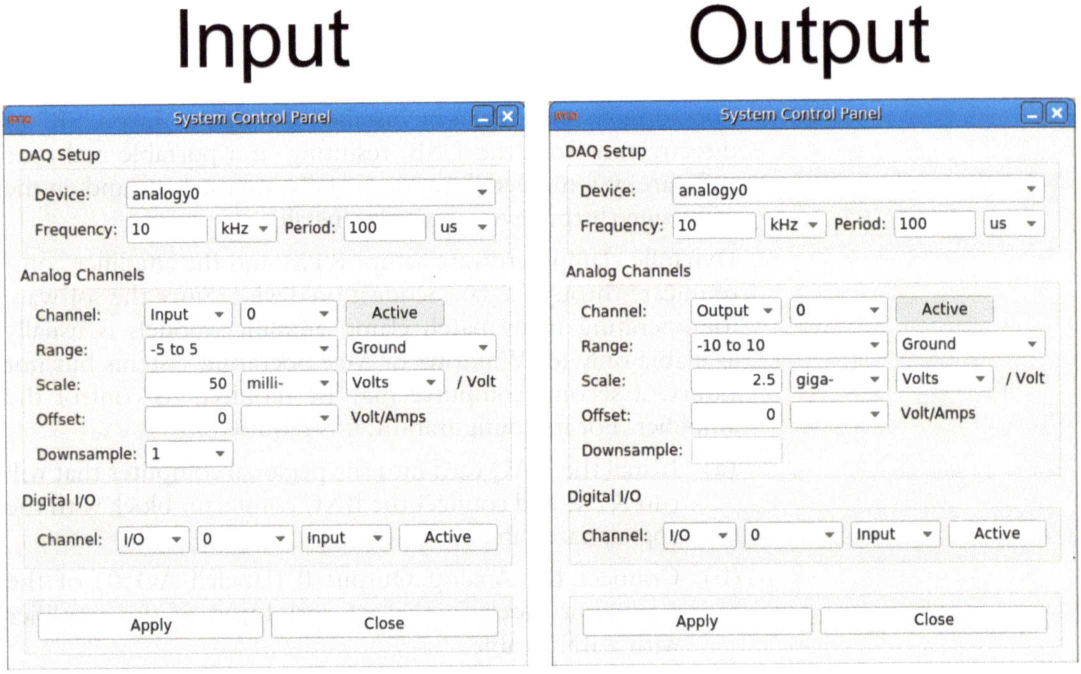

Fig. 2 System control panel. Representative system control panel interfaces, respectively, for the configuration of the input and output analog channels

amplifier is set to 10×, RTXI divides that signal by a factor of 10 by using the System Control Panel (Fig. 2, scale box) as described:

(a) Open RTXI.
(b) Navigate to the System tab, and open the Control Panel (System → Control Panel).
(c) Activate the channels for analog input and output.
(d) Adjust the scaling values accordingly.
(e) Save the setting by pressing Apply button.

3.3 Injection of an Endogenous Current

3.3.1 Choice and Creation of the Model (See Subheading 4.1.1)

1. RTXI works in modules. Based on the experiment and the cell types, the user should choose and write in the C++ programming language the computational model that adequately simulates the current in the target cell. The synthetic current is based on mathematical equations that describe the time course of the interest current in response to the AP recorded in real time in the patched cell. The current formulation can be taken from existing models with modifications, e.g., to mimic the effect of a genetic mutation or built ex novo by fitting the equations to voltage-clamp experiments. The required files are the following:
   (a) Header file. It contains the class definitions and variable declarations required for the module. These variables are divided into (i) model parameters (e.g., cycle length, membrane capacitance, stimulus setting), (ii) experimental parameters (e.g., electrolyte concentration, Nernst potential), or (iii) module control list functions (e.g., *void initialize()*; *void execute()*; *void update()*).
   (b) C++ source file. It contains (i) necessary inputs, outputs, parameters, and gating variables; (ii) real-time executed code in the "execute" function; (iii) and code executed during module initialization and update.
   (c) Make file for compiling the module.

3.3.2 Experiment Setup in RTXI

1. Open RTXI and the newly built module.
2. In the RTXI dashboard, select Modules → Load User Module → [[*Load current block*, e.g., *IK1.so*]].
3. Load additional modules required for dynamic clamp experiment (e.g., the Oscilloscope for the visualization (System → Oscilloscope) or Data Recorder for the data registration (System → Data Recorder)).
4. Connect the modules' input/output:
   (a) Open the Connector window (System → Connector, Fig. 3).
   (b) Set the Output block (Source): Current Block Module—Output Current.

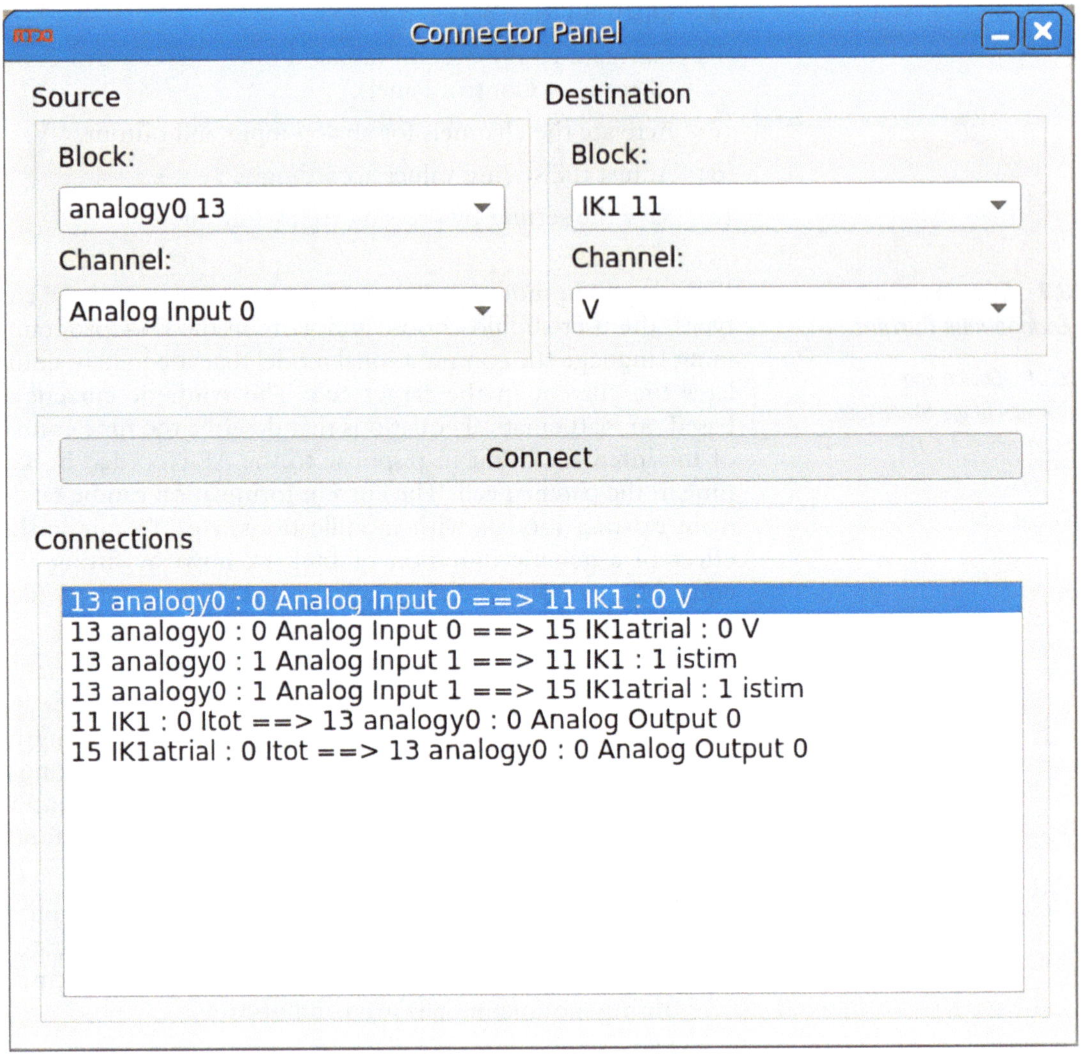

Fig. 3 DC connector panel: RTXI connector settings for I_{K1} dynamic clamp experiment. The connector panel showcases the signal and slot architecture of RTXI. The Analog Input 0 Channel of a DAQ device (signal), which contains the AP, is connected to the input of the I_{K1} module; the Analog Input 1 Channel, which contains the stimulus current, is connected to another input of the I_{K1} module. Conversely, the output of the module, the total current ($I_{K1} + I_{stim}$), is connected to the Analog Output 0 Channel. The Analog Output Channel is in turn connected to a patch clamp amplifier which will interact with the cell and stimulate it with these synthetic signals

(c) Set the Input block (Destination): */dev/comedi0*—Analog Output 0.

(d) Click the Connect button to complete the connection.

(e) Repeat for the following combinations:

- Block analogy0 13—Analog Input 0|IK1—Input Voltage.

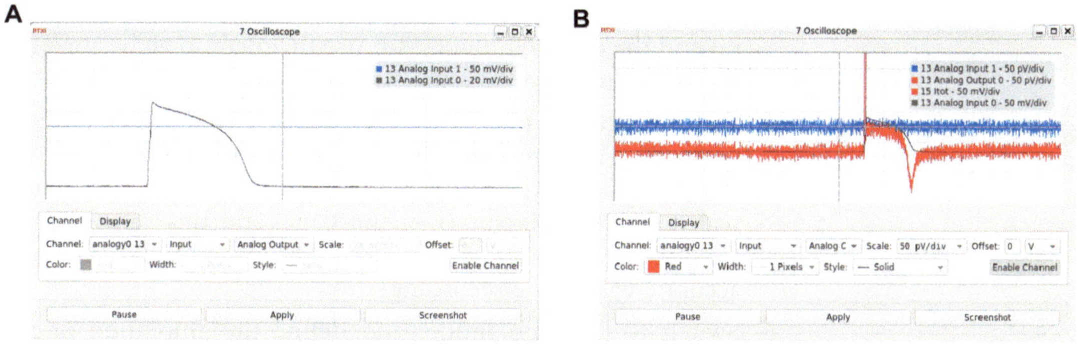

Fig. 4 Oscilloscope panel. (**a**) Oscilloscope window displaying only the voltage signal (AP, black) received from the cell. (**b**) Oscilloscope window displaying the AP received from the cell (black), the stimulus current required to trigger the AP and the total current injected in the cell (red), including both the stimulus current and the modelled current (inverted I_{K1} in this case). The traces have different offset values on the y-axis for better visualization during the experimental procedures

- Block analogy0 13—Analog Input 1|IK1—Input Istim.
- IK1—Itot Output|Block analogy0 13—Analog Output 0.

5. Set the Oscilloscope (Fig. 4) to display the CM membrane potential and currents.
   (a) Open Oscilloscope properties (right click on Oscilloscope → Properties).
   (b) Set Channel: IK1|Input|Input V.
   (c) Toggle the Active button.
   (d) Set Channel: IK1|Output|Output IK1.
   (e) Set Scale: 20 mV/V.
   (f) Click the Apply button.
   (g) Click Display tab.
6. Set Time Scale: 100 ms/div.
7. Save Workspace for future experiments.
   (a) File → Save Workspace.
   (b) Choose filename and click the Save button.

3.3.3 Patch Clamp Recording

1. Identify an isolated CM with a microscope, and approach it with a borosilicate glass pipette having a resistance of 1.5–2 MΩ filled with the intracellular solution.
2. Through a manual or automated micromanipulator, put the glass pipette in close proximity to the cell, and, upon touching the membrane, apply a gentle negative pressure with a syringe.

3. After the formation of a GΩ seal resistance between the pipette tip and the cell membrane, rupture the membrane by applying a stronger negative pressure with a syringe.
4. Calculate the cell membrane capacitance (Cm) and series resistance (Rs), and annotate the Cm value in the DC software for later use. The DC current magnitude has to be appropriately scaled for the cell size (i.e., the Cm), and this value will be included as a variable in the model. The lower the Rs, the more accurate will be the recording.
5. Switch to I-Clamp mode, and define the amplitude of the positive square pulse of 2–3 ms of duration and 0.5–1.5 nA of amplitude to evoke APs at a specific frequency.
6. Depending on the animal species used for the experiment (e.g., small rodents generally have a faster heart rate than large mammals) and on the protocol of choice, select one or more consecutive pacing frequencies. The most used ranges vary between 0.5 and 4 Hz.
7. Record triggered APs until a steady-state in the APD is reached. In some conditions as for some pacing frequencies, diseases, or drug exposures, an ideal APD stability may not be reached.
   (a) If interested in modifying and investigating the biophysical properties of a current, first, superfuse the cell with a specific ion channel blocker to remove the contribution of the respective current to the AP, and proceed with the activation of the DC mode.
   (b) Activate the DC mode if interested in injecting a current into a cell type characterized by a low or absent magnitude for that specific current.
8. In the classical I-Clamp configuration, the voltage profile of the cell is recorded. In DC mode, the voltage profile is recorded, but also included into a mathematical model which computes the appropriate current as output for each voltage point of the AP. The current is then injected into the cell within microseconds with a very rapid feedback loop, causing a dynamic modification of the AP contour.

3.4 Applications of DC on K+ Channels

3.4.1 Specifically Injection of a Current That Is Absent in the Target Cell

The following paragraph describes an experiment in which a current, normally absent in the targeted cell population, has been injected in primary guinea pig cardiomyocytes.

Injection of the transient-outward potassium current (I_{to}) into guinea pig (GP) adult ventricular CMs to clarify whether the presence of a phase-1 AP notch (i.e., an AP contour characterized by a spike-and-dome profile) may affect membrane current and repolarization response to β-adrenergic stimulation [17] (Fig. 5).

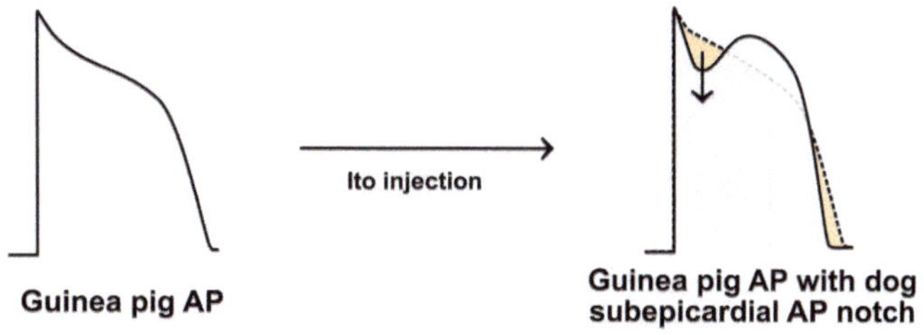

Fig. 5 Graphical representation of I_{to} injection. The impact of species-specific AP features on the response to specific stimuli, e.g., β-adrenergic stimulation, a numerical model of I_{to} is injected into ventricular CMs from adult guinea pigs to recreate the characteristic notched AP profile normally observed in canine subepicardial CMs.

(a) Select the I_{to} component from a canine numerical model, suitable to reproduce the notched AP profile observed in sub-epicardial canine CMs.
(b) Create the RTXI module with the implementation of equations describing the I_{to} current.
(c) Load the module and proceed with the configuration as described in Subheading 3.3.2 and with the patch clamp recording as in Subheading 3.3.3.
(d) Set the Cm in the Parameters menu.
(e) Upon reaching a steady state, activate the bottom FlagDC from the Parameters menu, and inject the I_{to} into the cell by using RTXI.
(f) Evaluate the impact of the injected current on the AP profile.
(g) If required, modify the biophysical parameters of the I_{to} model to obtain a suitable AP contour.

3.4.2 Injection of Currents Typically Exhibiting Low Magnitude in the Target Cell

The following paragraph describes an experiment in which the inward rectifier potassium current, I_{K1}, normally present with a low density in the targeted cell population, has been injected in hiPSC-derived cardiomyocytes (hiPSC-CMs) to discriminate chamber specificity [12] (*see* Fig. 6 and Subheading 4.1.2).

Injection of the inward rectifier potassium current (I_{K1}) in hiPSC-CMs to discriminate chamber specificity [12] (*see* Fig. 6 and Subheading 4.1.2).

(a) Select the I_{K1} formulation from human numerical models specific for atrial [18] or ventricular [19] cells.
(b) Create the RTXI modules by implementing the equations describing the atrial and ventricular I_{K1} profiles.
(c) Load the module (*see* Fig. 7, on the left the atrial module and on the right the ventricular one), and proceed with the

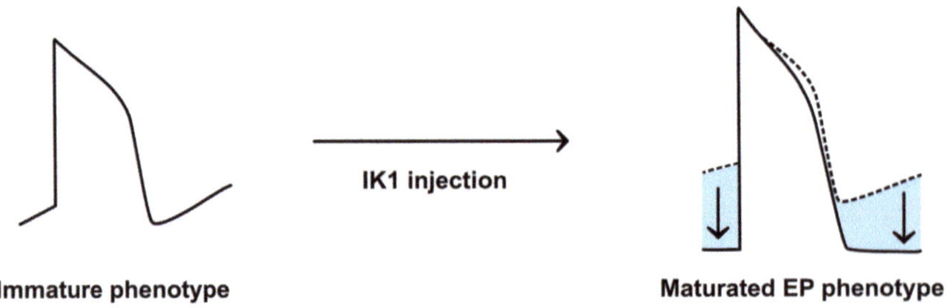

Fig. 6 Graphical representation of I_{K1} injection. The immature electrophysiological phenotype of hiPSC-CMs is corrected by injecting a virtual I_{K1} to hyperpolarize the resting membrane potentials to levels typical of adult human cardiomyocytes

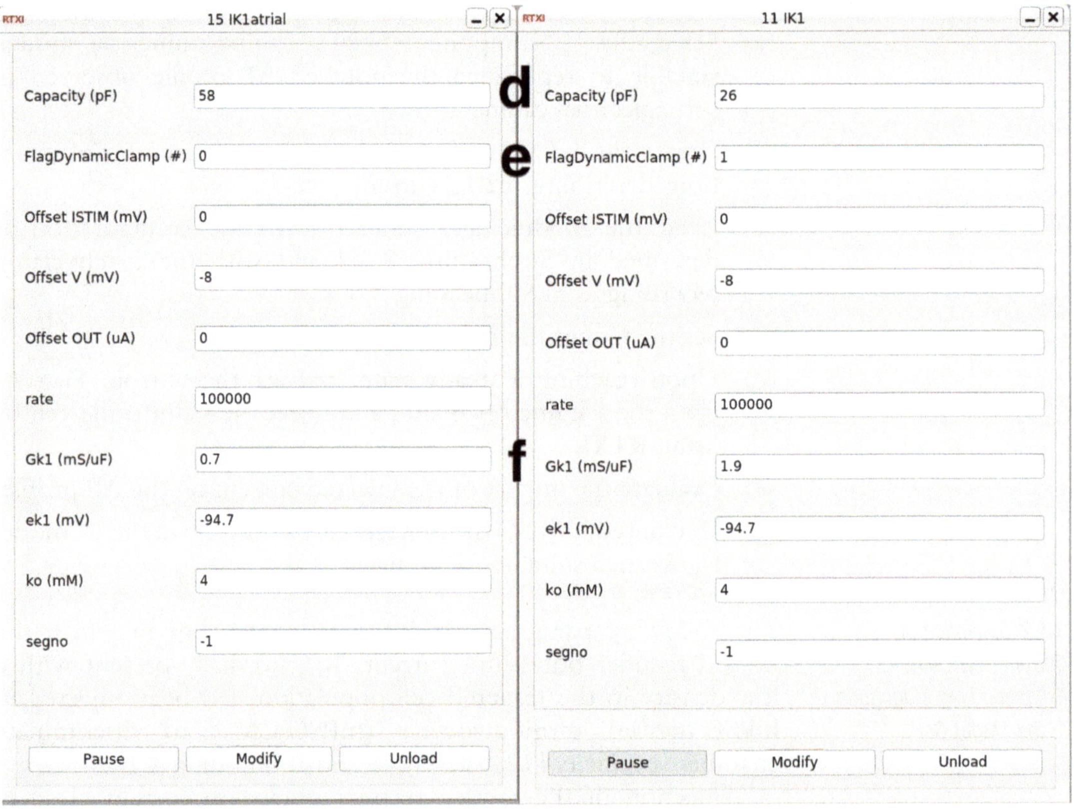

Fig. 7 Model parameter panels. Parameters menu refers to the I_{K1} models (atrial vs ventricular) loaded in RTXI. The letters beside each parameter refer to the respective letter in the protocol bullet point

configuration as described in Subheading 3.3.2 and with the patch clamp recording as in Subheading 3.3.3.

(d) Set the Cm in the Parameters menu.

(e) Upon reaching a steady state, activate the bottom FlagDC (set to 1, default is 0) from the Parameters menu, and inject the I_{K1} into the cell by using RTXI.

(f) Evaluate the impact of the injected current in terms of AP waveform changes. Observe whether the current is capable of repolarizing the membrane potential and whether it does not significantly impact the contour during repolarization, leading to abnormal AP contours. Parameters such as the conductance (G_{K1}) or the reversal potential (e_{K1}) can be modified to finely tune the result. However, it is important not to modify these parameters among different cells to properly compare the results.

(g) To evaluate the most appropriate chamber-specific I_{K1} profile, the comparison between the models should be performed in series within the same cell by turning on the atrial or ventricular I_{K1} module.

3.4.3 Injection of Currents Existing in the Target Cell to Study the Ion Channel Biophysics

This paragraph exploits DC to systematically analyze the effect of changes in I_{Kr} conductance and gating properties on GP AP duration (APD) and its time variability, an index of electric instability [20, 21] (Fig. 8).

(a) Select the I_{Kr} formulation from a GP numerical model.

(b) Compare the selected I_{Kr} model to in vitro data obtained by isolating the E4031-sensitive current (I_{Kr}) under AP-clamp.

(c) Should optimization be required, adapt the equations describing the I_{Kr} model to generate an optimized version.

(d) Create the RTXI modules by implementing the equations describing the optimized I_{Kr}.

(e) Load the module and proceed with the configuration as described in Subheading 3.3.2 and with the patch clamp recording as in Subheading 3.3.3.

(f) Superfuse the CM with 5 μM E-4031 dissolved in Tyrode's to block I_{Kr}.

(g) Record the AP during I_{Kr} blockade, simulating a genetic or drug-induced APD prolongation.

(h) Set the Cm in the Parameters menu.

(i) Upon reaching a steady state, activate the bottom FlagDC (set to 1) from the Parameters menu, and inject the I_{Kr} into the cell by using RTXI.

(j) Evaluate the capacity of the injected I_{Kr} formulation to restore the control AP profile.

(k) Change the biophysical parameters of the current (e.g., maximal conductance, half-maximal activation voltage, activation constant of decay, etc.) to simulate specific congenital or drug-induced alterations.

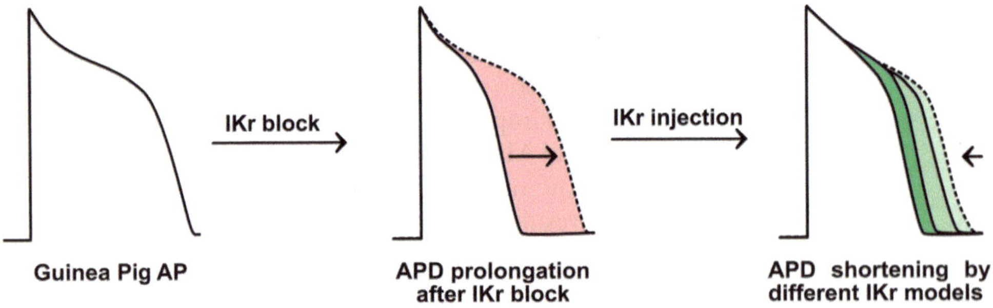

Fig. 8 Graphical representation of I_{Kr} injection. To understand the impact of I_{Kr} gating parameters on repolarization stability, multiple I_{Kr} numerical models are sequentially injected with DC to verify the major contributor to I_{Kr}-dependent repolarization abnormalities

4 Notes

4.1 Limitations

4.1.1 Lack of Ion Selectivity (See Subheading 3.3.1)

One of the inherent limitations of using DC to inject currents is the discrepancy existing between the current that is simulated and the corresponding ion flux across the membrane. Simulating a K^+ conductance with DC does not imply an equal flux of potassium ions across the membrane. This limitation becomes particularly evident for Ca^{2+} ions, when the carried ion also acts as a second messenger or when it triggers other functions such as contractility. In these settings, the ion flux across the membrane is limited to that flowing through selective ion channels.

4.1.2 Maturation-Specific Differences (See Subheading 3.4.2)

Other aspects that require caution when working with immature cell types are (i) the profile of the current and (ii) the magnitude of the current that needs to be injected.

Ion current models are built over specific experimental datasets, with the relative ratio between conductances often being species- and maturation-specific. This derives from the different ion channel isoforms and expression levels that each species expresses at each maturation stage. It is then crucial to verify that the species from which the computational model or ionic current model has been generated does share electrophysiological properties with the one used in the experiment.

The maturation status of hiPSC-CMs has been revealed crucial for the switch between fetal and adult isoforms of the sodium channel encoded by the SCN5A gene [22, 23]. If the mathematical model is not properly calibrated to take into account the SCN5A isoform switch, the injection of a conductance calibrated on a different isoform may lead to severe electrophysiological alterations, as emerged in vivo with gene editing [24]. The proper calibration of ion current conductance is essential for accurate simulations. In the early experiments with the injection of the I_{K1} current in hiPSC-CMs to compensate for their electrophysiological immaturity, the ion current magnitude injected was obtained from the O'Hara-Rudy model (ORd), which indicated an I_{K1} conductance of 0.19 nS/pF. Once applied to hiPSC-CMs, these values were insufficient to properly hyperpolarize the membrane potential, and the conductance had to be increased tenfolds to ~ 1.9–2 nS/pF [11, 13]. These differences may sometimes be subtle, thus not trivial to be identified and complex to be addressed.

4.2 Current and Future Advancements

The capacity of DC to artificially improve the electrophysiological phenotype of hiPSC-CMs has also been upscaled to benefit from the throughput of automated patch clamp systems [25]. This combination significantly expands the possibilities of using human-based models for pharmacological and toxicological studies, reducing the amount of manual work required and providing standardized and semi-automated procedures [26] that can become compatible with industry standards and requirements to assess drug-induced liabilities [27].

The concept of DC has also been coupled with the most recent advances in optogenetic actuators. This combination, termed optical dynamic clamp [28], harnesses the conductive properties of light-sensitive ion channels to alter the cell membrane potential and aims to remove the requirements for invasive, laborious, and single-use patch clamp pipettes of the standard DC approach. Current implementations still require an electrode to accurately measure the membrane potential required to calculate the I_{K1} model but represent the most advanced approach to date in the view for all-optical electrophysiological investigations [29, 30].

References

1. Wilders R (2006) Dynamic clamp: a powerful tool in cardiac electrophysiology. J Physiol 576: 349–359. https://doi.org/10.1113/JPHYSIOL.2006.115840
2. Sharp AA, Abbott LF, Marder E (1992) Artificial electrical synapses in oscillatory networks. J Neurophysiol 67:1691–1694. https://doi.org/10.1152/JN.1992.67.6.1691
3. Sharp AA, O'Neil MB, Abbott LF, Marder E (1993) Dynamic clamp: computer-generated conductances in real neurons. J Neurophysiol 69:992–995. https://doi.org/10.1152/JN.1993.69.3.992
4. Sharp AA, O'Neil MB, Abbott LF, Marder E (1993) The dynamic clamp: artificial conductances in biological neurons. Trends Neurosci 16:389–394. https://doi.org/10.1016/0166-2236(93)90004-6
5. Robinson HPC, Kawai N (1993) Injection of digitally synthesized synaptic conductance transients to measure the integrative properties of neurons. J Neurosci Methods 49:157–165. https://doi.org/10.1016/0165-0270(93)90119-C
6. Berecki G, Zegers JG, Bhuiyan ZA et al (2006) Long-QT syndrome-related sodium channel mutations probed by the dynamic action potential clamp technique. J Physiol 570: 237–250. https://doi.org/10.1113/JPHYSIOL.2005.096578
7. Berecki G, Zegers JG, Verkerk AO et al (2005) HERG channel (dys)function revealed by dynamic action potential clamp technique. Biophys J 88:566–578. https://doi.org/10.1529/BIOPHYSJ.104.047290
8. Dong M, Sun X, Prinz AA, Wang HS (2006) Effect of simulated I(to) on guinea pig and canine ventricular action potential morphology. Am J Physiol Heart Circ Physiol 291.

https://doi.org/10.1152/AJPHEART.00084.2006

9. Wang Y, Cheng J, Joyner RW et al (2006) Remodeling of early-phase repolarization: a mechanism of abnormal impulse conduction in heart failure. Circulation 113:1849–1856. https://doi.org/10.1161/CIRCULATIONAHA.106.615682
10. Lee YK, Sala L, Mura M et al (2021) MTMR4 SNVs modulate ion channel degradation and clinical severity in congenital long QT syndrome: insights in the mechanism of action of protective modifier genes. Cardiovasc Res 117:767–779. https://doi.org/10.1093/CVR/CVAA019
11. Rocchetti M, Sala L, Dreizehnter L et al (2017) Elucidating arrhythmogenic mechanisms of long-QT syndrome CALM1-F142L mutation in patient-specific induced pluripotent stem cell-derived cardiomyocytes. Cardiovasc Res 113:531–541. https://doi.org/10.1093/CVR/CVX006
12. Altomare C, Bartolucci C, Sala L et al (2023) A dynamic clamping approach using in silico IK1 current for discrimination of chamber-specific hiPSC-derived cardiomyocytes. Commun Biol 6:291. https://doi.org/10.1038/s42003-023-04674-9
13. Van Putten RMEM, Mengarelli I, Guan K et al (2015) Ion channelopathies in human induced pluripotent stem cell derived cardiomyocytes: a dynamic clamp study with virtual IK1. Front Physiol 6. https://doi.org/10.3389/FPHYS.2015.00007
14. Lian X, Zhang J, Azarin SM et al (2013) Directed cardiomyocyte differentiation from human pluripotent stem cells by modulating Wnt/β-catenin signaling under fully defined conditions. Nat Protoc 8:162–175. https://doi.org/10.1038/NPROT.2012.150
15. Sala L, Leonov V, Mura M et al (2022) Use of hiPSC-derived cardiomyocytes to rule out proarrhythmic effects of drugs: the case of hydroxychloroquine in COVID-19. Front Physiol 12. https://doi.org/10.3389/fphys.2021.730127
16. Feyen DAM, McKeithan WL, Bruyneel AAN et al (2020) Metabolic maturation media improve physiological function of human iPSC-derived cardiomyocytes. Cell Rep 32. https://doi.org/10.1016/J.CELREP.2020.107925
17. Sala L, Hegyi B, Bartolucci C et al (2018) Action potential contour contributes to species differences in repolarization response to b-adrenergic stimulation. Europace 20:1543–1552. https://doi.org/10.1093/europace/eux236
18. Koivumäki JT, Korhonen T, Tavi P (2011) Impact of sarcoplasmic reticulum calcium release on calcium dynamics and action potential morphology in human atrial myocytes: a computational study. PLoS Comput Biol 7. https://doi.org/10.1371/JOURNAL.PCBI.1001067
19. O'Hara T, Virág L, Varró A, Rudy Y (2011) Simulation of the undiseased human cardiac ventricular action potential: model formulation and experimental validation. PLoS Comput Biol:7
20. Altomare C, Bartolucci C, Sala L et al (2015) IKr impact on repolarization and its variability assessed by dynamic clamp. Circ Arrhythm Electrophysiol 8:1265–1275. https://doi.org/10.1161/CIRCEP.114.002572
21. Bartolucci C, Altomare C, Bennati M et al (2015) Combined action potential- and dynamic-clamp for accurate computational modelling of the cardiac IKr current. J Mol Cell Cardiol 79:187–194. https://doi.org/10.1016/j.yjmcc.2014.11.011
22. Campostrini G, Kosmidis G, Ward-Van Oostwaard D et al (2023) Maturation of hiPSC-derived cardiomyocytes promotes adult alternative splicing of SCN5A and reveals changes in sodium current associated with cardiac arrhythmia. Cardiovasc Res 119:167–182. https://doi.org/10.1093/CVR/CVAC059
23. Veerman CC, Mengarelli I, Lodder EM et al (2017) Switch from fetal to adult SCN5A isoform in human induced pluripotent stem cell-derived cardiomyocytes unmasks the cellular phenotype of a conduction disease-causing mutation. J Am Heart Assoc 6. https://doi.org/10.1161/JAHA.116.005135
24. Pang PD, Alsina KM, Cao S et al (2018) CRISPR-mediated expression of the fetal Scn5a isoform in adult mice causes conduction defects and arrhythmias. J Am Heart Assoc 7. https://doi.org/10.1161/JAHA.118.010393
25. Becker N, Horváth A, De Boer T et al (2020) Automated dynamic clamp for simulation of IK1 in human induced pluripotent stem cell–derived cardiomyocytes in real time using patchliner dynamite8. Curr Protoc Pharmacol 88: e70. https://doi.org/10.1002/CPPH.70
26. Clark AP, Wei S, Kalola D et al (2022) An in silico-in vitro pipeline for drug cardiotoxicity screening identifies ionic pro-arrhythmia mechanisms. Br J Pharmacol 179:4829–4843. https://doi.org/10.1111/BPH.15915
27. Kernik DC, Yang PC, Kurokawa J et al (2020) A computational model of induced pluripotent stem-cell derived cardiomyocytes for high throughput risk stratification of KCNQ1

genetic variants. PLoS Comput Biol 16: e1008109. https://doi.org/10.1371/JOURNAL.PCBI.1008109

28. Quach B, Krogh-Madsen T, Entcheva E, Christini DJ (2018) Light-activated dynamic clamp using iPSC-derived cardiomyocytes. Biophys J 115:2206–2217. https://doi.org/10.1016/J.BPJ.2018.10.018
29. Klimas A, Ambrosi CM, Yu J et al (2016) OptoDyCE as an automated system for high-throughput all-optical dynamic cardiac electrophysiology. Nat Commun 7(1):1–12. https://doi.org/10.1038/ncomms11542
30. Hart WL, Needham K, Richardson RT et al (2023) Dynamic optical clamp: a novel electrophysiology tool and a technique for closed-loop stimulation. Biomed Signal Process Control 85:105031. https://doi.org/10.1016/j.bspc.2023.105031

Chapter 14

Automated Patch Clamp Recordings of GPCR-Gated Ion Channels: Targeting the MC4-R/Kir7.1 Potassium Channel Complex

Ciria C. Hernandez, Luis E. Gimenez, and Roger D. Cone

Abstract

Automated patch clamp recording is a valuable technique in drug discovery and the study of ion channels. It allows for the precise measurement and manipulation of channel currents, providing insights into their function and modulation by drugs or other compounds. The melanocortin 4 receptor (MC4-R) is a G protein-coupled receptor (GPCR) crucial to appetite regulation, energy balance, and body weight. MC4-R signaling is complex and involves interactions with other receptors and neuropeptides in the appetite-regulating circuitry. MC4-Rs, like other GPCRs, are known to modulate ion channels such as Kir7.1, an inward rectifier potassium channel, in response to ligand binding. This modulation is critical for controlling ion flow across the cell membrane, which can influence membrane potential, excitability, and neurotransmission. The MC4-R is the target for the anti-obesity drug Imcivree. However, this drug is known to lack optimal potency and also has side effects. Using high-throughput techniques for studying the MC4-R/Kir7.1 complex allows researchers to rapidly screen many compounds or conditions, aiding the development of drugs that target this system. Additionally, automated patch clamp recording of this receptor–channel complex and its ligands can provide valuable functional and pharmacological insights supporting the development of novel therapeutic strategies. This approach can be generalized to other GPCR-gated ion channel functional complexes, potentially accelerating the pace of research in different fields with the promise to uncover previously unknown aspects of receptor–ion channel interactions.

Key words Automated patch clamp, High-throughput screening, Drug discovery, MC4-R, Kir7.1, APC, HTS, GPCR-gated ion channels

1 Introduction

Automated patch clamp (APC) recording, also known as high-throughput patch clamp, is a powerful technique used in drug discovery and basic science research to study ion channels and their role in various physiological processes [1, 2]. This technique allows for high-throughput screening (HTS) of ion channel activity from 384 cells simultaneously, making it particularly useful for sifting large libraries of compounds for drug discovery purposes.

Simone Furini (ed.), *Potassium Channels: Methods and Protocols*, Methods in Molecular Biology, vol. 2796,
https://doi.org/10.1007/978-1-0716-3818-7_14,

One exciting area of study in drug discovery is the investigation of the molecular mechanisms underlying G protein-coupled receptor (GPCR) biased signaling by directly or indirectly regulating ion channel gating [3–7]. GPCRs are among the most targeted proteins in drug discovery because of their central role in cellular signaling. Understanding the complex interactions between GPCRs and ion channels can aid in developing drugs that modulate these pathways for therapeutic purposes. The melanocortin 4 receptor (MC4-R) is a GPCR that regulates various physiological processes, primarily related to energy balance, appetite, and body weight [8–10]. Due to its central role in appetite regulation and energy balance, MC4-R have been exploited as a target for pharmacological interventions to treat obesity and related metabolic disorders [11–14].

The MC4-R is an interesting target for exploring biased signaling because activation of this receptor inhibits food intake, produces weight loss, and increases autonomic activity, creating an undesirable target-mediated pressor response [15–17]. MC4-R, like many GPCRs, can activate multiple intracellular signaling pathways upon ligand binding. Biased signaling, also known as biased agonism or functional selectivity, refers to the phenomenon by which a ligand (agonist) binding to a specific receptor, such as MC4-R, activates one signaling pathway over others, in contrast to balanced agonism where the receptor can activate multiple intracellular pathways at full efficacy. Biased signaling can have important implications for drug development, as it allows the design of ligands that selectively activate specific therapeutic pathways while avoiding the unwanted side effects mediated by the other signaling modalities.

MC4-R canonical pathways include the $G\alpha_s$/cAMP, $G\alpha_q$/Ca^{2+}, and β-arrestin-dependent pathways [18–21]. MC4-Rs can indirectly influence ion channel activity by modulating second messenger intracellular levels such as 3′,5′-cyclic adenosine monophosphate (cAMP) or calcium (Ca^{2+}) ion mobilization. Non-canonical pathways include the recent discovery regarding the direct coupling between the inward-rectifying potassium (K^+) channel Kir7.1 and MC4-R [22, 23]. The MC4-R/Kir7.1 complex represents an example of GPCR-gated ion channel crosstalk [24]. As a novel unexploited target, drugs targeting the MC4-R/Kir7.1 complex could potentially treat obesity or other metabolic disorders. This complex plays a crucial role in regulating neuronal excitability within the paraventricular nucleus of the hypothalamus (PVH), where Kir7.1 signaling plays a central role in the melanocortin-mediated regulation of energy homeostasis [22, 23, 25]. MC4-R-biased signaling research aims to identify or design ligands that act as selective functional agonists or antagonists of the MC4-R/Kir7.1 complex. These ligands can be used to modulate MC4-R activity in a highly targeted manner, potentially leading to more effective and safer anti-obesity drugs.

Automated patch clamp recording can be employed to develop drugs that target the MC4-R/Kir7.1 complex and assess the impact of potential drug candidates on its activity. This is how the methodology can be applied:

1. High-throughput screening: Automated patch clamp systems with 384 amplifier channels and a 384-pipetting head, enabling the parallel recording from an equal number of cells and allowing the screening of large compound libraries. With this capability, scientists can investigate the effect of different compounds on the activity of the MC4-R/Kir7.1 complex, identifying potential agonists, antagonists, or modulators that may have therapeutic relevance.
2. Characterization of compound potency and selectivity: Automated patch clamp recordings can generate concentration-response curves to determine the potency of compounds in modulating the MC4-R/Kir7.1 complex. This information helps assess the selectivity of compounds, ensuring they specifically target the desired molecular interaction without significant off-target effects.
3. Mechanistic studies: Researchers can study the mechanisms underlying the MC4-R/Kir7.1 complex by manipulating the experimental conditions. For example, they can investigate the effects of mutations in either protein component on complex formation and function, shedding light on the structure–function relationship and potential therapeutic targets.
4. Safety profiling: Automated patch clamp recordings can assess the impact of compounds on the electrophysiological properties of cells, providing insights into their safety profile. This information is crucial for identifying potential liabilities, such as off-target effects on other ion channels or adverse effects on cellular excitability.
5. Mode-of-action studies: Once promising compounds are identified, automated patch clamp recording can be used to further elucidate their mode of action. By studying the effects of compounds on specific ion currents associated with the MC4-R/Kir7.1 complex, scientists can gain insights into the underlying cellular signaling pathways and validate their proposed therapeutic targets.

Automated patch clamp recording is a valuable tool in drug discovery. It can be applied to investigate the MC4-R/Kir7.1 complex, providing insights into its function, identifying potential drug candidates, and facilitating the development of targeted therapies for metabolic disorders and related conditions.

Traditional drug discovery processes can be slow and costly. HTS accelerates the process by testing thousands of compounds simultaneously, which can significantly reduce the time and resources required to find potential therapies for melanocortin receptor-related metabolic disorders (Subheading 1.1). HTS can also determine the repurposing potential of approved drugs that may have unanticipated effects on melanocortin receptors or related pathways. By bypassing the taxing process of determining drug safety, repurposing can lead to the rapid development of new treatments by leveraging existing pharmaceuticals with established safety profiles (Subheading 1.2).

1.1 High-Throughput Electrophysiological Assays for MC4-R and Kir7.1 Channels Using the SyncroPatch 384PE

The use of APC to identify MC4-R agonists and antagonists targeting the MC4-R/Kir7.1 complex has yet to be documented. Therefore, we developed an assay to evaluate the coupling of the MC4-R/Kir7.1 complex based on real-time detection of the channel response and validated this assay for HTS using a pilot screen designed to detect MC4-R modulators. We employed the SyncroPatch 384PE (Fig. 1) platform to determine the voltage-dependent properties of mutant Kir7.1M125R (Kir7.1_{M125R}) channels. Replacement of Kir7.1M125 with the conserved arginine at the homologous position within the Kir family restored sensitivity to Ba^{2+} and Cs^{+} and increased pore conductance, yielding larger currents and improving the signal-to-noise ratio [23, 26, 27] without affecting MC4-R/Kir7.1 coupling [24].

We compared the macroscopic current-voltage properties (I/V curves) of Kir7.1_{M125R} obtained in three heterologous expression systems from whole-cell recordings using the traditional manual patch clamp setup and the APC (Fig. 3). HEK293T cells were transfected with Kir7.1_{M125R} for manual patch clamp experiments and recorded 36–48 h later, as previously described [24]. T-REx-HEK-293 cells were used in the automated system, with Kir7.1_{M125R} and the MC4-R/Kir7.1_{M125R} complex. Representative SyncroPatch 384PE recordings extracted using the DataControl software (Nanion Technologies) are shown in Fig. 2b. The DataControl screen capture displays all 384 wells in the whole-cell voltage-clamp recording configuration. The central panel shows individual I/V curves for all wells acquired during the voltage step protocol. Green color-coded wells passed the quality control parameters (*see* **Note 1**); failed wells are shown in blue. The red-highlighted well corresponds to the expanded I/V curves at the bottom, and the black squares correspond to the I/V curves on the right. Kir7.1_{M125R} currents were evoked by 500 ms voltage steps from −180 mV to 0 mV (Fig. 3a, for manual patch clamp) or +40 mV at a holding potential of −60 mV (Fig. 3b, c, for APC recordings). We found that regardless of the whole-cell configuration system used, the macroscopic properties of Kir7.1_{M125R} were unchanged. Kir7.1_{M125R} channels exhibited maximum inward currents at −180 mV with a reversal potential (V_{rev}) close to −60 mV. The mean I/V relationship was not significantly different between

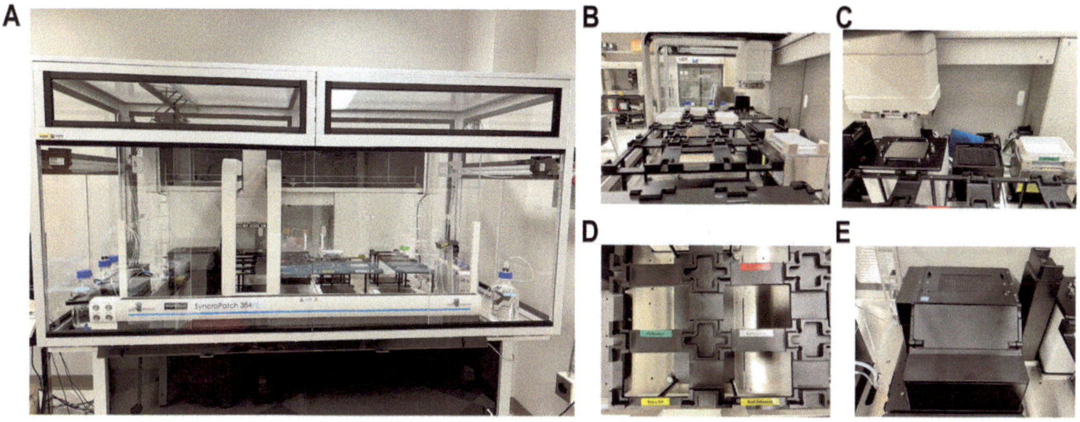

Fig. 1 The SyncroPatch384PE instrument. (**a**) General view of the instrument. As shown, the internal configuration holds different parts that integrate the Biomek robot and the Nanion. (**b–e**) Different details of the main components of the instrument. (**b**) Side and (**d**). top-down view of the Biomek deck. (**c**) Detail of the cell hotel (Teflon reservoir on the right), tip wash station (center), and Biomek pipette head (left). (**e**) Nanion module

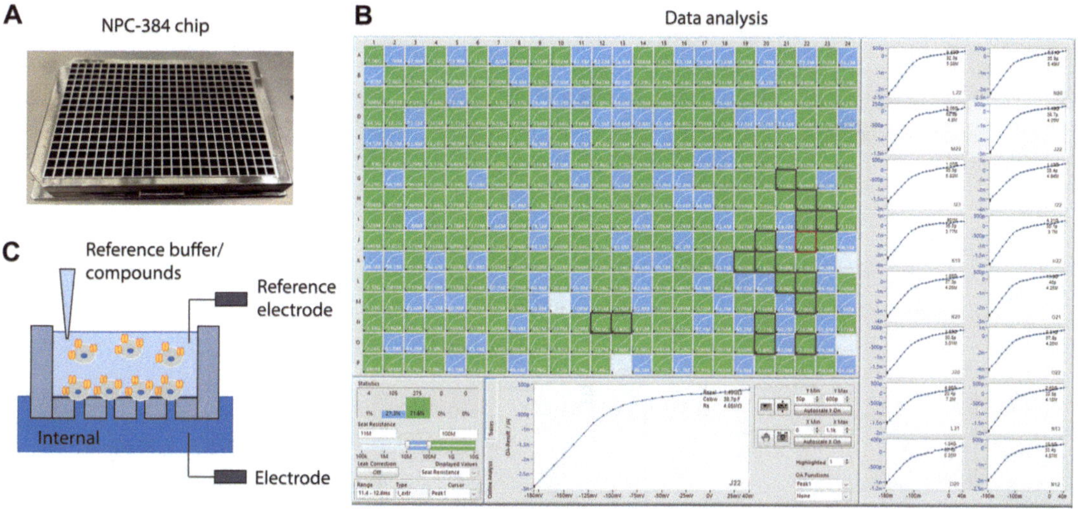

Fig. 2 High-throughput recording of 384 cells in parallel planar patches. (**a**) NPC-384 Nanion patch clamp chip. (**b**) DataControl384 main analysis screen capture. (**c**) Schematic depicting the different components of each NPC-384 chip well. This includes the well fenestrations, internal and external solution milieu, and electrode placement

the three conditions, indicating that the inward rectification that characterized these channels is the same, regardless of the method used to obtain the responses (i.e., manual vs. automated) and the presence or absence of the receptor (i.e., Kir7.1 alone or in complex with MC4-R), demonstrating that it is possible to obtain reliable and reproducible currents from Kir7.1-expressing cells regardless of the method or the condition used.

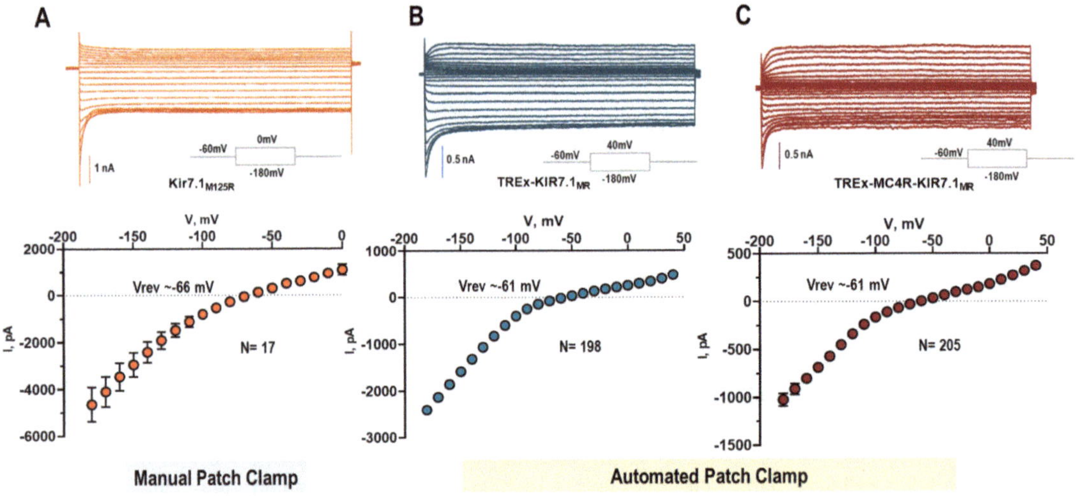

Fig. 3 Manual and APC acquired voltage-current curves yield similar results. Comparison of voltage–current relationship curves recorded from cells expressing the Kir7.1$_{M125R}$ after transient transfection of HEK-293 cells recorded using a manual patch clamp rig (**a**), from T-REx cells stably expressing the same channel induced by doxycycline and recorded by APC (**b**), or T-REx cells expressing the channel and MC4-R also acquired through APC (**c**)

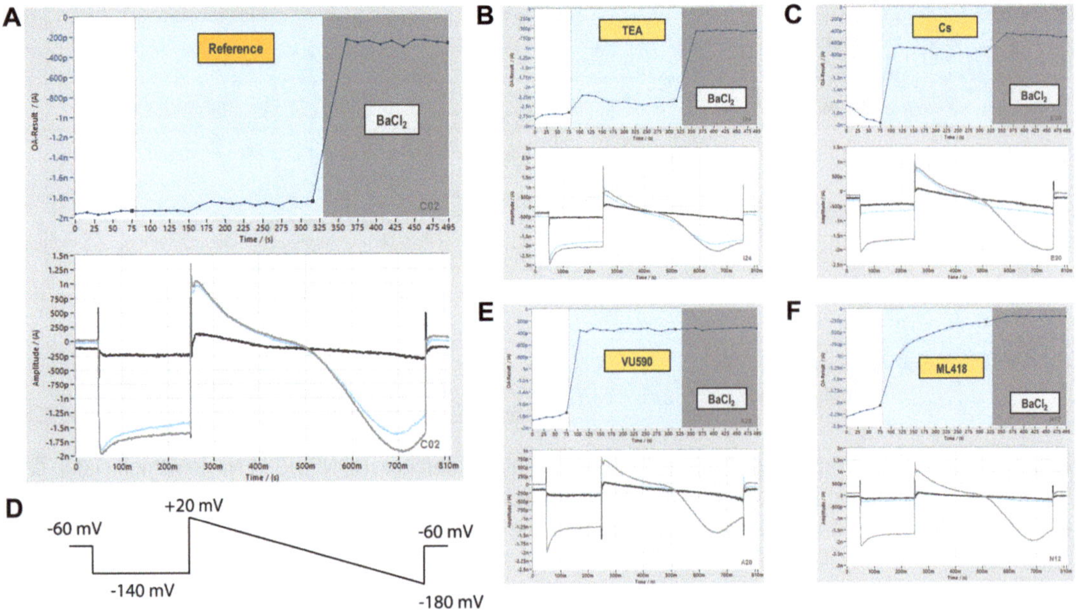

Fig. 4 Biophysical properties of Kir7.1$_{M125R}$ channels recorded in APC. Each panel shows the time course (upper panel) and maximal peak currents (lower panel) evoked after applying a two-stage voltage protocol (D) during different experimental conditions

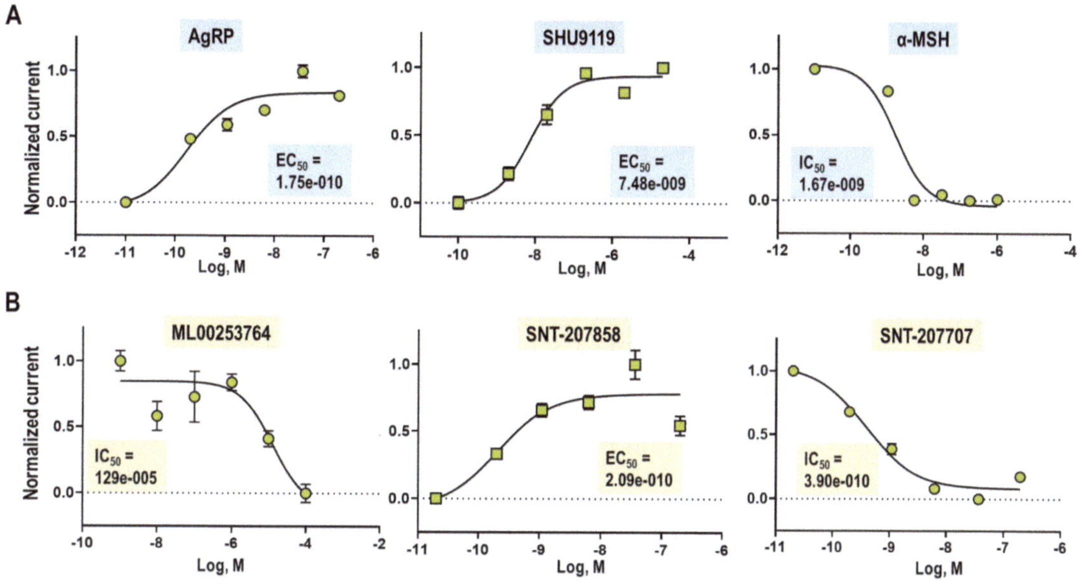

Fig. 5 Positive or negative receptor-channel coupling can be characterized by concentration response curves. (**a**) Concentration–response curves for three prototypical melanocortin ligands: Agouti-related peptide (AgRP), SHU9119 (channel activators, but G protein coupling antagonists), and α melanocyte-stimulating hormone (agonist). (**b**) Small-molecule MC4-R-specific antagonist channel coupling responses

The next step was to determine the biophysical properties of these channels against known selective and non-selective potassium channel blockers (Fig. 4). For this, we conceived a two-stage voltage protocol consisting of a hyperpolarization step to −140 mV for 200 ms, followed by a ramp (−1.2 V/s) from +20 mV to −180 mV for 500 ms (holding potential −60 mV) and repeated every 5 s (Fig. 4d). This protocol allows measuring changes in the maximum inward current response. Therefore, the peak amplitude of Kir7.1$_{M125R}$ inward tail currents was measured during the hyperpolarizing step and normalized to the full block current (i.e., in the presence of 10 mM $BaCl_2$) after applying a specific concentration of the blocker/compound at the end of each experiment. The full block current was determined in the presence of 10 mM $BaCl_2$. Figure 4a shows an example of an APC whole-cell recording from Kir7.1$_{M125R}$ channel-expressing cells during the perfusion of the reference solution. The top panel shows the time course of the two-stage voltage application protocol. This voltage protocol is applied repeatedly at each time point. The bottom panel shows a single time point recording used to evaluate the shape of the measured currents. Correspondingly, in agreement with the upper panel, the gray lines match the beginning of the recording (white area), the blue line compound or drug (blue area) addition, and the black line pairs to the full block (gray area). As demonstrated in Fig. 4a, Kir7.1$_{M125R}$ currents remained stable and showed no measurable decrease throughout the experiment, and as expected, Ba^{2+} almost completely blocked the currents.

Further, Kir7.1_{M125R} had increased sensitivity to 10 mM Cs^+, an open channel blocker, with a current decrease of ~80% (Fig. 4c). In comparison, it was found to be insensitive to the application of 10 mM TEA^+ (~14% block, Fig. 4b), a non-selective K^+ channel blocker, as previously reported [26, 27]. We also evaluated the effects of two known specific Kir7.1 pore blockers, VU590 [28] and ML418 [29]. As shown in Fig. 4e, f, the two blockers used at a concentration of 50 μM reduced Kir7.1 currents by up to 90%, in agreement with the literature.

To further validate the MC4-R/Kir7.1 coupling in the SyncroPatch platform, a set of reference melanocortin peptide ligands [13, 14, 30] (Fig. 5a) and known small-molecule MC4-R antagonist [31, 32] (Fig. 5b) were used (*see* **Note 2**). We performed concentration-response curves (CRCs) of AgRP, SHU9119, α-MSH, ML00253764, SNT-207858, and SNT-207707 on T-REx-HEK-293 cells expressing Kir7.1_{M125R} channels and human MC4-R. We found that peptide ligand-induced coupling responses of MC4-R opened (via AgRP and SHU9119) or closed (via α-MSH) the channel in a comparable fashion to previous reports [22, 24] (Fig. 5a). On the other hand, the effects of small-molecule MC4-R antagonists were variable. ML00253764 performed poorly by shutting the channel with a high IC_{50} (Fig. 5b). The SNT compounds produced opposite results on channel coupling, perhaps due to differences in bias between Gs-coupling antagonism and Kir7.1 coupling.

1.2 Application of APC in Primary Drug Screening: The High-Throughput Screening of the Drug Repurposing Library

To validate the applicability of this methodology for primary drug screening, we evaluated a set of 1,280 small molecule compounds from a drug repurposing set from the Center for Chemical Genomics (CCG) at the University of Michigan (Fig. 6). This collection is composed of FDA-approved compounds. In a preliminary trial, using a "checkerboard" distribution of AgRP at an EC_{80} concentration, we assessed the uniformity and variability across a 384-well plate. By contrasting the signal-to-noise ratio with the response variability (measured as the standard deviations across vehicle or AgRP), we calculated a Z factor value of 0.60 [33] (Fig. 6a). These parameters indicated that we had developed a high-throughput-ready assay to screen agonists and antagonists for the MC4-R/Kir7.1 complex. Compounds were screened at a final assay concentration of 20 mM (0.2% DMSO), and each plate included vehicle and positive and negative controls alongside probe compounds (Fig. 6c). High-throughput performance statistics were adequate, with mean peak current densities of ~1.5 nA and a whole-cell patch success rate of ~83% across all plates. Two hundred and forty-three compounds (19%) were excluded due to failure to pass the Qc parameters (Fig. 6b). Accordingly, a total of 1037 compounds exceeded the Qc parameters.

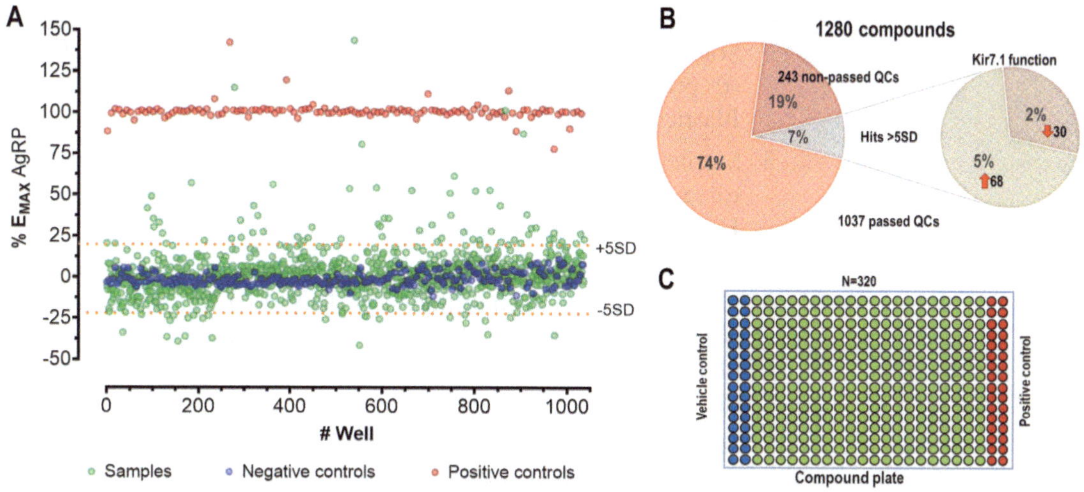

Fig. 6 APC is amenable to high-throughput screening. (**a**) Repurposing library screening campaign results. Test compounds are shown in green, positive control (AgRP) in red, and negative control (vehicle) in blue. (**b**) Proportion of hits identified and breakdown by channel activators and inhibitors. (**c**) Screening compound plate compound distribution map

Changes in peak current density were recorded during a continuous two-stage voltage protocol to identify either inhibitors or enhancers. After applying a cutoff value of >5 standard deviations (5SD), 68 compounds showed various degrees of potentiation of the MC4-R/Kir7.1-mediated effect, while 30 compounds decreased the channel function. Thus, our pilot screening showed a positive identification rate of possible lead compounds by high-throughput electrophysiological assays for MC4-R/Kir7.1 complexes using the SyncroPatch384.

In summary, the HTS of the MC4-R/Kir7.1 complex serves as an example of the broader importance of investigating GPCR-ion channel interactions. It boosts our understanding of fundamental cellular processes, contributes to developing new therapeutic strategies, and highlights the significance of high-throughput techniques in advancing biomedical research. This knowledge could lead to more targeted and effective treatments for GPCR dysfunction and ion channel dysregulation conditions.

2 Materials

2.1 Cell Culture and Harvesting

1. Cell lines: Tetracycline-inducible stable HEK293 cell lines (T-REx-HEK-293) expressing human MC4-R and Kir7.1 (M125R) or solely Kir7.1(M125R) were generated as previously described [34].
2. Fetal bovine serum (FBS): 10% dilution in growth media.
3. TrypLE™ Express Enzyme (1X).

4. Prophylactic and selection antibiotics: Antibiotic Antimycotic Solution (100X) containing 10,000 units penicillin, 10 mg streptomycin, and 25 μg amphotericin B per mL, 10 mg/mL Blasticidin S HCl, and 1 mg/mL doxycycline hyclate.
5. Sterile water and buffers: UltraPure™ DNase/RNase-Free distilled water, phosphate-buffered saline (PBS) pH 7.4 (1X), Hanks' balanced salt solution (HBSS) (1X).
6. Growth media: Dulbecco's modified Eagle's medium (DMEM) (1×) containing 10% dialyzed FBS, 4.5 g/L Glucose, L-Glutamine, 1% Antibiotic Antimycotic Solution (100X), 0.25 mg/mL Hygromycin B, and 5 μg/mL Blasticidin S HCl.
7. Induced-expression media: Dulbecco's modified Eagle's medium (DMEM) (1X) containing 10% dialyzed FBS, 4.5 g/L Glucose, L-Glutamine, 1% Antibiotic Antimycotic Solution (100X), 0.25 mg/mL Hygromycin B, 5 μg/mL Blasticidin S HCl, and 1 μg/mL Doxycycline.
8. Dissociation media: Dulbecco's modified Eagle's medium (DMEM) (1X) containing 10% dialyzed FBS, 4.5 g/L Glucose, L-Glutamine, and 1% Antibiotic Antimycotic Solution (100X).
9. Cell culture vessels: vent cap, sterile, tissue culture-treated 75 cm^2 flasks, and 150 mm cell culture-treated surface dishes.
10. Pipettes and pipette tips: 2 mL, 5 mL, 10 mL, 25 mL, and 50 mL non-pyrogenic serological pipets and 10 μL, 200 μL, 1000 μL filtered pipette tips.
11. Sterile plasticware: 15 mL and 50 mL centrifuge tubes, conical tubes, and other containers required for cell handling and storage. Sterility is crucial to prevent contamination.
12. Cell culture equipment: 37 °C 5% CO_2 humidified incubator, class II laminar flow hoods, table-top centrifuge, brightfield or phase-contrast microscope, automated cell counting equipment, and water bath.
13. Cryopreservation supplies: Cryovials, Recovery Cell Culture Freezing Medium, and controlled rate freezing equipment to freeze cell lines for long-term storage.
14. Liquid nitrogen freezer: used for storing frozen cell stocks at ultra-low temperatures in liquid nitrogen tanks.
15. Safety equipment: Personal protective equipment (PPE), including gloves, lab coats, and safety goggles, to ensure safe handling of cell cultures and minimize contamination risks.

2.2 Automated Patch Clamp (APC) Solutions

1. Extracellular buffer solution (Reference): 140 mM NaCl, 4 mM KCl, 1 mM $MgCl_2$, 2 mM $CaCl_2$, 5 mM glucose, 10 mM HEPES, pH 7.4 (NaOH), 298 mOsm.
2. Extracellular buffer solution plus cations (seal enhancer): 140 mM NaCl, 4 mM KCl, 1 mM $MgCl_2$, 6 mM $CaCl_2$, 5 mM glucose, 10 mM HEPES, pH 7.4 (NaOH).
3. Extracellular buffer solution without cations (extra fill): 140 mM NaCl, 4 mM KCl, 5 mM glucose, 10 mM HEPES, pH 7.4 (NaOH), 291 mOsm.
4. Intracellular buffer solution (KF internal): 110 mM KF, 10 mM NaCl, 10 mM KCl, 10 mM EGTA, 10 mM HEPES, pH 7.2 (KOH), 280 mOsm.
5. 20 mM $BaCl_2$ in extracellular buffer solution.
6. 1 M $BaCl_2$.
7. 1 M $CaCl_2$.
8. 1 M $MgCl_2$.

2.3 Automated Patch Clamp (APC) Components

The SyncroPatch 384PE (Nanion Technologies) is an automated patch clamp system that combines advanced hardware components, automation capabilities, and software control to enable high-throughput patch clamp recordings [35]. It allows parallel recordings from multiple cells, efficient liquid handling, and accurate data acquisition and analysis (*see* **Note 3**). Here are the main system components of the SyncroPatch 384PE (Fig. 1a):

1. Robotic Liquid Handling System (Fig. 1b–d): The SyncroPatch 384 incorporates a Biomek FX pipetting robot (Beckman Coulter) with 384 pipettors and software for control of the robot (Biomek Software) that automates the process of preparing and delivering solutions to the cells being patched. The liquid handling system accurately dispenses solutions, such as drug compounds or test substances, to the wells containing the cells in the multiwell plates.
2. Patch Clamp Headstage Array (Fig. 1e): The APC recording module is a headstage array containing 384 individual digital amplifier channels. Each channel can patch-clamp a single cell, simultaneously allowing for parallel recordings from multiple cells (Fig. 2c). The headstage array amplifies and filters the electrical signals generated by ion channels in the patched cells.
3. Multiwell Plates (Fig. 2a): The APC system utilizes a 384-well plate (NPC-384 chip) with a borosilicate glass slide containing 384 recording chambers. The APC recording module can individually access each well. We used NPC-384 4x medium resistance (four holes per well) plates for cell capture and recording (Fig. 2c).

4. Data Acquisition and Analysis (Fig. 2b): The APC system includes two specialized software application components for accurate data acquisition (PatchControl 384, Nanion Technologies) and online analysis (DataControl 384, Nanion Technologies). The data analysis software provides tools for analyzing and interpreting recorded data, such as generating current-voltage relationships and analyzing concentration–response curves.

3 Methods

The general approach for APC recording involves the following steps.

3.1 Cell Preparation for APC Recordings

Cells of interest are isolated or cultured for experimentation. These can be native cells, established cell lines, or cells derived from stem cells.

1. T-REx-HEK-293 cells are cultured in 150 mm culture dishes in growth media, kept in a 37 °C 5% CO_2 incubator, and passage twice a week when they are 80–90% confluent (*see* **Note 4**).
2. For APC recordings, cells are plated in 150 mm dishes at a density of 2,000,000–3,000,000 cells per dish in growth media, and let settle for 24 h to allow the cells to attach to the bottom of the dish.
3. The next day, aspirate off the medium, and replace it with fresh, induced-expression media containing doxycycline. Incubate for at least 24 h before the APC recordings (*see* **Note 5**).
4. On day 3, cells should be approximately 80% confluent before harvesting (*see* **Note 6**).
5. Aspirate off the medium, and wash once with 10 mL HBSS (*see* **Note** 7) for cell harvesting.
6. Add 3 mL of TrypLE™ (1X), gently swirl the dish sideways until the enzyme covers the entire surface, and incubate for 5 min in a humidified CO_2 incubator at 37 °C.
7. Complete the detachment of the cells by tapping the plates on the sides. Add 7 mL of dissociation medium to stop trypsinization. Cover the entire dish surface, and pipet the solution up and down with a 10 mL pipet to break up clumps of cells, taking care not to introduce bubbles.
8. Centrifuge the cells at 200× *g* for 5 min in a table-top centrifuge at room temperature, and carefully aspirate off the supernatant medium.

9. Resuspend the cells at a density of approximately 1,000,000 cells/mL in the Reference recording solution (*see* **Note 8**).
10. Transfer the cell suspension to the cell hotel of the SyncroPatch, where cells are continuously shaken to maintain the suspension as single cells and maintain viability.

3.2 Plate Setup and Deck Preparation

Cells and solutions are plated in the Biomek deck (Fig. 1b, d).

1. Prepare the recording solutions: Make sure the recording solution is ready. This includes buffer solutions, salts, and compounds required for the experiment (Subheading 2.2).
2. Prepare the cells: Ensure that cells are ready. The cells should be suspended in the Reference recording solution immediately after dissociation (Subheading 3.1).
3. Prepare the recording plate: Select the appropriate NPC-384 plate type that fits the experimental needs. The SyncroPatch typically uses single-hole or multi-hole 384-well plates with low, medium, or high resistance (*see* **Note 9**).
4. Load cells into the cell hotel reservoir: Carefully load the cells into the Teflon vat placed at the cell hotel station of the Biomek deck (Fig. 1c).
5. Position the NPC-384 plate on the SyncroPatch: Turn on the SyncroPatch, and ensure it is properly calibrated and ready for use (*see* **Note 10**). Carefully place the NPC-384 plate into the plate holder of the APC recording module. Ensure the plate is securely positioned and aligned with the placement guides (Fig. 1e).

3.3 Recording and Compound Application in the SyncroPatch

Once a stable giga-ohm seal is established, the automated system can rupture the cell membrane to obtain access to the cytoplasm (whole-cell configuration) and apply voltage protocols to the cell through the recording wells. In this patch configuration, it is possible to record the resulting ion channel-mediated electrical signals. These signals are amplified, digitized, and stored for further analysis. The SyncroPatch works in unison with a robotic liquid handling system (Biomek and control software application) compounds to be automatically applied to cells.

1. Power on the Nanion and the Biomek platform that integrates the Biomek robot, Biomek control, and Nanion computers (*see* **Note 11**).
2. Select the Experimental Protocol: Open the Biomek software on the Biomek computer, where the methods are programmed and stored. The Biomek software allows to define the voltage or current settings, recording parameters, and analysis options.

3. At this experiment stage, the compound solutions should be ready at the desired concentrations and equilibrated at room temperature. Compound plates are prepared using a highly precise and accurate low-volume acoustic dispensing instrument.
4. Set the compound addition parameters in the compound dialog box: Within the Biomek control software, open the compound dialog box, and edit the compound positions. Define the compound parameters, such as the concentration, duration, volume, and compound dispensing timing. The software will control the pipette system to apply the compound per the defined settings.
5. Running the experiment: Start the experiment using the Biomek control software application interface. According to the defined protocol, PatchControl will perform automated electrophysiological recordings from cells in the plate wells (*see* **Note 12**).
6. Apply compounds: Initiate the compound addition process. The Biomek robot will accurately add the compounds to all wells as defined in the compound dialog box on the Biomek software.
7. Monitor data acquisition: During the experiment, you can monitor the data acquisition process in real time using the PatchControl software. This includes observing voltage-clamp recordings and current traces (Fig. 2b).
8. Data collection during compound addition: Once the Biomek software initiates a given protocol, data will continue to be recorded during and after the compound addition to capture the compound effects on ion channel function.

3.4 Data Analysis

The recorded currents are analyzed to extract relevant information about ion channel behavior, such as current–voltage relationships, activation kinetics, or drug effects. Data analysis can be performed using the DataControl software and other specialized software to assist in quantifying and interpreting the results.

1. Current–Voltage (IV) Curve Analysis: IV curves are crucial for understanding ion channel functionality. Analyze the IV curves to extract parameters such as current amplitudes, reversal potentials, activation/inactivation kinetics, and conductance (Fig. 3).
2. Time Course Analysis: For time-dependent compound experiments (Fig. 4), analyze the effects of the compound additions on the channel peak current response latency.
3. Concentration–Response Curve Analysis: The determination of concentration–response curves is necessary to determine

compound potency and efficacy (Fig. 5). If standard non-biased full agonist compounds for the targets are known and included in the experiment, operational model-based determinations of affinity and efficacy can be determined [36] (*see* **Note 13**). Fit the data by non-linear regression to the appropriate models (e.g., Hill equation) to calculate EC_{50} or IC_{50} and Hill slope values and compound efficacy as a result of ion channel activity (*see* **Note 14**).

4. Statistical Analysis: Perform statistical tests (e.g., t-tests, ANOVA) to assess the significance of differences between control and experimental groups or between different experimental conditions (*see* **Note 15**).
5. Data Visualization: Create clear and informative plots and graphs to visualize your results. Visualization methods include voltage-clamp traces, concentration–response curves, bar charts, and scatter plots (*see* **Note 15**).
6. Data Export and Reporting: Export analyzed data in a suitable format for further analysis or inclusion in research reports, presentations, or publications (*see* **Note 15**).
7. Quality Control (Qc): Perform quality control checks to ensure the data is reliable and free from artifacts or errors. Identify and exclude any problematic data points (*see* **Note 1**).

4 Notes

1. Qc parameters: Recordings are analyzed from cells with a 500 MΩ or greater resistance seal, <10 MΩ series resistance, and stable cell capacitance throughout the recording session. Under optimized cell patching parameters, the success rate is expected to be ≥90%.
2. α-MSH, a tridecapeptide, is one of the endogenous melanocortin agonist peptides. Agouti-related peptide (AgRP) is an endogenous antagonist or biased agonist for the different receptor signaling modalities (including β-arrestin-dependent signaling and channel coupling) [13, 14, 30]. The structure of SHU9119, a cyclic lactam macrocyclic peptide, is based on the core His^6-Phe^7-Arg^8-Trp^9 pharmacophore of α-MSH. ML-00253764 is an MC4-R antagonist [31]. It prevents the loss of lean mass and enhances light-phase food consumption. SNT-207858 and SNT-207707 are selective, blood-brain barrier penetrating, potent, and orally active MC4-R antagonists that induced weight loss in mice [32].
3. Automated patch clamp technology has advanced significantly in recent years, and several options are available in addition to the Nanion SyncroPatch384PE. Some alternative APC

platforms include the following: (a) QPatch and Qube 384 from Sophion Bioscience, which offer simultaneous recordings from up to 384 cells, allow perforated and whole-cell patch configurations and single- and multi-hole patch clamping. (b) The Fluxion Biosciences IonFlux systems are offered in different models, such as the IonFlux 16, the IonFlux 48, and the IonFlux HT, each with matching features and capacities. (c) The Molecular Devices PatchXpress and IonWorks Barracuda. (d) The Multi-Channel Systems Roboocyte. (e) The CytoPatch from Cytocentrics as alternative APC systems with various high-throughput capabilities. Further, Nanion Technologies offers a high-quality portable APC system called Port-a-Patch, while the Patchliner from the same company offers voltage and current clamp modes. Nanion Technologies also offers an updated SyncroPatch variant called SyncroPatch 384i and 768PE, optimized for recording in 384- and 768-well formats. While not a traditional APC system, the Nanion Technologies Vesicle Prep Pro is used for recording from proteoliposome preparations, which can also find application in HTS ion channel research. These are examples of APC systems available on the market. The system choice may depend on specific research requirements, throughput needs, and budget constraints. Researchers typically evaluate each system's features, performance, and support to determine the best suited for their experiments.

4. Cells in culture should be split regularly to ensure the culture remains healthy by avoiding 100% confluency. To this effect, the cells were subcultured into T75 flasks for maintenance between APC experiments and transferred to 150 mm dishes before APC recordings to expand and obtain the needed cell number. If cells are split at a 1:10 dilution ratio, they will reach 80–90% confluence in 3–4 days. To successfully grow and maintain cells, it is strongly recommended that all solutions and equipment that come in contact with the cell culture be sterile. Always use proper sterile technique and work in a laminar flow hood.

5. We used doxycycline as an alternative gene expression-inducing agent for the T-REx system. Doxycycline and tetracycline share a similar mechanism of action, showing similar induction and dose–response characteristics in T-REx cells. One advantage is that doxycycline has a longer half-life than tetracycline (48 h vs. 24 h, respectively).

6. It is generally recommended that all solutions and buffers used for harvesting the cells and APC recordings be at room temperature before the start of the experiment.

7. Ensure that cell culture procedures are performed in a laminar flow hood. Depending on the number of dishes and the different experimental conditions, this step lasts at least 30 min. It is essential to avoid any risk of contamination, as even a small amount can affect the outcome of the experiment.
8. Cells must be completely dissociated to obtain stable APC recordings without clumps or cell debris. Visual control of the cell suspension under the microscope should reveal single, round cells with smooth membrane edges and few cell clusters. After pipetting up and down five times, look at the cells under a microscope. No further pipetting is needed if the cells are completely dissociated (80–90%). If cells still form clusters, gently pipette cells up and down an additional 10 times. Repeat this step until cells are isolated (80–90%). However, rough and excessive pipetting should be avoided to ensure healthy cells for recording. Nanion Technologies suggests a wide range of 33,000 to 5,000,000 cells/mL for successful recordings. The cell density should be optimized before embarking on a drug discovery campaign.
9. The selection of NPC-384 plate resistance depends on cell size to maximize the successful formation of giga-ohm seals. Typically, recording from HEK cells is successfully achieved using four-hole plates with an average resistance of ~4–5 MΩ.
10. Always follow the SyncroPatch manufacturer's specific guidelines, and adjust the procedure based on the experimental design and objectives to ensure accurate and reliable results during recordings and compound applications in electrophysiology experiments.
11. This step includes opening two software applications, the PatchControl 384 and Biomek control software, which integrate the two systems. The experimenter should launch the Biomek control application to gain access to one of many PatchControl startup helper routines. For example, a helper routine that cleans the fluid system and prepares the system at the start of the experiment is available.
12. PatchControl initiates with the cell capture and sealing procedure, where the chip is first filled with external solution (Reference) and then with internal K^+ solution. This step occurs simultaneously as the internal solution fills beneath the chip. Once the chip has been fully primed with both solutions, the cell suspension is applied to all 384 wells by the Biomek robotic head. Once the cells settle to the bottom of the chip, PatchControl applies gentle suction steps to attract one cell to each opening (default = −150 mBar). The pressure is changed back to −50 mBar at the end of the cell capture. After this, a high Ca^{2+} solution is added to facilitate the formation of the

giga-ohm seal and is removed and replaced with Reference solution once the giga-ohm seal is achieved.

13. Moreover, full agonist (if known or available) concentration-response curve dextral shifts in the presence of several antagonist concentrations can be used to determine pA_2 values, a systems-independent parameter for said antagonists [37].
14. As an initial "quick/first look at the data," analysis of compound EC_{50} (for potentiators or activators) and IC_{50} (for blockers) values can be extracted using the DataControl software. Thus, concentration–response curves (CRCs) are calculated by fitting the data to a four-parameter sigmoid equation with constrained bottom to 0 and top to 1. For example, in data extracted from time-course experiments, compound inhibition is calculated as the percentage of peak current (I) decrease from before compound application (I_reference) to the end of compound application (I_end) and both being normalized to the end of experiment full block current (I_full-block) using this following equation: Block = {1 − (I_end − I_fullblock/ I_reference − I_fullblock)}.
15. The data generated in DataControl can be extracted for more complete and detailed analysis offline using more specialized software applications. In our routine, raw data is exported to Microsoft Excel and graphed in GraphPad Prism after transformation and formatting.

Acknowledgments

We thank the Center for Chemical Genomics at the University of Michigan Life Sciences Institute for the collaboration and for providing the compounds from the Drug Repurposing Library. This work was partly funded by the Michigan Drug Discovery Initiative Grant No. U076275-22204S (to C. C. Hernandez and L. E. Gimenez), the Klatskin-Sutker Discovery Fund Award No. G026085 (to C. C. Hernandez), and NIH grant 1S10OD025203 to the Center for Chemical Genomics at the University of Michigan.

References

1. Hernandez CC, Tarfa RA, Miguel ILJ, Liu R, Mondal P, Hill C et al (2022) Development of an automated screen for Kv7.2 potassium channels and discovery of a new agonist chemotype. Bioorg Med Chem Lett 71:128841. https://doi.org/10.1016/j.bmcl.2022.128841
2. Hernandez CC, Shen Y, Hu N, Shen W, Narayanan V, Ramsey K et al (2023) GABRG2 variants associated with febrile seizures. Biomolecules 13(3):414. https://doi.org/10.3390/biom13030414
3. Buch TR, Heling D, Damm E, Gudermann T, Breit A (2009) Pertussis toxin-sensitive

signaling of melanocortin-4 receptors in hypothalamic GT1-7 cells defines agouti-related protein as a biased agonist. J Biol Chem 284(39):26411–26420. https://doi.org/10.1074/jbc.M109.039339

4. Breit A, Buch TR, Boekhoff I, Solinski HJ, Damm E, Gudermann T (2011) Alternative G protein coupling and biased agonism: new insights into melanocortin-4 receptor signalling. Mol Cell Endocrinol 331(2):232–240. https://doi.org/10.1016/j.mce.2010.07.007
5. Yang LK, Tao YX (2017) Biased signaling at neural melanocortin receptors in regulation of energy homeostasis. Biochim Biophys Acta Mol basis Dis 1863(10 Pt A):2486–2495. https://doi.org/10.1016/j.bbadis.2017.04.010
6. Lotta LA, Mokrosiński J, Mendes de Oliveira E, Li C, Sharp SJ, Luan J et al (2019) Human gain-of-function MC4R variants show signaling bias and protect against obesity. Cell 177(3):597–607.e9. https://doi.org/10.1016/j.cell.2019.03.044
7. Carrington SJ, Hernandez CC, Swale DR, Aluko OA, Denton JS, Cone RD (2018) G protein-coupled receptors differentially regulate glycosylation and activity of the inwardly rectifying potassium channel Kir7.1. J Biol Chem 293(46):17739–17753. https://doi.org/10.1074/jbc.RA118.003238
8. Farooqi IS, Keogh JM, Yeo GS, Lank EJ, Cheetham T, O'Rahilly S (2003) Clinical spectrum of obesity and mutations in the melanocortin 4 receptor gene. N Engl J Med 348(12):1085–1095. https://doi.org/10.1056/NEJMoa022050
9. Butler AA, Marks DL, Fan W, Kuhn CM, Bartolome M, Cone RD (2001) Melanocortin-4 receptor is required for acute homeostatic responses to increased dietary fat. Nat Neurosci 4(6):605–611. https://doi.org/10.1038/88423
10. Fan W, Boston BA, Kesterson RA, Hruby VJ, Cone RD (1997) Role of melanocortinergic neurons in feeding and the agouti obesity syndrome. Nature 385(6612):165–168. https://doi.org/10.1038/385165a0
11. Ericson MD, Lensing CJ, Fleming KA, Schlasner KN, Doering SR, Haskell-Luevano C (2017) Bench-top to clinical therapies: a review of melanocortin ligands from 1954 to 2016. Biochim Biophys Acta Mol basis Dis 1863(10 Pt A):2414–2435. https://doi.org/10.1016/j.bbadis.2017.03.020
12. Collet TH, Dubern B, Mokrosinski J, Connors H, Keogh JM, Mendes de Oliveira E et al (2017) Evaluation of a melanocortin-4 receptor (MC4R) agonist (Setmelanotide) in MC4R deficiency. Mol Metab 6(10):1321–1329. https://doi.org/10.1016/j.molmet.2017.06.015
13. Hruby VJ, Lu D, Sharma SD, Castrucci AL, Kesterson RA, al-Obeidi FA, et al. (1995) Cyclic lactam alpha-melanotropin analogues of Ac-Nle4-cyclo[Asp5, D-Phe7,Lys10] alpha-melanocyte-stimulating hormone-(4-10)-NH2 with bulky aromatic amino acids at position 7 show high antagonist potency and selectivity at specific melanocortin receptors. J Med Chem 38(18):3454-3461. doi: https://doi.org/10.1021/jm00018a005.
14. Sutton GM, Josephine Babin M, Gu X, Hruby VJ, Butler AA (2008) A derivative of the melanocortin receptor antagonist SHU9119 (PG932) increases food intake when administered peripherally. Peptides 29(1):104–111. https://doi.org/10.1016/j.peptides.2007.10.014
15. Greenfield JR, Miller JW, Keogh JM, Henning E, Satterwhite JH, Cameron GS et al (2009) Modulation of blood pressure by central melanocortinergic pathways. N Engl J Med 360(1):44–52. https://doi.org/10.1056/NEJMoa0803085
16. Li P, Cui BP, Zhang LL, Sun HJ, Liu TY, Zhu GQ (2013) Melanocortin 3/4 receptors in paraventricular nucleus modulate sympathetic outflow and blood pressure. Exp Physiol 98(2):435–443. https://doi.org/10.1113/expphysiol.2012.067256
17. da Silva AA, do Carmo JM, Wang Z, Hall JE (2019) Melanocortin-4 receptors and sympathetic nervous system activation in hypertension. Curr Hypertens Rep 21(6):46. https://doi.org/10.1007/s11906-019-0951-x
18. Podyma B, Sun H, Wilson EA, Carlson B, Pritikin E, Gavrilova O et al (2018) The stimulatory G protein Gsalpha is required in melanocortin 4 receptor-expressing cells for normal energy balance, thermogenesis, and glucose metabolism. J Biol Chem 293(28):10993–11005. https://doi.org/10.1074/jbc.RA118.003450
19. Li Y-Q, Shrestha Y, Pandey M, Chen M, Kablan A, Gavrilova O et al (2016) Gq/11α and Gsα mediate distinct physiological responses to central melanocortins. J Clin Invest 126(1):40–49. https://doi.org/10.1172/JCI76348
20. Sohn JW, Harris LE, Berglund ED, Liu T, Vong L, Lowell BB et al (2013) Melanocortin 4 receptors reciprocally regulate sympathetic and parasympathetic preganglionic neurons. Cell 152(3):612–619. https://doi.org/10.1016/j.cell.2012.12.022

21. Sharma S, Garfield AS, Shah B, Kleyn P, Ichetovkin I, Moeller IH et al (2019) Current mechanistic and pharmacodynamic understanding of melanocortin-4 receptor activation. Molecules 24(10). https://doi.org/10.3390/molecules24101892
22. Ghamari-Langroudi M, Digby GJ, Sebag JA, Millhauser GL, Palomino R, Matthews R et al (2015) G-protein-independent coupling of MC4R to Kir7.1 in hypothalamic neurons. Nature 520(7545):94–98. https://doi.org/10.1038/nature14051
23. Hernandez CC, Gimenez LE, Dahir NS, Peisley A, Cone RD (2023) The unique structural characteristics of the Kir 7.1 inward rectifier potassium channel: a novel player in energy homeostasis control. Am J Phys Cell Physiol 324(3):C694–c706. https://doi.org/10.1152/ajpcell.00335.2022
24. Yu J, Gimenez LE, Hernandez CC, Wu Y, Wein AH, Han GW et al (2020) Determination of the melanocortin-4 receptor structure identifies Ca(2+) as a cofactor for ligand binding. Science 368(6489):428–433. https://doi.org/10.1126/science.aaz8995
25. Anderson EJP, Ghamari-Langroudi M, Cakir I, Litt MJ, Chen V, Reggiardo RE et al (2019) Late onset obesity in mice with targeted deletion of potassium inward rectifier Kir7.1 from cells expressing the melanocortin-4 receptor. J Neuroendocrinol 31(1):e12670. https://doi.org/10.1111/jne.12670
26. Doring F, Derst C, Wischmeyer E, Karschin C, Schneggenburger R, Daut J et al (1998) The epithelial inward rectifier channel Kir7.1 displays unusual K+ permeation properties. J Neurosci 18(21):8625–8636
27. Krapivinsky G, Medina I, Eng L, Krapivinsky L, Yang Y, Clapham DE (1998) A novel inward rectifier K+ channel with unique pore properties. Neuron 20(5):995–1005. https://doi.org/10.1016/s0896-6273(00)80480-8
28. Lewis LM, Bhave G, Chauder BA, Banerjee S, Lornsen KA, Redha R et al (2009) High-throughput screening reveals a small-molecule inhibitor of the renal outer medullary potassium channel and Kir7.1. Mol Pharmacol 76(5):1094–1103. https://doi.org/10.1124/mol.109.059840
29. Swale DR, Kurata H, Kharade SV, Sheehan J, Raphemot R, Voigtritter KR et al (2016) ML418: the first selective, sub-micromolar pore blocker of Kir7.1 potassium channels. ACS Chem Neurosci 7(7):1013–1023. https://doi.org/10.1021/acschemneuro.6b00111
30. Haskell-Luevano C, Cone RD, Monck EK, Wan YP (2001) Structure activity studies of the melanocortin-4 receptor by in vitro mutagenesis: identification of agouti-related protein (AGRP), melanocortin agonist and synthetic peptide antagonist interaction determinants. Biochemistry 40(20):6164–6179. https://doi.org/10.1021/bi010025q
31. Vos TJ, Caracoti A, Che JL, Dai M, Farrer CA, Forsyth NE et al (2004) Identification of 2-[2-[2-(5-bromo-2- methoxyphenyl)-ethyl]-3-fluorophenyl]-4,5-dihydro-1H-imidazole (ML00253764), a small molecule melanocortin 4 receptor antagonist that effectively reduces tumor-induced weight loss in a mouse model. J Med Chem 47(7):1602–1604. https://doi.org/10.1021/jm034244g
32. Weyermann P, Dallmann R, Magyar J, Anklin C, Hufschmid M, Dubach-Powell J et al (2009) Orally available selective melanocortin-4 receptor antagonists stimulate food intake and reduce cancer-induced cachexia in mice. PLoS One 4(3):e4774. https://doi.org/10.1371/journal.pone.0004774
33. Zhang JH, Chung TD, Oldenburg KR (1999) A simple statistical parameter for use in evaluation and validation of high throughput screening assays. J Biomol Screen 4(2):67–73. https://doi.org/10.1177/108705719900400206
34. Raphemot R, Lonergan DF, Nguyen TT, Utley T, Lewis LM, Kadakia R et al (2011) Discovery, characterization, and structure-activity relationships of an inhibitor of inward rectifier potassium (Kir) channels with preference for Kir2.3, Kir3.x, and Kir7.1. Front Pharmacol 2:75. https://doi.org/10.3389/fphar.2011.00075
35. Obergrussberger A, Rinke-Weiß I, Goetze TA, Rapedius M, Brinkwirth N, Becker N et al (2022) The suitability of high throughput automated patch clamp for physiological applications. J Physiol 600(2):277–297. https://doi.org/10.1113/jp282107
36. Kenakin T (2011) Functional selectivity and biased receptor signaling. J Pharmacol Exp Ther 336(2):296–302. https://doi.org/10.1124/jpet.110.173948
37. Schild HO (1947) pA, a new scale for the measurement of drug antagonism. Br J Pharmacol Chemother 2(3):189–206. https://doi.org/10.1111/j.1476-5381.1947.tb00336.x

Chapter 15

Machine Learning Methods for the Analysis of the Patch-Clamp Signals

Monika Richter-Laskowska, Agata Wawrzkiewicz-Jalowiecka, Aleksander Bies, and Paulina Trybek

Abstract

Patch-clamp technique provides a unique possibility to record the ion channels' activity. This method enables tracking the changes in their functional states at controlled conditions on a real-time scale. Kinetic parameters evaluated for the patch-clamp signals form the fundamentals of electrophysiological characteristics of the channel functioning. Nevertheless, the noisy series of ionic currents flowing through the channel protein(s) seem to be bountiful of information, and the standard data processing techniques likely unravel only its part. Rapid development of artificial intelligence (AI) techniques, especially machine learning (ML), gives new prospects for whole channelology. Here we consider the question of the AI applications in the patch-clamp signal analysis. It turns out that the AI methods may not only enable for automatizing of signal analysis, but also they can be used in finding inherent patterns of channel gating and allow the researchers to uncover the details of gating machinery, which had been never considered before. In this work, we outline the currently known AI methods that turned out to be utilizable and useful in the analysis of patch-clamp signals. This chapter can be considered an introductory guide to the application of AI methods in the analysis of the time series of channel currents (together with its advantages, disadvantages, and limitations), but we also propose new possible directions in this field.

Key words Patch-clamp, Artificial intelligence, Ion channels, Machine learning, Time series analysis, Time series classification

1 Introduction

Ion channels are transmembrane proteins responsible for the fast and selective transport of ions through biological membranes. They are expressed ubiquitously in living cells, including viruses, and regulate fundamental biological processes such as cell growth, motility, excitability, and apoptosis [1]. Therefore, they are considered a class of membrane proteins widely targeted by drugs [2–

Authors Agata Wawrzkiewicz-Jalowiecka and Paulina Trybek have contributed equally to this work.

Simone Furini (ed.), *Potassium Channels: Methods and Protocols*, Methods in Molecular Biology, vol. 2796, https://doi.org/10.1007/978-1-0716-3818-7_15,

4]. Understanding the molecular mechanisms that control the functioning of ion channels is a prerequisite for progress in channel-oriented pharmacology. For this reason, experimental studies in this field continue to generate significant interest within the scientific community.

The patch-clamp method is the most powerful experimental approach that enables researchers to obtain unique real-time information about the activity of ion channels [5]. Invented by Sakmann and Neher over four decades ago, it remains the primary empirical source of knowledge concerning the conformational switching between the conducting and non-conducting states of these channels under various external conditions. Patch-clamp experiments can be conducted in different modes, allowing experimenters to record ion channel activity at various levels, such as whole-cell or individual channels. This method provides flexibility to alter the surrounding media on either side of the membrane and to "clamp" either voltage or current. One of the most popular patch-clamp modes involves establishing a mechanical and electrical connection between a small "patch" of biological membrane and the measuring equipment, typically via a micropipette. In this setup, the membrane potential is fixed at a certain level. In such cases, the recorded signals take the form of time series of currents, often with a single channel responsible for the ionic flow. These single-molecule level recordings provide valuable insights into the conformational dynamics of the investigated transport protein.

Over the years, patch-clamp electrophysiologists have established standards for data preprocessing and analysis. These standards include the development of numerous case-specific algorithms for event identification, such as distinguishing conducting and non-conducting events, correcting "missed events," and estimating single-channel kinetic parameters. These parameters encompass open state probability, dwell-time distributions, estimations of the number of conducting and non-conducting states, as well as transition rates between them [6–8]. These characteristics offer valuable insights into the functional changes of channel activity under controlled conditions, including variations in ligand concentration, membrane potential, temperature, and mechanical stimulation. It is worth noting, however, that the analysis of patch-clamp signals is a labor-intensive and human-supervised process that requires highly experienced personnel. Furthermore, all existing data processing algorithms are susceptible to subjectivity, especially when dealing with signals that contain pronounced noise components, have low amplitudes, or involve the gating of multiple channel proteins [9]. From this perspective, the advent of the universal application of artificial intelligence methods may bring new possibilities and conveniences to channel-oriented research.

The application of AI methods in the context of ion channels is still a rapidly expanding field [10–14]. To date, AI techniques have found utility in various areas such as sequence-based recognition of channel mutants, deducing structure or topology from genetic information, quantitative structure–activity relationship analysis, mechanistic studies, and drug design. However, only a few studies have explored the utilization of AI in the analysis of single-channel patch-clamp signals. One of the simplest reasons for this lies in the fact that the conventional approach for ion channel data analysis primarily centers around kinetic analysis, following established procedures.

Many biologists have yet to fully recognize the potential benefits of AI methods. Ion channel data demands different preprocessing strategies due to varying signal-to-noise ratios among different types of ion channels. For example, highly conductive BK channels and low-conductive channels like GABA receptors may not require the same noise elimination techniques. Consequently, implementing automated analysis supported by machine learning can be challenging, particularly when dealing with complex system activity that requires detailed interpretation.

On the other hand, the key advantages of the AI approach in the context of functional analysis of ion channels lie in its direct reliance on the characteristics of the signals. It does not require in-depth knowledge of channel gating mechanisms or expertise in statistical descriptions of gating kinetics. In this work, we evaluate the potential of artificial intelligence methods in the analysis of ion channel activity, primarily based on previously published results for high-conductivity BK channels [15, 16].

2 Methods

In this section, we introduce classification, pattern recognition, and idealization methods of the patch-clamp recordings.

2.1 Classification Methods

2.1.1 The Dataset

Our dataset comprises patch-clamp recordings that reflect the single-channel activity patterns of mitoBK channels from three different cell types: the human endothelial cell line (EA.hy926), primary human dermal fibroblasts cell line (HDFa), and embryonic rat hippocampal neurons. Detailed electrophysiological analyses related to the acquired traces can be found in [17–19]. To ensure a high level of activation, the solutions of $CaCl_2$ were appropriately adjusted, with 100 μM $CaCl_2$ for endothelium and hippocampus and 200 μM $CaCl_2$ for fibroblasts. All types of single-channel currents from these three cell types were recorded at six different membrane potential levels, namely, −60, −40, −20, +20, +40, and +60 mV, covering a wide range of voltage activation.

As indicated by the findings reported in [17–19], the most significant disparity in activity patterns among the three analyzed groups of mitoBK channels arises from their association with distinct β subunits: β2 in endothelium, β3 in fibroblasts, and β4 in hippocampus. This hypothesis received confirmation through Western blot analysis and immunohisto-chemical labeling [17–19]. Naturally, there may also be other factors unique to various cell lines, such as varying mixtures of BK channel splice variants in inner mitochondrial membranes across different cell types, as well as disparities in the lipid and protein composition of the membranes under investigation.

The exemplary data samples analyzed are depicted in Fig. 1. Regardless of the experimental conditions, they all exhibit a typical, highly fluctuating conductance pattern. Initially, distinguishing between recordings corresponding to the same voltage activation level but taken from different cells is challenging. Conversely, some variations between the analyzed traces can be observed in the activation curves shown in Fig. 2. The most notable difference is observed in the endothelial mitoBK channel, where the sigmoid curve exhibits a rightward shift compared to other cell types. On the contrary, a strong resemblance is noted in the dependence on p_{op} between hippocampal and fibroblast patches. However, these distinctions are solely quantitative in nature and merely reflect global disparities. They do not provide any insight into variations in channel gating and switching patterns.

To address this gap, we introduce classification algorithms grounded in machine learning (ML) techniques. These algorithms are designed to identify distinctions in mitoBK channel activation, not only on a global scale but also at a local level. In the subsequent sections, we provide a detailed description of the preprocessing stage, encompassing anomaly detection technique, as well as an evaluation of the efficiency of various ML approaches in classifying distinct traces.

2.1.2 The Outlier Detection and Preprocessing

Preprocessing the acquired data is a crucial step in every data analysis. Typically, experimental data tends to have low quality and may contain numerous artifacts that can hinder sophisticated data mining techniques. Consequently, unclean datasets can make it impossible to uncover hidden characteristics that are of interest to domain experts. They may also contain serious errors such as outliers. Moreover, some algorithms necessitate proper data normalization and dimensionality reduction beforehand. In general, data preprocessing comprises four essential steps [20]: data cleaning; data reduction; data transformation; and data integration. Each of these steps holds equal significance and can be executed independently. The choice of the preprocessing procedure is strictly data-dependent and may require the intervention of a domain expert who can effectively interpret the analyzed data.

In this subsection, our goal is to introduce popular classification techniques applied to patch-clamp data. As emphasized in the previous paragraphs, data analysis hinges on the processing stage.

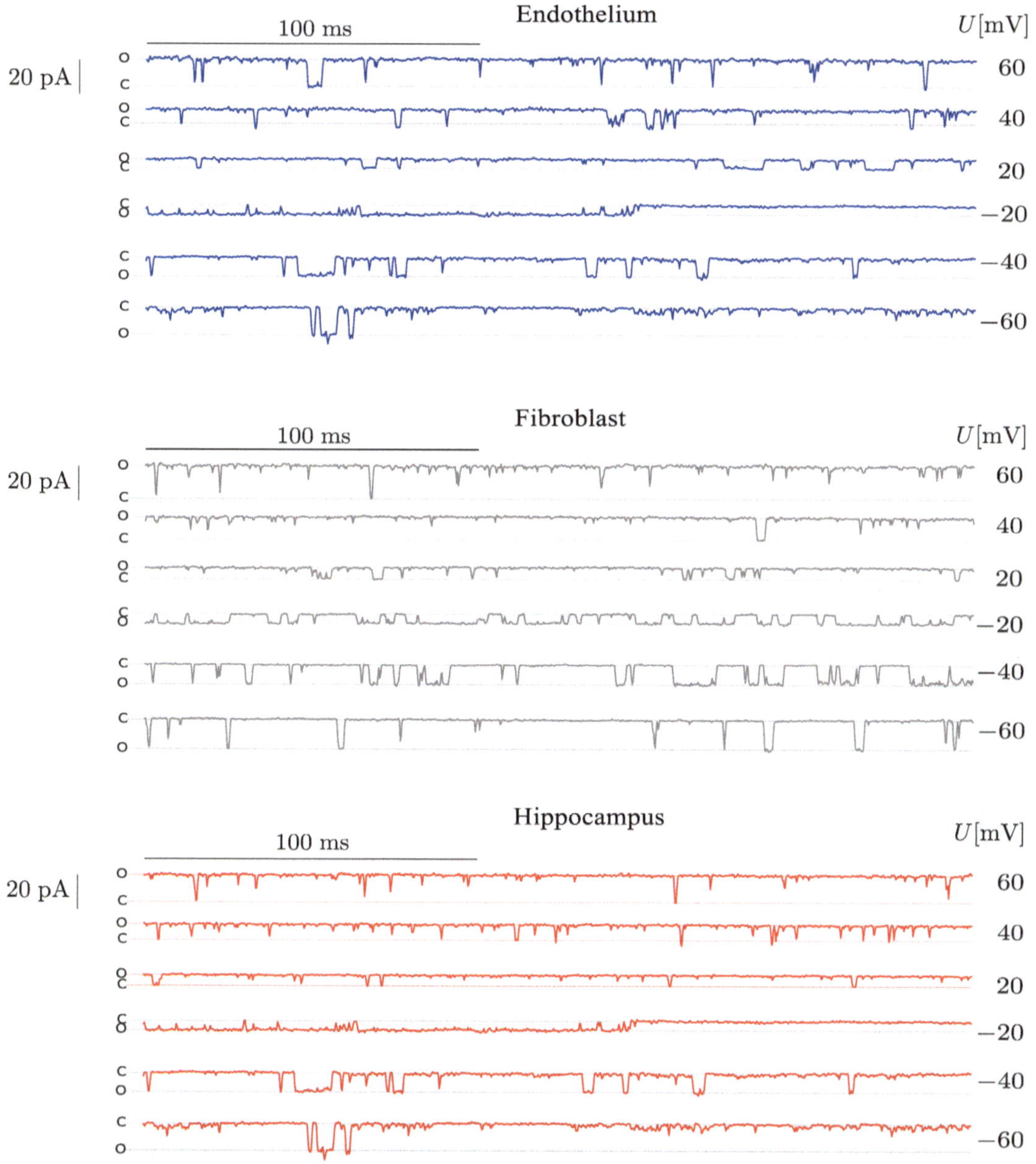

Fig. 1 A set of recordings acquired from the patch-clamp experiment performed on different cell types: endothelium, fibroblast, and hippocampus at different values of pipette potentials U expressed in mV. The value of 20 pA denotes the amplitude of the obtained ion current. The symbols C and O stand for the closed and open states of the ion channel, respectively

Consequently, before we delve into demonstrating the various data mining methods, the algorithms to prepare the dataset are discussed.

The general steps for preparing the raw data for classification are depicted in Fig. 3. The original data obtained from the experiment consists of multiple traces for each cell type. Given the lengthy nature of these recordings, we divide them into smaller sub-series,

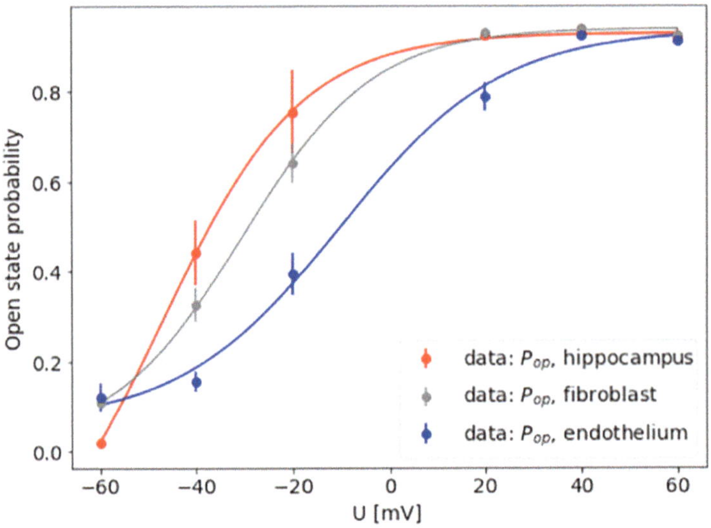

Fig. 2 The probability of channel opening pop averaged over all recordings obtained at the same experimental conditions (Um, cell type). The resulting probabilities are fitted by the sigmoidal curves

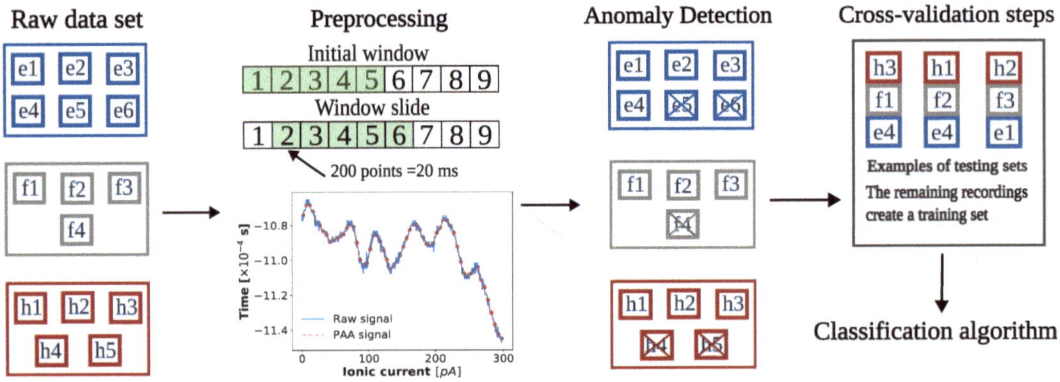

Fig. 3 Data preprocessing stage. The preprocessing of the raw data involves several general steps. Initially, the raw data input comprises multiple traces of each type, corresponding to different cellular lines. These recordings are then segmented into smaller subseries, each with a length of 1000 data points (equivalent to 0.1 s) using the sliding window method, with a stride parameter set to $r = 200$. Subsequently, the length of these subseries is further reduced from 1000 to 200 points through the application of the PAA method. In the subsequent step, any outlier recordings are removed from the dataset. The remaining traces are systematically divided into training and testing sets. This division is carried out iteratively as follows: one recording from each type is chosen and assigned to the testing set, while the remaining recordings form the training set. The classification algorithm is then evaluated for each of these cases. Finally, the ultimate result is averaged across all such defined cross-validation steps

each comprising 1000 data points (equivalent to 0.1 s), using the sliding window method with a stride of $r = 200$ points (20 ms). It is important to note that this division ensures that each sample is sufficiently long to capture the characteristic activity of the ion channel in a given cell type. However, due to the limited size of the dataset, these long traces can potentially lead to overfitting during the classification process. To mitigate this risk, we further condense the samples using the PAA (Piecewise Aggregate Algorithm) method as described in [21]. Consequently, the samples originally of length 1000 are reduced to 200 points.

In the next step, we eliminate the outliers from the dataset. Outliers are defined as observations that significantly deviate from the typical data distribution. Such deviations may arise from various sources such as erroneous data entry or faulty measuring devices. When dealing with traces obtained from patch-clamp experiments, occurrence of the anomalous recordings may be a consequence of some rarely expressed splice variants of the mitoBK channels or atypical protein–protein interactions. These events occur rather rarely and cannot be considered representative of the general characteristics of mitoBK channel activity in various tissues. Therefore, it is necessary to, prior to the data classification task, remove the outliers. One approach to achieve to remove outliers is to employ techniques such as the autoencoder neural network.

The autoencoder, as described in [22], is a neural network designed to map an input to an output. Because it does not rely on data labeling, it falls under the category of unsupervised learning methods, as outlined by [23]. During the reconstruction process, the autoencoder reduces the dimensionality of the input data while simultaneously removing noise and redundancy. A standard autoencoder comprises two main components: the encoder and the decoder, which are illustrated in Fig. 4. Generally, the encoder

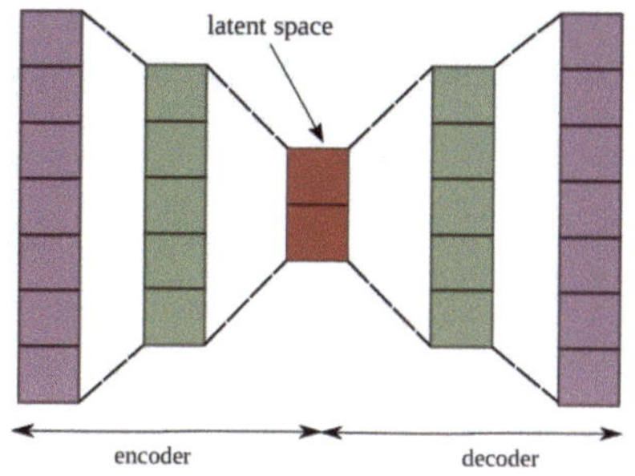

Fig. 4 Autoencoder neural network. The diagram illustrates an autoencoder neural network. The input and output layers of the neural network are represented by purple blocks, while the intermediate layer, which produces the reduced (latent) representation of the input data, is depicted in green. In the center of the diagram, two red squares symbolize neurons that generate a two-dimensional latent representation for each data point

takes the input data x and maps it to a lower-dimensional space known as the latent space L. Subsequently, the decoder's role is to reconstruct x based on the information contained within L.

Regarding the data derived from the patch-clamp experiment, our encoder is configured with a 200-dimensional input, featuring two hidden layers consisting of 100 and 50 neurons, culminating in a two-dimensional latent space. The decoder's architecture is the reverse of the encoder's. We apply the sigmoid activation function in all layers.

2.1.3 The Ion Channel Recordings' Classification

Recordings attributed to distinct cell types can be categorized using a range of contemporary data mining algorithms. In our previous work [24], we employed the K-Nearest Neighbors Classifier (KNN) method for this purpose. The key benefit of this approach lies in its simplicity, low computational demands, and ease of interpretation. Moreover, its efficacy in time series classification tasks has been demonstrated, as indicated in previous studies [25–27]. KNN algorithm can be also adapted for complex sequences data by replacing the Euclidean distance metric with the Dynamic Time Warping (DTW).

DTW has demonstrated its effectiveness in classifying non-stationary and nonlinear sequences of data, such as ECG and EEG signals. However, it is important to note that DTW may come with the drawback of being computationally intensive and resource-demanding, particularly when dealing with large datasets, as highlighted in [28]. Additionally, the choice of the appropriate K parameter in KNN and the parameters in DTW can significantly impact the quality and accuracy of classification. In situations involving high-dimensional data, KNN + DTW may not perform optimally, as the DTW metric may not adapt well to the variability between dimensions.

The working principle of this algorithm is straightforward and can be summarized in a few steps:

1. Determine the number of neighbors K and the distance metric d.
2. Calculate the distances between the new sample to be classified and its K nearest points.
3. Assign the sample to the category that is most frequently represented among these nearest neighbors as illustrated in Fig. 5.

In [24], we achieved remarkably high classification accuracy using this technique supported with the anomaly detection algorithm. Specifically, in that study, we utilized the Euclidean distance metric. Although the obtained results were satisfactory, it is possible that we could achieve even more accurate predictions by substituting the Euclidean metric with the DTW distance metric.

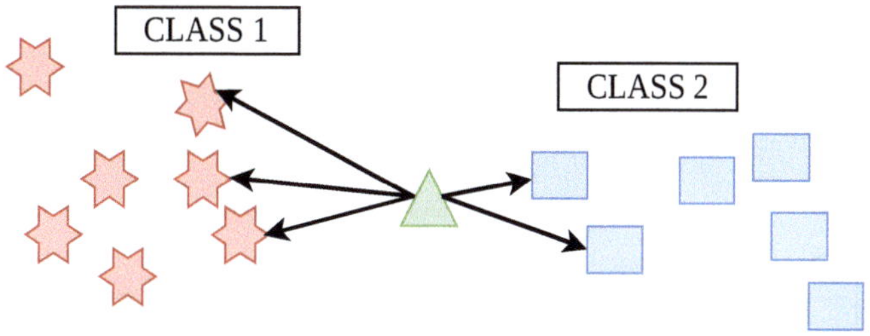

Fig. 5 Illustration of the K-Nearest Neighbors (KNN) algorithm. It involves the process of assigning a new sample, indicated by a green triangle, to one of the predefined categories. This assignment is determined by calculating the distances to the five nearest data points (referred to as neighbors), using specific distance metrics. The decision for assignment is made through majority voting. In this particular instance, the sample will be allocated to the first class

In [29], you can access a Python script containing basic machine learning classifiers designed for distinguishing between different traces. We also provide a set of mitoBK channel recordings corresponding to various cell lines, which can serve as a basis for conducting classification experiments. The brief description of the implemented code can be found in Appendix 1.

2.2 The Pattern Recognition

In addition to the classification task, another crucial data mining method in the context of patch-clamp data analysis involves the recognition of characteristic patterns present in the recordings. This type of analysis can offer valuable insights into the switching dynamics of ion channels under various experimental conditions. It allows for the identification of significant patterns and behaviors, which can provide important information for understanding the underlying mechanisms and responses of ion channels to different experimental conditions. The next subsections delve into possible application of ML algorithms in the analysis of mitoBK patch-clamp data in the context of the local modification of experimental traces by external stimuli. In the subsequent analyses, we will focus on mitoBK channel data activated by both membrane depolarization and naringenin, as described in [16].

2.2.1 The Dataset

Our dataset comprises the mitoBK channel patch-clamp recordings obtained at various values of the pipette potential U and naringenin concentration [Nar]. Some representative sequences are presented in Fig. 6. For the purpose of our analysis, we partition the data into two groups characterized by either the same value of the membrane potential U or the naringenin concentration [Nar].

1. First group: $U = 60$ mV and [Nar] $= 10$ μM, or $U = 60$ mV and [Nar] $= 0$ μM.

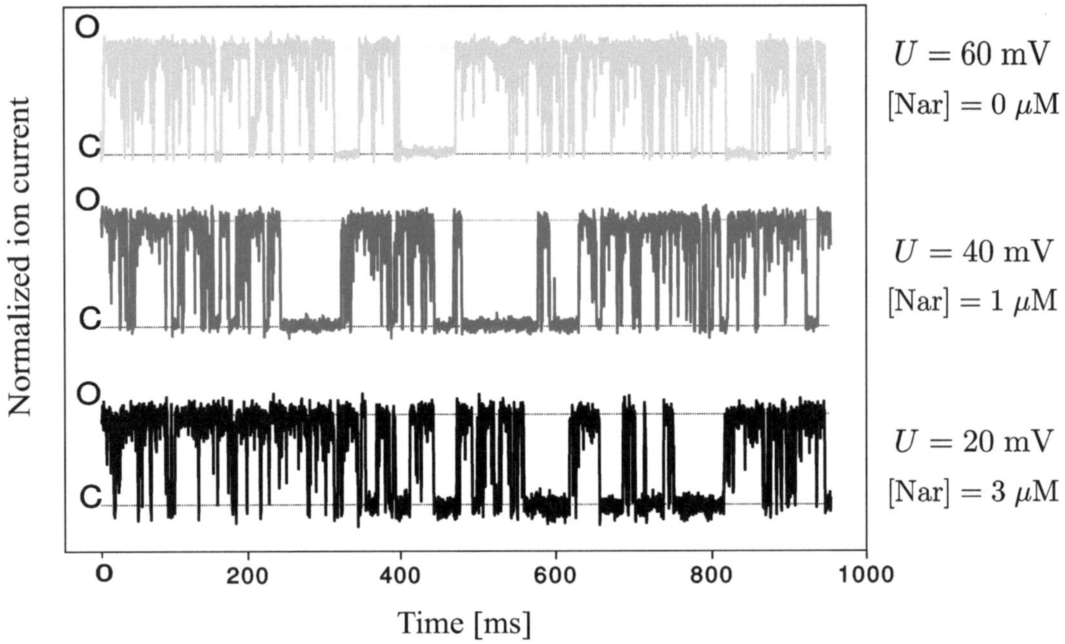

Fig. 6 Examples of normalized patch-clamp recordings acquired from patch-clamp experiments, showcasing various values of naringenin concentration [Nar] and pipette potentials U. In the visual representation, the letters "O" and "C" denote the open and closed states of the ion channels, respectively

2. Second group: $U = 60$ mV and [Nar] = 10 μM, or $U = 20$ mV and [Nar] = 0 μM.

The primary objective of this study is to investigate how different activators have an influence on the local shaping of the recorded mitoBK channel traces. In other words, our primary interest is not to ascertain the effects of external stimuli on the opening probability of the ion channel (a task achievable through straightforward statistical methods). Instead, our aim is to uncover the distinct patterns and features within the analyzed data corresponding to different stimuli. To accomplish this, we employ the novel data mining technique known as the shapelet method detailed in the following subsection.

2.2.2 The Shapelet Method

The shapelet method is a novel data mining technique employed for time series classification [30]. In contrast to the KNN classifier, the technique is devoted to time series data and places emphasis on extracting local representative features of the time series, as opposed to global characteristics. Another advantage of this method is its ability to generate easily interpretable results which facilitates a more insightful analysis of the dataset, allowing a more insightful study of the dataset. This approach has found success in the classification of biological time series and applications in bioinformatics [31–33].

The fundamental concept behind the shapelet algorithm is to identify a representative sequence (or a set of sequences) within a time series that maximizes its classifiability. In other words, the chosen subsequence, denoted as S, should best represent a given class. To determine the optimal candidate for a time series pattern, we begin by considering all subseries S' of a fixed length l (where the value of l can be chosen arbitrarily). We then calculate the Euclidean distances of S' to the time subsequences attributed to different classes according to the following equation:

$$M_{i,k} = \min_{j=1,\ldots,J}\left[\frac{1}{l}\sum_{i=1}^{l}\left(T_{i,j+l-1} - S'_{k,l}\right)^2\right] \tag{1}$$

A subsequence for which the distance to the time series representing one of the classes is notably greater than the distance to other classes is referred to as the *shapelet* denoted as S. Then the distance of S to the time subseries can be viewed as a discrimination measure between two or more classes. In essence, the *shapelet S* serves as a discriminative feature that effectively distinguishes one class from the others in the classification task.

The brute force approach to finding optimal shapelets is time-consuming and, for more challenging classification problems, demands a lot of computational power. Therefore, in practical applications, it is advisable to determine the optimal set of shapelets using the algorithm outlined in [34] which significantly accelerates the computations.

To gain a better understanding of the applied technique, let us focus on the scenario involving only two classes of subsequences for which the labels y_i are either 0 or 1. Then the predicted labels $\hat{y}_i$ predicted by the algorithm can be calculated with the logistic regression model as [34]:

$$\hat{y}_i = \sum_{k=1}^{K} M_{i,k} W_k \tag{2}$$

where W_k (weights) are free parameters learned during the training process. The coefficients $M_{i,k}$ indicate the distance of a time series T_i to a shapelet S_k calculated according to Eq. 1. The weights and, indirectly, the shapelets S incorporated in the definition of M are found by minimizing the loss function $L(y, \hat{y})$:

$$L(y, \hat{y}) = -\sum_{i=1}^{N}\left[y_i \ln(\sigma(\hat{y}_i)) + (1 - y_i)\ln(1 - \sigma(\hat{y}_i))\right] \tag{3}$$

where $\sigma(x)$ denotes the sigmoid function. It is important to emphasize that the above-mentioned form of the objective function is not ultimate and can be extended by additional regularization terms.

One of the advantages of this method is the possibility to visualize the outcomes of applying the shapelet method in two-dimensional space. To achieve it, it is essential to select two most discriminative shapelets, denoted as s_1 and s_2, and compute their distances to all samples in a dataset k. In the following step, we can create the feature vectors $d(s_1, k)$ and $d(s_2, k)$ defined as:

$$\begin{aligned} D(s_1, k) &= [D(s_1, k_1), D(s_1, k_2), \ldots, D(s_1, k_n)] \\ D(s_2, k) &= [D(s_2, k_1), D(s_2, k_2), \ldots, D(s_2, k_n)] \end{aligned} \tag{4}$$

where $D(s_j, k_i)$ stands for the distance of the subsequence k_i to the shapelet s_j. These distances serve as the basis for creating a visualization that allows for a more intuitive and insightful understanding of the data.

2.2.3 Determination of the Characteristic Subsequences

In the case of the pattern recognition analysis with the shapelet method, as opposed to the classification task described in Subheading 2.1.3, we consider, instead of the raw patch-clamp data, the dwell-time series. The dwell-time series are defined as a normalized series of subsequent durations of open (O) and closed (C) states (Fig. 7). They can be extracted from the raw experimental data using, for instance, idealization method presented in Subheading 2.3. The example of such a subsequence is presented in Fig. 8.

In order to feed the data into the ML model, we need, at first, to divide into a set of smaller subseries. Our experiments show that the optimal length of a single dwell-time subsequence is $N = 50$. Such a choice strikes a balance between the dimension of the training dataset and the number of samples (the larger the lengths of individual samples, the smaller their number). Moreover, to

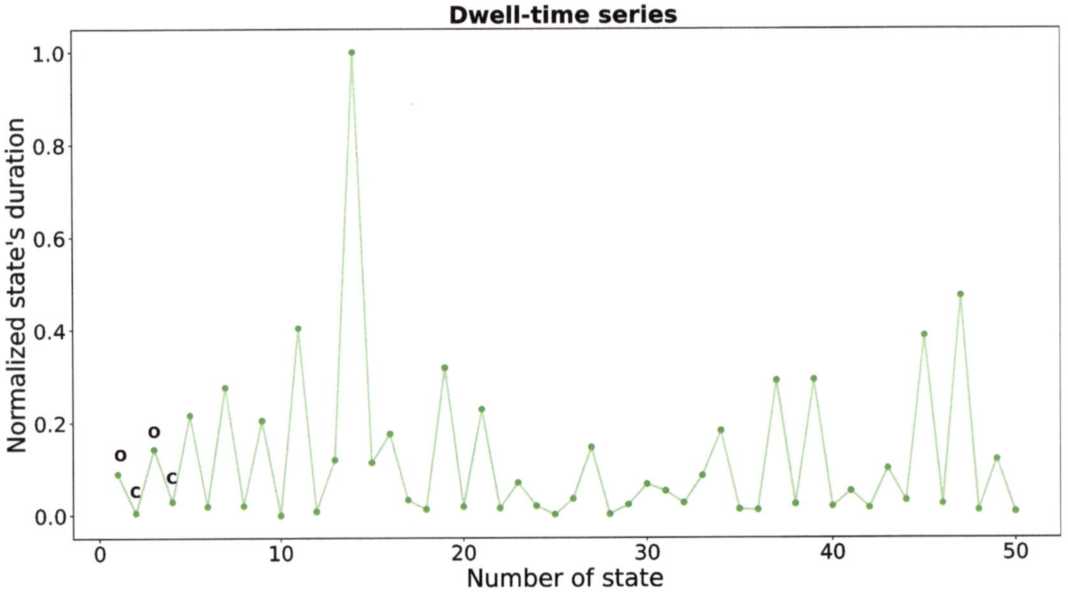

Fig. 7 An example of a dwell-time series with a length of 50, representing the normalized values of the durations of consecutive open (O) and closed (C) ion channel states

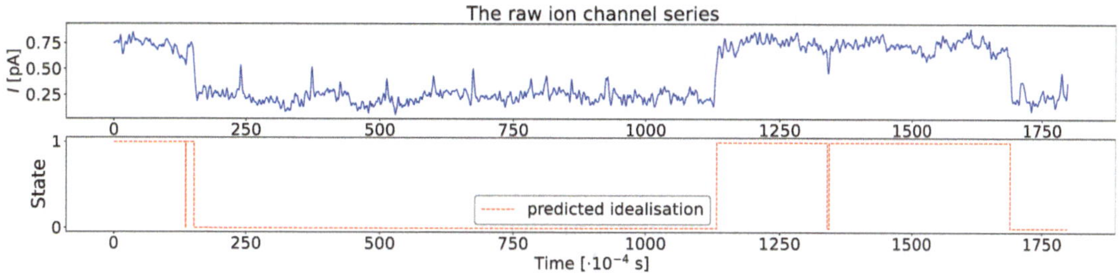

Fig. 8 Dwell-time series acquired from the Deep Channel model. The upper panel of the figure represents the fragment of the raw mitoBK channel recording. The lower panel demonstrates the predicted (by the Deep Channel model) signal's idealization

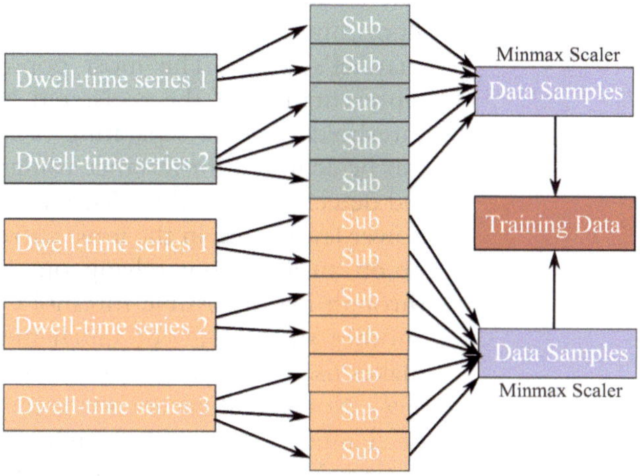

Fig. 9 The illustration of the data preprocessing stage. Raw dwell-time recordings from various classes are initially divided into smaller subseries. Subseries belonging to the same category are then globally normalized into the range [0,1] using the MinMax Scaler. The final training (and testing) dataset is formed by combining all these prepared subsequences. This process ensures that the data is properly preprocessed and ready for further analysis and modeling

ensure the independence of the data points, we apply no overlap between consecutive subseries. Such extracted subsequences are then globally (separately for each class) normalized into the range [0,1]. The general steps taken during preprocessing stage are presented in Fig. 9.

In our calculations supported by shapelet, we compared the dwell-time sequences of BK channel activity. It turned out that for most cases related to classification according to membrane potential, the shapelet technique gains a slight advantage over the KNN method, but it was more difficult to classify the data due to the concentration level of the channel's modulator. It is worth emphasizing that the shapelet method let us to select the potentially interpretative channel activation pattern. Thus, it can bring us closer to the nature of the phenomenon under study and has greater potential in the context of channel data classification. This opens up greater perspectives, not only in the process of the channel data classification but also at the level of the comprehensive analysis of ion channel gating dynamics.

In Appendix 2, you will find a concise explanation of the code that was utilized to generate the results featured in [16], as well as illustrative outcomes attainable through the shapelet analysis. This code is accessible to the public via the following link [35].

2.3 The Idealization Method

The machine learning methods can be also used to idealize the patch-clamp series [36]. In the literature, we can find a couple of algorithms aiming to distinct conducting from non-conducting ion channel's states. The most popular techniques find the binary representation of the registered traces by determining a current threshold value *Id* separating between closed and open states. It can be achieved for instance using a simple half-amplitude threshold detection algorithm [5], a method basing on the analysis of the power law scaling in the ionic current probability distribution [37] or on an algorithm relying on the evaluation of the transitions through many possible threshold currents [38]. There exists also another successful, but much more complicated method which makes use of the Minimum Description Length (MDL) Principle [39]. Not all of them are however applicable to an arbitrary number of channels, N_{ch}, in the patch. Even if they are, the user has to declare N_{ch} manually and sometimes to manually correct a baseline drift.

In order to resolve this problem, the authors of [36] proposed a hybrid recurrent CNN model (RCNN) which is able to idealize the single molecule events in ion channel records with up to five ion channels in a patch. The main advantage of the proposed algorithm is that it does not require any human supervision: it is necessary neither to declare the number of ion channels, provide information about the signal's amplitude, nor to initially preprocess the signal with regard to a baseline drift.

The architecture of the proposed deep learning model (called by the authors Deep Channel) is depicted in Fig. 10. A more detailed description of the applied architecture can be found in the caption to the abovementioned figure.

A simplified script based on the codes provided by the authors can be found in [40]. Basing on it, we have created a simplified script written in Jupyter Notebook allowing to process the experimental data with only one channel in a patch. In order to test the code, we added a couple of experimental recordings. The code and the abovementioned dataset can be found on the website [41], and it is described in more detail in Appendix 3.

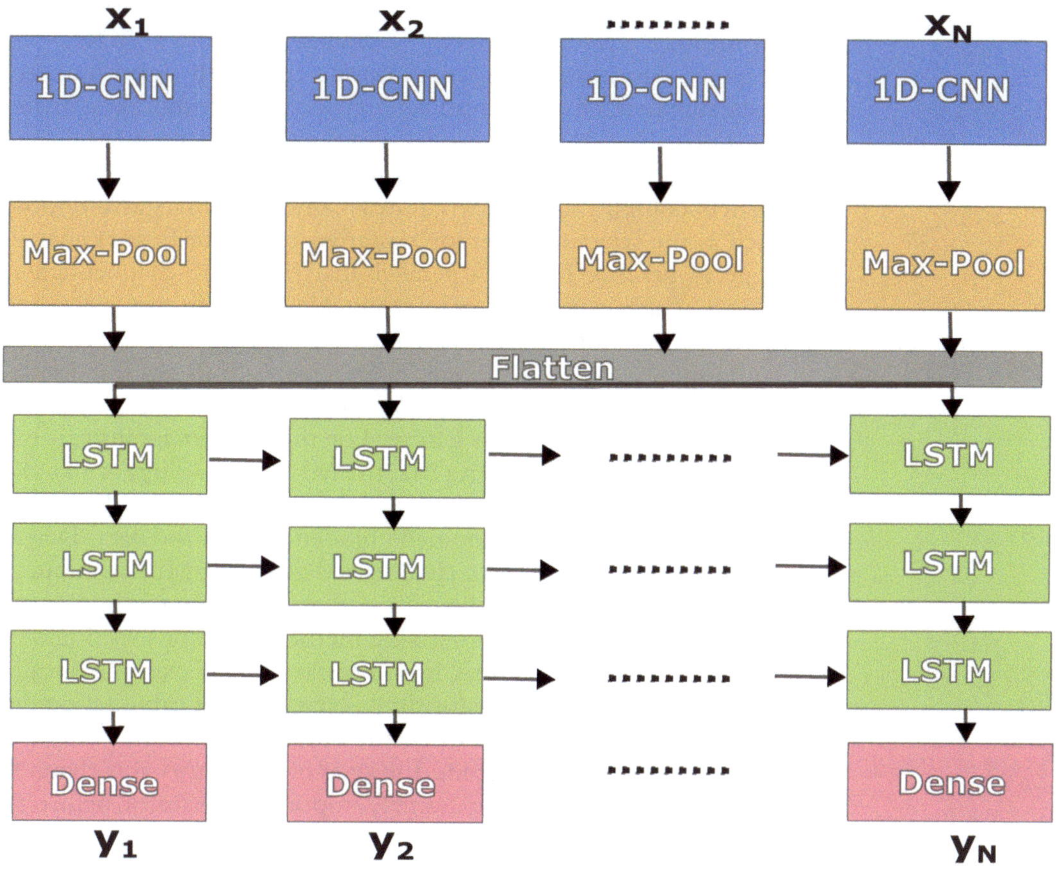

Fig. 10 The architecture of the Deep Channel model. The training data (x_1, x_2, ..., x_N) acquired from simulations of Markovian models were fed to one-dimensional Convolutional layer (1D–CNN) with 64 filters activated with Relu function. The dimension of the data was then reduced with the Max–Pooling layer (Max–Pool). Subsequently, the output of the Max–Pooling layer was flattened and fed into three LSTM layers, each composed of 256 units and activated with Relu function. In order to prevent the overfitting, each of the LSTM layers was accompanied by the Dropout layer with the probability value $p = 0.2$. The signal's features returned by the LSTM layers were in the next step forwarded to a dense layer activated with the Softmax activation function returning the probability yN of a Nth channel to be open at each time step. In post-network processing, the most likely number of channels open at a given time is calculated simply as the class with the highest probability at a given instant (Argmax)

3 Discussion

The application of artificial intelligence methods has already given unprecedented progress in many fields. The ion channel-oriented research is not an exception here. In this chapter, we characterized the applicability and usefulness of machine learning techniques in the analysis of ion channel sequences. Machine learning was primarily used for the classification of ion channel activity according to the specific condition: the different values of membrane potential or the impact of channel's modulators. In these two cases, promising results were obtained, using KNN or shapelet algorithms. It is worth noting that both algorithms have been supported by the advanced preprocessing techniques, including normalization, dimensionality reduction, or the identification of anomalous sequences by an Autoencoder Network. Data in the form of complex sequences requires many preliminary assumptions, such as the selection of a representative length for further analysis. A good practice here is a parallel analysis of the channel activity model, which may be better interpretable than sensitive experimental data. In addition, the patch clamp signals consist of stochastic components, and some of unwanted perturbations like seal instabilities, leak currents, electronic noise, etc. may affect the final representation of the channel's gating. For the experimentalists, it may be additionally discouraging that the ML methodology is a kind of black box that does not necessarily bring us closer to the correct interpretation of the phenomena hidden behind the data. Despite the fact that the analysis of channel activity using ML techniques is not an easy task, due to the choice of an adequate method, attentiveness in the use of preprocessing methodology, or selection of an appropriate activity pattern, it is necessary to mention the unquestionable potential of AI algorithms. New fields of exploitation may concern the identification of channel mutations, or the computer-aided system for recognizing the type of regulatory ion channel's units, which may be an extremely important indicator before the use of immunological tests. An important aspect also concerns automatic methods of idealization of channel's signal in order to avoid subjective assessment of experimentalists and significantly support their daily work.

4 Conclusions

The implementation of AI techniques in the analysis of ion channel activity is not a trivial task and requires a lot of engagement in both aspects—the selection of the appropriate methodology and the preparation of data. However, the benefits that can be gained in the form of establishing non-obvious relationships that are omitted in standard kinetic analysis gain an advantage and should be an incentive for scientists to support the ion channel-oriented research by the AI potentiality.

Appendices

The simple script written in the Python language containing different classification algorithms of the patch-clamp recordings can be found under the link [29]. It is divided in several sections:

Appendix 1: The Classification of Different Patch-Clamp Recordings

1. The data preprocessing stage

   The data preprocessing stage is written in the class data preprocess including the following functions:

   (a) *read_files(nr_dir, nr_pot)* reads the files from the directory corresponding to a given cell type (*nr_dir*) and the membrane potential value (*nr_pot*).

   (b) *sliding_window* divides the recording into a set of smaller subseries of length *ts_length* and stride *stride*.

   (c) *PAA_transformation* transforms the set of subsequences of length *ts_length* to a set of subseries of length *ts_length/ window_size*.

   (d) *all_files* returns the set of preprocessed subsequences from all recordings (files) corresponding to the given cell type and the membrane potential value.

2. The anomaly detection algorithm

   The anomaly detection algorithm is written in the Anomaly detection class containing two functions:

   (a) *model_autoencoder* returns two neural network models: autoencoder and encoder. The architecture of the autoencoder is structured according to the specifications outlined in [24], but it is also adaptable and can be customized as needed.

   (b) *is_anomaly* determines whether a particular recording associated with a specific cell type and membrane potential significantly deviates from other traces based on the analysis of the histogram of the autoencoder reconstruction loss as presented in Fig. 11.

3. The ML classifiers

   This section evaluates the performance of various machine learning classifiers available as part of the Scikit-learn module library [42] in the context of classifying patch-clamp data. These are:

   (a) Support Vector Machine (linear kernel) classifier
   (b) Support Vector Machine (non-linear kernel) classifier
   (c) KNN classifier with the Euclidean distance metric
   (d) Naive Bayes classifier
   (e) Quadratic Discriminant Analysis classifier
   (f) Random Forest classifier
   (g) AdaBoost classifier
   (h) Gaussian Process classifier
   (i) Linear Discriminant Analysis classifier

4. KNN + DTW

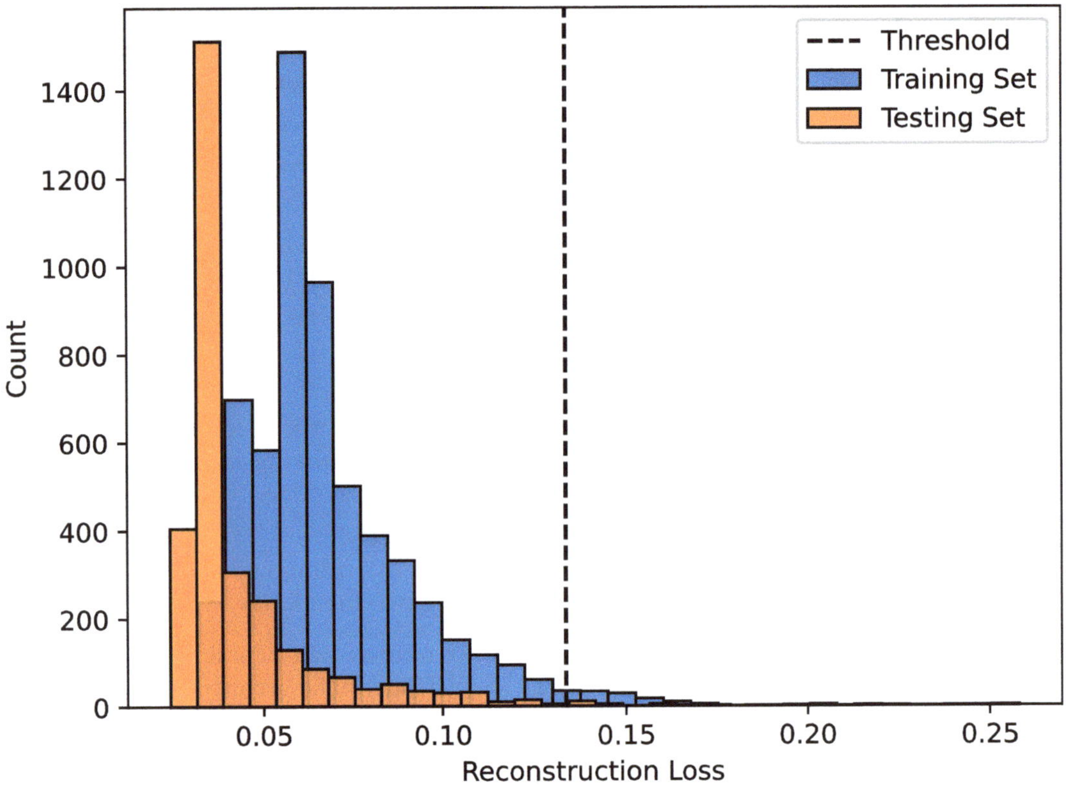

Fig. 11 Histogram of the reconstruction loss generated by the autoencoder neural network. It is calculated for both the training and testing datasets. In the training set, a threshold for identifying abnormal observations is established at the 98th percentile of the loss distribution. Any data points in the testing set that exceed this threshold are considered outliers and are treated as such

In the provided script, there is also implementation of the KNN algorithm with the DTW distance metric.

Appendix 2: The Pattern Recognition with the Shapelet Method

Under the link [35] we provide a simple code allowing to generate shapelets corresponding to different patch-clamp experimental conditions. As described in Subheading 2.2.1, we partition the dataset into two groups. We then preprocess the dwell-time data according to the scheme depicted in Fig. 9 with the following functions:

1. *divide(data, length, stride)*—divides the dwell-time series data into subseries of length length and stride stride
2. *load(dir, length, stride)*—concatenates dwell-times series from different recordings and split them into subsequences using divide function
3. *data_preprocessing(dirs, length, stride)*—employs the global MinMax Scaler to the data prepared with function load

In the next step, we feed the preprocessed data into the shapelet algorithm. We first choose which group of the data we

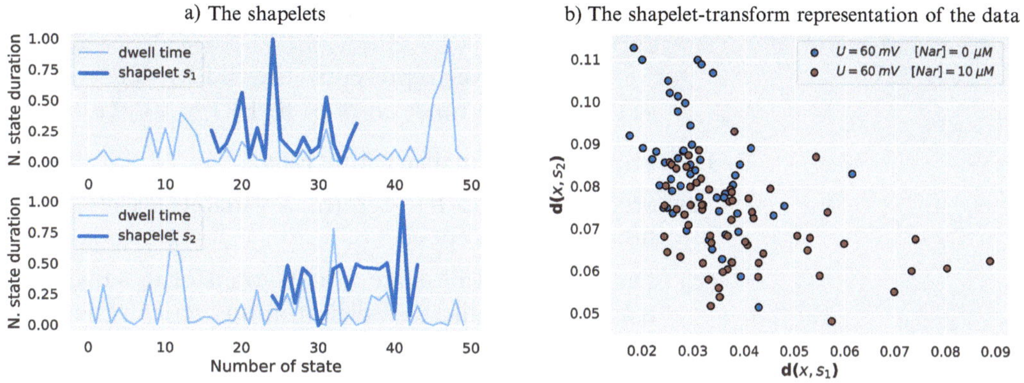

Fig. 12 Shapelet representation. The left panel (**a**) illustrates the examples of shapelets imposed on the normalized dwell-time sample in their best-matching locations. The right panel (**b**) illustrates the shapelet—transform representation of the input data. $d(x,s_1)$ and $d(x,s_2)$ denote the distances of the individual data points to the shapelets s_1 and s_2 depicted in the left panel

want to analyze: naringenina (first group) or voltage (second group). Subsequently, the data is split randomly into the training and testing datasets and fitted with the classificator *LearningShapelets*, which is part of the *tslearn* Python package [43]. The code returns the classification accuracy and a collection of discovered shapelets.

Subsequently, you can visualize the obtained shapelets in their best matching locations by overlaying some of the found shapelets on the original dwell-time subsequences where the pattern matching is the strongest. It is also possible to visualize the distances to the selected shapelets in a two-dimensional feature space as defined by Eq. 4. An example of such visualization is provided in Fig. 12.

Appendix 3: The Patch-Clamp Data Idealization: Deep Channel Algorithm

The simplified code (in comparison to a fuller version provided by the authors of [36]) allowing to idealize the patch-clamp series (with one ion channel in the patch) is publicly available and can be found under the link [41]. It is written in the Python Jupyter Notebook and divided in five sections:

1. Definition of metrics evaluating the neural network performance
   (a) Matthews correlation function (MCC)

$$\mathrm{MCC} = \frac{\mathrm{TP} \times \mathrm{TN} - \mathrm{FP} \times \mathrm{FN}}{\sqrt{(\mathrm{TP}+\mathrm{FP})(\mathrm{TP}+\mathrm{FN})(\mathrm{TN}+\mathrm{FP})(\mathrm{TN}+\mathrm{FN})}}$$

where TP = True Positives, number of samples that are labeled as 1 (True) and are classified as such; TN = True Negatives, number of samples that are labeled as 0 (False) and are classified as such; FP = False Positives, number of samples that are labeled as 0 (False) and are classified as 1 (True); FN = False Negatives, number of samples that are labeled as 1 (True) and are classified as 0 (False).

   (b) AUC–ROC metrics defined as the Area Under ROC curve.

   ROC curve represents the TPR, TPR = TP/(TP + FP), as a function of FPR, FPT = FP/(FP + TN).

2. Definition of some auxiliary functions
   (a) Step decay (epoch)—returns a variable learning rate in function of the epoch.
   (b) Dichotomous (file, state = 'o')—transforms a dwell-times read from the file to a dichotomous series. The variable state can take two values: "o" if the first state in the dwell-time series is open or "c" if it is closed.
3. Reading of the training dataset
   (a) Read of the data acquired from the Markovian model included in the file outfinaltest10.csv. The file has the following structure:

	0	1	1
1	t1	I1	s1
2	t2	I2	s2
3	t3	I3	s3
...	...	...	...

   where the first (denoted as 0), second (denoted as 1), and third (denoted as 2) columns include the timestep t, current value at a given time I, and state of the ion channel (open (1) or closed (0)), respectively.

   (b) Transformation of the training dataset to the feature range (0, 1) and its visualization.
   (c) Splitting of the data into training and testing datasets in proportion 80% to 20%.
   (d) Upsampling the dataset in order to acquire the balanced number of samples (number of samples corresponding to the closed state should be equal to the number of samples representing open state) with the SMOTE (Synthetic Minority Oversampling Technique) strategy.
   (e) Reshaping the data to match the neural network input.
4. The training process.
   (a) DeepChannel()—returns the Deep Channel neural network architecture presented in Fig. 10.
   (b) The training of the Deep Channel neural network.
5. Validation of the training process

(a) Dichotomous dwell time (dichotomous)—returns the opening probability of ion channel based on the dichotomous representation of the patch-clamp recording.

(b) plot(val set, class predict val, filename)—returns the plot of the original patch-clamp sample val_set and its corresponding dwell-time representation (class_predict_-val). The plot is saved as an image file with the given filename.

(c) Read of all mitoBK channel recordings in the dataset and calculation, for each of them, the opening probability with the trained Deep Channel neural network.

References

1. Hille B (1978) Ionic channels in excitable membranes. Current problems and biophysical approaches. Biophys J 22(2):283–294
2. Santos R, Ursu O, Gaulton A, Bento AP et al (2017) A comprehensive map of molecular drug targets. Nat Rev Drug Discov 16(1): 19–34
3. Hutchings CJ, Colussi P, Clark TG (2019) Ion channels as therapeutic antibody targets. MAbs 11:265–296
4. Zhang Y, Wang K, Yu Z (2020) Drug development in channelopathies: allosteric modulation of ligand-gated and voltage-gated ion channels. J Med Chem 63(24):15258–15278
5. Sakmann B, Neher E (2013) Single-channel recording. Springer, Boston
6. Qin F, Auerbach A, Sachs F (1996) Estimating single-channel kinetic parameters from idealized patch-clamp data containing missed events. Biophys J 70(1):264–280
7. Colquhoun D, Hawkes A, Srodzinski K (1996) Joint distributions of apparent open and shut times of single-ion channels and maximum likelihood fitting of mechanisms. Philos Trans R Soc London, Ser A 354(1718):2555–2590
8. Colquhoun D, Sigworth FJ (1983) Fitting and statistical analysis of single-channel records. In: Sakmann B, Neher E (eds) Single-channel recording. Springer, Boston
9. Sivilotti L, Colquhoun D (2016) In praise of single channel kinetics. J Gen Physiol 148(2): 79–88
10. Menke J, Maskri S, Koch O (2021) Computational ion channel research: from the application of artificial intelligence to molecular dynamics simulations. Cell Physiol Biochem 55(S3):14–45
11. Zhu Z, Deng Z, Wang Q, Wang Y et al (2022) Simulation and machine learning methods for ion-channel structure determination, mechanistic studies and drug design. Front Pharmacol 13:939555
12. Han K, Wang M, Zhang L, Wang Y et al (2019) Predicting ion channels genes and their types with machine learning techniques. Front Genet 10:399
13. Ashrafuzzaman M (2021) Artificial intelligence, machine learning and deep learning in ion channel bioinformatics. Membranes 11(9): 672
14. Singh A, Tiwari AK (2023) Machine learning-based approach for pre- diction of ion channels and their subclasses. J Cell Biochem 124(1): 72–88
15. Richter-Laskowska M, Trybek P, Bednarczyk P, Wawrzkiewicz-Jalowiecka A (2021) Application of machine-learning methods to recognize mitoBK channels from different cell types based on the experimental patch-clamp results. Int J Mol Sci 22(2):840
16. Richter-Laskowska M, Trybek P, Bednarczyk P, Wawrzkiewicz-Jałowiecka A (2022) To what extent naringenin binding and membrane depolarization shape mitoBK channel gating – a machine learning approach. PLoS Comput Biol 18(7):e1010315
17. Piwonska M, Szewczyk A, Schröder U, Bednarczyk K (2016) Effectors of large-conductance calcium-activated potassium channel modulate glutamate excitotoxicity in organotypic hippocampal slice cultures. Acta Neurobiol Exp 76(1):20–31
18. Kicinska A, Augustynek B, Kulawiak B, Jarmuszkiewicz W et al (2016) A large-conductance calcium-regulated k+ channel in

human dermal fibroblast mitochondria. Biochem J 473(23):4457–4471
19. Bednarczyk P, Koziel A, Jarmuszkiewicz W, Szewczyk A (2013) Large-conductance ca2+-activated potassium channel in mitochondria of endothelial EA.hy926 cells. Am J Physi Heart Circ Physiol 304(11):1415–1427
20. Roy S, Sharma P, Nath K, Bhattacharyya DK, Kalita JK (2019) Pre-processing: a data preparation step. In: Ranganathan S, Gribskov M, Nakai K, Schönbach C (eds) Encyclopedia of bioinformatics and computational biology. Academic Press, Oxford, pp 463–471
21. Lin J, Keogh E, Wei L, Lonardi S (2007) Experiencing sax: a novel symbolic representation of time series. Data Min Knowl Disc 15: 107–144
22. Goodfellow I, Bengio Y, Courville A (2016) Deep learning. MIT Press, Cambridge, MA
23. Kramer MA (1991) Nonlinear principal component analysis using autoassociative neural networks. AICHE J 37(2):233–243
24. Richter-Laskowska M, Trybek P, Bednarczyk P, Wawrzkiewicz-Jalowiecka A (2021) Application of machine-learning methods to recognize mitobk channels from different cell types based on the experimental patch-clamp results. Int J Mol Sci 22(2):840
25. Yang K, Shahabi C (2007) An efficient k nearest neighbor search for multivariate time series. Inf Comput 205(1):65–98
26. Chaovalitwongse WA, Fan Y-J, Sachdeo RC (2007) On the time series k-nearest neighbor classification of abnormal brain activity. IEEE Trans Syst Man Cybern Part A Syst Hum 37(6):1005–1016
27. Lee Y-H, Wei C-P, Cheng T-H, Yang C-T (2012) Nearest-neighbor-based approach to time-series classification. Decis Support Syst 53(1):207–217
28. Chatzigeorgakidis G, Skoutas D, Patroumpas K, Palpanas T, Athanasiou S, Skiadopoulos S (2023) Efficient range and kNN twin sub-sequence search in time series. IEEE Trans Knowl Data Eng 35(6):5794–5807
29. Richter-Laskowska M, Wawrzkiewicz-Jalowiecka A, Trybek P (2023) The classification of ion channel recordings. https://www.kaggle.com/code/monialisa91/ion-classification
30. Ye L, Keogh E (2009) Time series shapelets: a new primitive for data mining. In: Proceedings of the 15th ACM SIGKDD international conference on knowledge discovery and data mining, pp 947–956
31. Dai C, Pi D, Becker SI (2020) Shapelet-transformed multi-channel EEG channel selection. ACM Trans Intell Syst Technol 11(5): 1–27
32. Zorko A, Frühwirth M, Goswami N, Moser M, Levnajić Z (2020) Heart rhythm analyzed via shapelets distinguishes sleep from awake. Front Physiol 10:1554
33. Ghalwash MF, Obradovic Z (2012) Early classification of multivariate temporal observations by extraction of interpretable shapelets. BMC Bioinf 13:1–12
34. Grabocka J, Schilling N, Wistuba M, Schmidt-Thieme L (2014) Learning time-series shapelets. In: Proceedings of the 20th ACM SIGKDD international conference on knowledge discovery and data mining, pp 392–401
35. Richter-Laskowska M, Wawrzkiewicz-Jalowiecka A, Trybek P (2023) DeepChannel-samplified. https://www.kaggle.com/code/monialisa91/shapelet-method
36. Celik N, O'Brien F, Brennan S, Rainbow RD et al (2020) Deep-Channel uses deep neural networks to detect single-molecule events from patch-clamp data. Commun Biol 3(1):3
37. Mercik S, Weron K, Siwy Z (1999) Statistical analysis of ionic current fluctuations in membrane channels. Phys Rev E 60(6):7343
38. Fulinski A, Grzywna Z, Mellor I, Siwy Z, Usherwood P (1998) Non-Markovian character of ionic current fluctuations in membrane channels. Phys Rev E 58(1):919
39. Gnanasambandam R, Nielsen MS, Nicolai C, Sachs F et al (2017) Unsupervised idealization of ion channel recordings by minimum description length: application to human piezo1-channels. Front Neuroinform 11:31
40. Celik N, O'Brien F, Brennan S, Rainbow RD, Dart C, Zheng Y, Coenen F, Barrett-Jolley R (2020) Deep-Channel. https://github.com/RichardBJ/Deep-Channel
41. Richter-Laskowska M, Wawrzkiewicz-Jalowiecka A, Trybek P (2023) DeepChannel-samplified. https://www.kaggle.com/code/monialisa91/deepchannel-simplified
42. Pedregosa F, Varoquaux G, Gramfort A, Michel V et al (2011) Scikit-learn: machine learning in python. J Mach Learn Res 12: 2825–2830
43. Tavenard R, Faouzi J, Vandewiele G, Divo F et al (2020) Tslearn, a machine learning toolkit for time series data. J Mach Learn Res 21(118): 1–6

Chapter 16

Isothermal Titration Calorimetry for Fragment-Based Analysis of Ion Channel Interactions

Crystal Archer

Abstract

Ion channels are membrane proteins that may also have intracellular and extracellular domains that interact with other ligands. In many cases, these interaction sites are highly mobile and may undergo changes in the configuration upon binding with regulatory signaling molecules. Isothermal titration calorimetry (ITC) is a powerful technique to quantify protein–ligand interactions of purified samples in solution. This chapter describes a fragment-based analysis method using ITC to quantify the interactions between a domain of the voltage-gated Kv7 channel and the calcium-regulated protein calmodulin. This example can be used to quantify the interactions between specific domains of other ion channels and their regulatory signaling proteins.

Key words Isothermal titration calorimetry, Fragment-based analysis, Kv7, Calmodulin, G-protein, Peptides

1 Introduction

The activity of ion channels is regulated by several factors. In particular, their interactions with other signaling molecules such as proteins, phospholipids, and nucleotides play a key role in fine-tuning ion channel activity in determining whether the channel is fully open or closed, desensitized or inactivated, or partially open [1–6]. While CryoEM and X-ray crystallography allows for visualization of the structure of many ion channels and complexes, those fixed structures represent the most energetically favorable structure and do not show the many different configurations that ion channels can adopt [7, 8]. For example, 15 different open and closed or inhibited configurations have been indicated for Kv7 channels [9–13]. A common feature of many ion channels is high mobility or disordered regions within their intracellular domain [7, 14]. This feature may allow the channels to fine-tune their gating properties or respond to many different signaling events that underlie the

Simone Furini (ed.), *Potassium Channels: Methods and Protocols*, Methods in Molecular Biology, vol. 2796,
https://doi.org/10.1007/978-1-0716-3818-7_16,

different gating configurations. Studying proteins in their mobile and unbound state, then comparing to their bound states and the fixed structures, can provide rich insights into the mechanisms of how ion channels are regulated by other signaling molecules. Therefore, methods to study protein interactions in solution are important for a more comprehensive understanding of the dynamic mechanisms of ion channel activity.

Isothermal titration calorimetry (ITC) is a solution-based method to measure molecular interactions. In an ITC experiment, one ligand is titrated into another molecule [15, 16]. The materials of interest can be small molecules, proteins, and macromolecular complexes. The heat of the reaction is measured, and the binding curve is plotted to the molar ratio, and this is done automatically by the instrument software. This provides the binding affinity and stoichiometry data based on the thermodynamic profile, using the Gibbs free energy formulas [15, 17]. Since the detection method is the heat of the reaction, there is no need to label or conjugate the protein or ligand. It might be possible to measure ligand or protein affinity to a full channel purified using specialized techniques. Another approach is to examine binding using purified protein fragments or synthetic peptides that correspond to key interaction domains, which is the focus of this chapter. This fragment-based approach allows us to observe protein interactions with specific domains of the channel. In the event there are multiple binding sites between two or more proteins, this approach could provide opportunity to build a model for the sequence of binding during complex formation. This chapter focuses on special factors to consider when designing fragment-based ITC experiments for ion channels. The example is based on the calcium-binding protein calmodulin (CaM) interacting with a fragment of the Kv7.2 channel protein subunit.

Kv7 channels are voltage-gated channels that serve as a "brake" to prevent excessive cellular excitability [18]. They are heterotetramers comprised of protein subunits Kv7.1-5. Each subunit has six transmembrane segments (TMS) that surround a pore domain between TMS-5 and TMS-6. TMS-4 is the voltage sensor that is regulated by possible interactions with CaM. Each segment is connected by short linkers that dip into the extracellular region and intracellular cytosol [19]. Kv7 subunits have a short intracellular N terminus and a long intracellular C terminus. The proximal half of the C terminus contains two CaM binding sites called the A helix and the B helix (Fig. 1a). Due to the challenge of expressing purified Kv7 channels in the absence of co-factors, CryoEM and X-ray crystallography are performed using co-purified CaM-Kv7 complex [20–22]. These structures show that a single CaM embraces the A and B helices together (Fig. 1b, c). Kv7 channels are open in low or zero Ca^{2+} environment and become closed when Ca^{2+} levels rise in the presence of CaM [23]. This suggests that

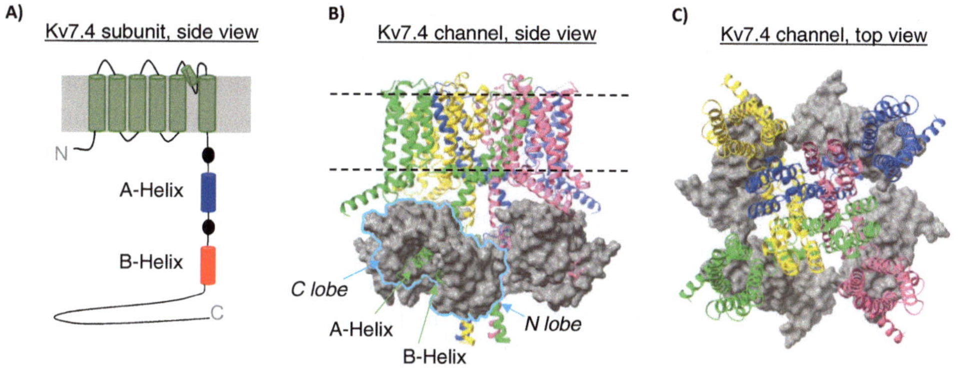

Fig. 1 Structure of Kv7.4 channels and protein subunits. (**a**) A schematic of the basic architecture of each Kv7.4 subunit. Each subunit has a short, intracellular N terminus followed by six transmembrane segments and a long intracellular C terminus. The distal end of the C terminus is responsible for joining four subunits together in a coiled-coil domain, to form the tetrameric channel. The proximal half of the C terminus makes up the "Regulatory Domain," comprised of several interaction sites for GPCR signaling cofactors. Specifically, the A helix (blue cylinder) and B helix (red cylinder) contain binding motifs for CaM. (**b**) The side view of the cryoEM structure of Kv7.4 is shown, with each subunit shown as blue, pink, green, and gold ribbons and each CaM shown as a grey surface structure outlined in light blue (PDB: 7VNQ). The dotted black lines represent the transmembrane domain. The A and B helices, indicated by yellow arrows, are folded into a hairpin structure by a single CaM, with the N lobe of CaM primarily interacting with the B helix and the C lobe of CaM interacting primarily with the A helix. (**c**) A top view from the extracellular-facing side of the of Kv7.4-CaM cryoEM structure, showing that a single CaM binds the A and B helix of the same subunit

Ca^{2+}-free CaM (apoCaM) interacts with Kv7 channels in a configuration that is much different from Ca^{2+}-bound CaM (Ca^{2+}/CaM) [14, 24, 25]. We hypothesized that the A helix interacts with CaM in both high and low [Ca^{2+}]. To test this hypothesis, we designed the synthetic peptides corresponding to the A helix of Kv7.4, and then we ordered the commercially prepared peptides at 95% purity. We purified CaM in the lab using phenyl sepharose matrix and size exclusion chromatography, following bacterial expression. CaM was dialyzed to HBS buffer (20 mM HEPES, 150 mM NaCl, pH 7.4). The peptides were dissolved in the same HBS solution as CaM. We found that the traces of contaminants from synthesizing the peptides did not interfere with the background signal and therefore did not need to dialyze the small peptides as long as they were in the same buffer as CaM. We added either $CaCl_2$ or EGTA to a final concentration of 500 μM or 1 mM, respectively, prior to the ITC experiment, in order to test CaM binding in the presence or absence of Ca^{2+}. CaM has four Ca^{2+}-binding sites, referred to as "EF-hands." This ensured that enough Ca^{2+} would be present to bind to 5 μM CaM [14]. More details of this method are described below.

A major benefit to ITC is that no special labeling is needed to monitor the interactions, eliminating the need to prepare sample in special media or the extra step for conjugation to a fluorophore

[26]. ITC has been most useful for this fragment-based approach to studying Kv7-CaM interactions when the dissociation constant (K_d) is ~5 μM or less. This is primarily due to limited solubility and protein quantities of some proteins we use. Lower-affinity reactions can be studied using ITC if higher concentrations of proteins can be attained [27]. We use other methods such as microscale thermophoresis or fluorescence polarization to determine lower-affinity interactions in low volumes and lower concentrations. The primary risk is that conjugating a fluorophore to the protein of interest might interfere with the reaction. Other methods can also be used to measure the affinities and evaluate the specific binding sites. Ultimately, comparing complementary validates the findings but also provides powerful insights to residues that are critical to the binding interaction.

Before starting ITC, we recommend reading the manuals and tutorials specific to the instrument and software you are using. Many of these manuals provide valuable instructions for properly caring for the instrument and running an experiment, and they may offer a guide to analyze the data using the accompanying software. The guides may also provide more extensive background on the equations used for each binding model. This is recommended even if in-person instruction is provided.

The next recommendation is to practice on the ITC instrument, technique, and analysis software by practicing with two simple reagents: $CaCl_2$ and EGTA. Observe the changes in the titration peaks and shifts in the binding isotherms by testing different concentrations, temperatures, and reference powers. Use this practice to learn how to subtract reference background data and analyze the data. Your software may also come with sample data and a tutorial to practice analysis methods.

Once you are confident with ITC, you may find many uses for it afterward. For example, you may want to test how a mutation affects the binding of a compound to your protein, compared to a wild-type protein. With proper planning and sample handling, ITC can be a valuable tool for your research.

1.1 Principles of Isothermal Titration Calorimetry Technique

There are two compartments in the instrument that hold the samples: the sample cell and the reference cell. The instrument measures the amount of power in μcal/sec that is needed to maintain a constant temperature that is close to zero difference between the sample cell and the reference cell [15, 17, 26, 28]. One protein or ligand (protein B) is loaded into the titration syringe (Fig. 2a), and the other protein or molecular target is placed in the sample cell (protein A) (Fig. 2b, d). When the syringe injects protein B into the cell, depending on the affinity, complexes between protein A and protein B are formed, illustrated by the blue and gold complexes in Fig. 2b. This is shown as a raw data trace of the heat change over time, which appears as peaks that come back to baseline at the end

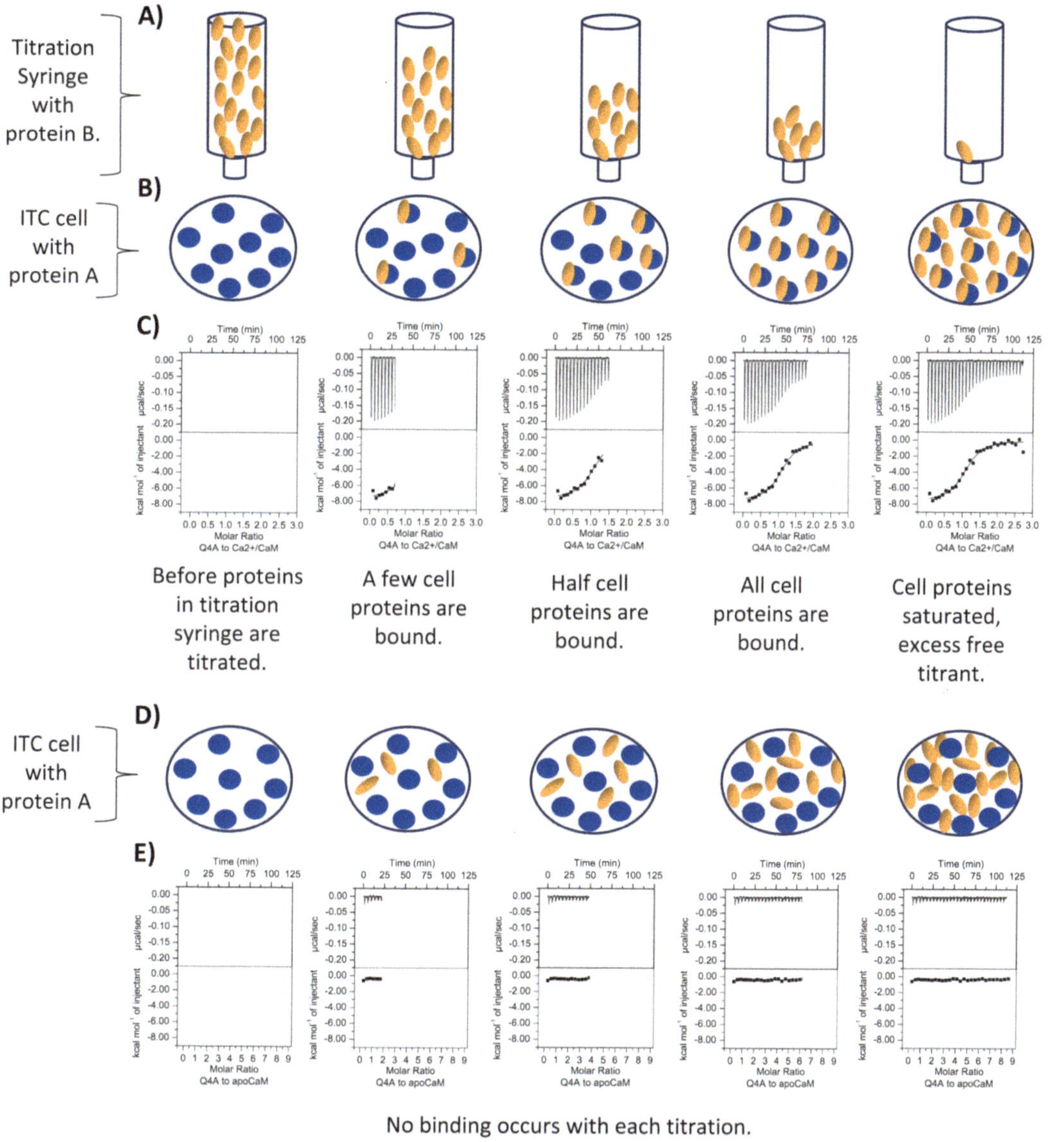

Fig. 2 Principles of ITC. (**a**) A schematic of the titration syringe holding a sample colored by tan ovals. As the experiment progresses from left to right, the sample is pushed out of the titration syringe into the cell. (**b**) A schematic of the sample cell holding a sample (blue circles) that is 10× lower concentration than the sample in the syringe. As the experiment progresses from left to right, more sample from the syringe is injected (from **a**) into the cell. Each injected molecule binds with a molecule inside the cell until all molecules in the cell are bound (shown in the fourth panel). The last panel illustrates saturation of binding, where all blue molecules in the cell are bound and the tan oval molecules from the syringe are now in excess. (**c**) An example of an experiment starting with 5 μM CaM in the cell and 70 μM Kv7.4-A helix peptide in the titration syringe, in the presence of Ca^{2+}. The raw heat measurements (μcal/s vs time) are shown as downward peaks in the upper box, indicating that this is an exothermic reaction. A small spacing at the baseline in between each peak indicates that the reaction for that injection has reached equilibrium. The lower box shows a binding curve (isotherm) plotted in kcal/mol of injected amount vs molar ratio injected. As the experiment progresses from left to right, the heat of the reaction begins to get smaller, reflected by the smaller peaks in the top box and the slope of the isotherm steepening toward the zero baseline in the lower box. In the last panel, the peaks are smaller and closer to zero, and the isotherm has reached a plateau, consistent with binding saturation. (**d**) A

of the titration. The larger higher heat change results in higher peaks. If there is very weak affinity or no binding, then no complex will form in the cell which will appear as very small peaks as an effect of the heat of protein diluted into the cell (Fig. 2d, e).

The formation of a complex results in a heat absorbed (endothermic reaction) or heat released (exothermic reaction) in the sample cell that is different from the temperature in the reference cell [29]. The instrument then adjusts the power to match the temperature between the two cells. After each injection of sample, the system reaches equilibrium, and temperature difference between the cells returns to zero. This is seen when each peak (upper panels of Fig. 2c) returns to the baseline and stabilizes at zero. The software automatically integrates the area under the peak using the baseline as the reference and reports that value as the amount of heat associated with the reaction from the injected protein. As the sample continues to be injected, more protein complex will form, and less free protein A in the cell will be available for protein B to bind, until all of protein A in the cell becomes saturated. This results in less heat signal per injection over time, until only the background heat is observed.

The binding enthalpy (ΔH) is a result of molecular interactions such as van der Waals, hydrogen bonds, and electrostatic interactions, which is related to the size of the raw peaks. The entropy (ΔS) is a result from the solvent rearrangement around the complex as it forms and the conformational changes of the proteins as they form the complex. An endothermic reaction has positive peaks relative to the baseline, and an exothermic reaction displays negative peaks. The reaction is characterized by the Gibbs free energy equations (Eqs. 1 and 2), where a spontaneous reaction corresponds to a negative ΔG [15, 26, 28, 29].

$$\Delta G = RT \ln K_{\mathrm{d}} \qquad (1)$$

where $R = 1.98$ cal mol^{-1} K^{-1}; $T = 273.2$ K, and $RT = 0.62$ kcal/mol at 37 °C

$$\Delta G = \Delta H - T\,\Delta S \text{ (and therefore)}, \text{-}RT\ln K_{\mathrm{a}} = \Delta H - T\,\Delta S \qquad (2)$$

The ΔH value is automatically determined based on the amount of power the instrument needed to put into the system to maintain the constant temperature (by increasing or decreasing the temperature). The area under the curve of the injected peaks is

Fig. 2 (continued) schematic of the sample cell holding a sample (blue circles) illustrating that no binding occurred as more sample (brown ovals) was injected into the cell sample (blue circles). (**e**) 5 μM apoCaM is loaded in the cell and 200 μM Kv7.4-A helix peptide in the titration syringe, with 1 mM EGTA added to the buffer. The data shows that no significant change in heat occurred over the course of the experiment, indicated by each peak staying near the baseline and no change in the data to model a binding curve. Both experiments were performed at 37 °C

integrated and then plotted as molar ratio vs ΔH to result in a binding curve. When binding occurs, a *sigmoidal* curve is generated because more protein A (target) is in the cell at the start of the experiment, which become saturated with protein B at the end of the experiment. N is the binding ratio, which can equal the stoichiometry if the concentrations are accurate and all proteins are active. N can be visualized as the inflection point of the ITC sigmoidal curve, when plotted to molar ratio [30]. The association constant (K_a) is based on the starting concentrations of protein A and protein B [15, 26, 28].

Molecule or protein B is gradually added to protein A, and the amount of free A and B changes as the complex [AB] increases. Therefore, the binding constant is determined based on the fractional saturation and total heat content fitted by a quadratic equation to obtain the association constant K_a. The software offers different fitting equations to determine the affinity and stoichiometry based on whether the system is a 1:1, sequential binding, or multiple binding partners. These equations used for fitting ITC data are described in the manual that comes with the software. With the estimated K_a, the measured ΔH, and known T and R, are plugged into Eq. (2) to obtain the K_a (association constant). To report the dissociation constant (K_d), use Eq. (3) [26]:

$$K_d = 1/K_a. \tag{3}$$

For a simple 1:1 reaction, the typical result is $N \sim 1$. In our studies, a single CaM protein binds a single peptide, so N will be close to 1. If N is expected to be 1 but the result is greater or less than 1, then this usually indicates that a protein in the syringe or cell is not correctly determined or that there is less active protein than calculated. There are now many resources to dive more deeply into the analysis of ITC [31]. This chapter shows a general method for preparing protein sample concentrations and running an ITC experiment. The goal of this text is to provide a general guide to test any ion channel fragment-ligand system in ITC, using the interaction between CaM and the A helix of Kv7.4 channels as an example.

2 Materials

1. ITC instrument (e.g., Microcal VP-ITC).
2. HBS buffer: 20 mM HEPES, 150 mM NaCl, pH 7.4.
3. Dialysis tubing (*see* **Note 4.1.1**).
4. Protein concentrators. We use centrifuge concentrators such as Vivaspin, with MWCO that is less than half the molecular weight of the protein being concentrated.

5. De-gassing apparatus (*see* **Note 4.1.2**).
6. Microcentrifuge tubes (2.0 mL and 0.6 mL) and pipette tips.
7. Calibrated pipettes.
8. Weighing station.
9. pH meter.
10. pH papers.
11. 6 × 50 mm borosilicate glass culture tubes.
12. 5 mL plastic test tube with cap.
13. Hamilton syringe, 2.5ML, 18GA, 8.5IN, PT3.

3 Methods

The ITC method to test the interaction between CaM and the Kv7.4-A helix is described below and summarized in the flow chart in Fig. 3. Although the actual ITC experiment only takes ~2 h, a considerable amount of time and effort is spent on planning the experiment and preparing the samples (*see* **Notes 4.1** and **4.2**).

1. Design and order the peptide fragments. The Kv7.4-A helix peptide (Q4A) with sequence EKRRMPAANLIQAAWR LYSTDMSRAYL was based on previously published structures and other experimental data [14, 32, 33]. 10 mg peptide was ordered at 95% from an external company.
2. Express and purify calmodulin. The CaM plasmid pET-GQ.H-CaM was a gift from William N. Zagotta, PhD. This protein is not tagged, and we purify using high-performance phenyl sepharose following a method adapted from an earlier published method [34].
3. Buffer exchange CaM and concentrate. Prepare 3 L HBS buffer and dialyze CaM to that buffer by placing in a dialysis tubing less than 10 MWCO and soaking 4 h in 1 L HBS and then

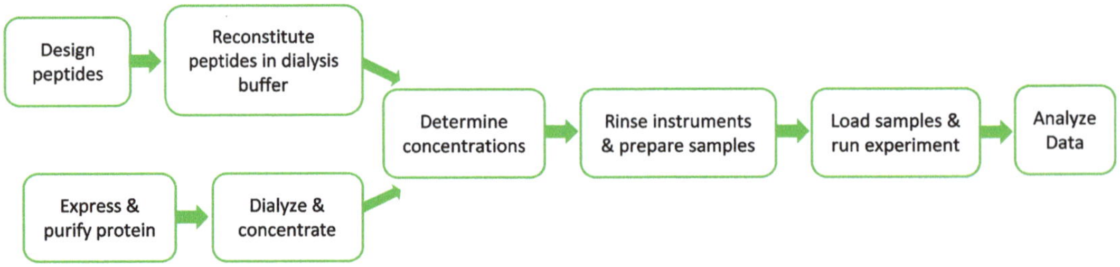

Fig. 3 A typical workflow for collecting ITC data. Most of the time is spent preparing the protein and peptide stocks, which need to be pure and have accurate concentrations. Once the samples are loaded, the experiment only takes a couple of hours to run. Data analysis may indicate that the experimental parameters may need to be adjusted or the process repeated to further optimize the experiment

transferring to second liter of HBS, and soak overnight at 4 °C. Concentrate the protein using a centrifuge concentrator or vacuum concentrator device. Sterile-filter and de-gas the remaining 1 L of that buffer through a 0.2 μM filter. This sterile HBS buffer will be used for the experiment and reconstituting the peptides.

4. Reconstitute the peptides in the HBS. Weigh a small amount of lyophilized peptide (~1–2 mg) into a low-binding microcentrifuge tube. Add ~250 μL HBS and examine whether the peptide is soluble. The easiest way to test this is to measure the protein on a NanoDrop, and centrifuge and measure the supernatant on the nanodrop. The Q4A peptides are soluble in the HBS up to 1.0 mM.
5. Determine accurate concentrations. Since the Q4A peptide contains a tryptophan, we determine its concentration at 280 nm. CaM and other peptides we work with do not have a tryptophan. This creates more possibility for errors when using absorbance at 280 nm to determine the concentration. It helps to have the concentrated stock of CaM and peptides analyzed by amino acid analysis (AAA) to obtain the most accurate concentration. From that AAA-analyzed concentration, it is helpful to prepare a standard curve of those AAA-quantified stocks (0.05, 0.1, 0.25, . . .2.0 mg/mL) to have accurate and equivalent protein concentration for testing the concentration using densitometry, BCA assay, or even an absorbance method. Quantifying proteins to their equivalent protein standard curve results in more accurate concentrations in the lab. When calculation is taken from the absorbance, be sure to use the calculated extinction coefficient of the protein.
6. Clean and rinse the ITC instrument cell and titration syringe. Prior to running the experiment, make sure the titration syringe and ITC cell are clean of any residual contaminants by rinsing both the cell and titration syringe with autoclaved, deionized water. Make sure the reference cell has clean deionized, autoclaved, de-gassed water loaded according to the manufacturer's recommendation. One way to check that the system is clean and working properly is to titrate water-to-water or buffer-to-buffer before running your experiment, and check that there are no baseline shifts and/or peaks or bubbles.

   For regular cleaning of the cell and titration syringe, prepare 5% Contrad detergent, and load that into the cell with the Hamilton syringe. Then rinse both the cell and syringe with autoclaved, deionized water, and follow that by flowing 200 mL clean water through the cell using the vacuum pump, following the manufacturer's protocol. If a sticky protein was used, soak the cell and syringe in 20% Contrad for a few minutes, and then rinse thoroughly with water. Frequent use of detergent to clean the titration syringe seems to weaken the white plunger tip, so do check for leaks in the syringe regularly in case the tip needs to be replaced. For regular cleaning of the syringe, do a final rinse with methanol, and then dry the syringe by attaching the tubing from the vacuum filter to pull air through for 10 min. Cleaning (and drying of the syringe) should be done at the end of each day and at the end of each experiment if working with proteins that tend to stick to surfaces. Rinsing with water should be done at the beginning of each experiment.

7. Prepare samples. Prepare 475 μL Q4A peptide at 75 μM and 2 mL CaM at 5 μM, in de-gassed HBS in 1.5 mL and 2.0 mL microcentrifuge tubes. Add $CaCl_2$ to each protein to a final concentration of 0.5 mM. If any other additives such as DTT are present in one sample, then also add it to the other sample in equal concentrations. If you are testing the binding with apoCaM, then add EGTA to a final concentration of 1.0 mM instead of $CaCl_2$ to both samples, or if the buffer has no contaminant levels of Ca^{2+}, then there may be no need to add EGTA. Check the pH of ~0.1 μL of each prepared sample on pH paper. Adjust the pH, if necessary. Centrifuge the samples for 1 min at 8000×*g* to remove aggregates, dust, or other ntaminants. Transfer the samples to the glass culture tubes and Thermovac tubes, and place samples in the Thermovac or other sample de-gasser. De-gas the samples and buffer for approximately 10 min. Rinse the cell and titration syringe with buffer.

8. Load samples into the titration syringe and ITC cell. Examine the titration syringe to be sure there is no leakage—if so, then replace the plug before starting the experiment. Remove the buffer from the cell and the titration syringe. For the Ca^{2+}/CaM binding experiment, load 70 μM Q4A peptide into the syringe and 5 μM CaM into the cell in the Ca^{2+}-HBS buffer. For the apoCaM binding experiment, load 200 μM Q4A peptide into the syringe and 5 μM CaM into the cell in the EGTA-HBS buffer. We increased the Q4A concentration in EGTA because we expected a weaker affinity, if there was binding between Q4A and CaM. In subsequent experiments we tested other concentrations and parameters in EGTA to confirm there is no binding. Make sure there is only a small gap of space between the tip of the titration syringe and the bottom of the glass tube, to minimize the possibility of air being pulled into the syringe. Connect the plastic syringe tubing slowly and carefully to the fill port and fill titration syringe with sample, while making sure that the tip of the syringe paddle is still immersed in a small amount of sample in the culture tube. This will minimize the chance of air and bubbles introduced

to your sample. Close the fill port, remove the tubing, and then press "purge and refill." Make sure there still is liquid at the bottom of the culture tube when the syringe is fully loaded.

While the titration syringe is purging and refilling, load the de-gassed CaM into the cell without introducing any bubbles. To do this, slowly draw up the 2 mL sample into the cleaned Hamilton syringe. Move or tap the syringe to loosen air bubbles, then turn the syringe upside down, and pull back until a small amount of air enters the syringe. Slowly move the air out of the syringe until liquid begins to be pushed out the tip, and then slowly add the sample into the cell while wiggling the syringe tip to dislodge any bubbles that may form in the cell. Once most of the liquid has been added and begins to spill over the top of the cell opening, pull the syringe needle to the top of the opening, and rest on the ledge, and then remove the excess volume from the cell. Place the remaining sample back into the vial, and store the remaining sample for future use, if needed.

9. Run the experiment. Make sure the Jacket temperature is set to the running temperature. The experimental parameters used to obtain the binding data for CaM-Q4A in Fig. 2 are as follows: 25 injections; 37 °C; reference power 15; initial delay of 60 s; and stirring speed of 307. For our injection parameters, the first injection volume is 2.0 μL, and subsequent injections are each 10 μL; spacing = 240 s (to allow enough time for the reaction to reach equilibrium before the next titration). Perform the experiment with the samples and then with peptide titrated into buffer or buffer titrated into CaM to obtain a reference experiment. Although ITC can provide a robust thermodynamic description of the reaction, it is still important to test controls to properly answer your overarching questions. Some suggestions are reported in **Note 4.3**.

10. Analyze the data. For simple binding, we use Origin 7. After collecting the experimental data and reference experiments, subtract the reference from the main experiment data. If the stoichiometry is not close to $N = 1$, this indicates that the concentration of either the CaM or the peptide is not accurate or that a certain amount of protein is inactive. Figure 2 shows the ITC binding curves using this described method for Q4A binding CaM. The dissociation constant (K_d) between Ca^{2+}/CaM and the Kv7.4-A helix is approximately 440 nM with a stoichiometry close to 1:1, and there is no binding between apoCaM and the Kv7.4-A helix. A few suggestions are offered as a starting place for troubleshooting in **Note 4.4**.

ITC is a powerful technique to quantify and understand the interactions between ion channels and other ligands or signaling proteins that act on the channels. We have used this fragment-based approach to build a model to explain the dynamic movement of ion channels. When used as a complementary technique combined with other biophysical and cellular methods, the quantitative results from ITC can strengthen understanding of how ion channels are regulated by their interactions with other signaling molecules.

4 Notes

4.1 Sample and Buffer Preparation

The key elements of sample preparation are ensuring that the buffer in the cell matches the titration syringe and to have accurate measurement of concentrations. Sample handling and buffer selection have a great impact on the quality of ITC data. Plan to dedicate time to prepare samples with care before each ITC run. Many resources are available online and journal articles offer guidance in preparing buffers and protein samples for ITC. Here are a few specific issues that we have encountered for ion channel peptides:

1. We use float-a-lyzers which are excellent for small volume samples and small proteins and use a molecular weight cutoff (MWCO) that is less than half the size of the protein being dialyzed.
2. If possible, use an apparatus to de-gas small volume samples. If one is not accessible, another option is to de-gas larger volumes in bottles overnight using vacuum filtration, and then use that de-gassed buffer to dilute the protein samples, and centrifuge the sample prior to the experiment.
3. It is important that the buffer in the titration syringe matches exactly the buffer in the cell. An ITC experiment with a buffer mismatch, such as different pH values, usually exhibits large peaks of the background throughout the experiment that hides the reaction binding curve. Whenever possible, dialyze the proteins together in the same buffer, and use the last dialysis buffer as the ITC buffer. Another tip is to check the pH of each sample prior to running the experiment. We do this by placing ~0.2 μL each sample on pH papers and doing a visual comparison of the results.
4. Buffer selection is important. Consider buffer compatibility issues with downstream experiments. If you wish to use the BCA or Bradford to estimate concentrations, then consider the buffer compatibility with those assays. HEPES is known to absorb at low UV ranges which impacts the readings for some proteins.
5. For Ca^{2+} binding proteins such as CaM, it is critical to control the Ca^{2+} levels in the buffer. EGTA or EDTA in the buffer reacting with Ca^{2+} can sometimes interfere with the measurements. To avoid this issue, we dialyze or buffer exchange to the

target buffer to remove unwanted components that can react. Sodium phosphate buffer should be avoided if using $CaCl_2$ for Ca^{2+} binding proteins because the two chemicals react and form a precipitate.

6. To perform experiments in the absence of Ca^{2+}, you can either use high concentration of EGTA or use buffer that has been confirmed to have all traces of Ca^{2+} removed. Be sure that enough EGTA is in the buffer to ensure there is no background binding between EGTA and Ca^{2+}.
7. Protein purity: Pure, single protein samples are essential for ITC experiments. Protein lysates and protein mixtures are not recommended. We have used synthetic peptides by directly reconstituting them in the target buffer. However, if residual chemicals do affect buffer composition, it is recommended to dialyze synthetic proteins and peptides in the target buffer.
8. Protein activity: It is important to know if the proteins change under different conditions such as concentration or pH. Protein aggregation, precipitation, or dimer formations could greatly affect the stoichiometry.
9. Protein amount: ITC requires a relatively large volume and potentially high concentration of pure protein. If this is the first time to test a binding reaction, plan to have enough protein at high enough concentration to be able to test a few practice experiments. Sometimes the temperature, pH, or concentrations may need to be adjusted in order to obtain the accurate binding data.
10. Accurate concentrations are critical for determining the correct stoichiometry. For example, $N = 0.8$ indicates that the concentration of one of your proteins may not be accurate. To this point, ITC can be a valuable tool for verifying the concentration of one protein if the precise concentration of the other protein is known. For example, if CaM is only estimated, but the other protein is known, and the stoichiometry is known to be 1:1, then $N = 0.8$ indicates that the CaM concentration is off by approximately 20%. Alternatively, this could indicate that only 80% of one of the proteins is active, and the other 20% is inactive.
11. When estimating protein concentration using the A_{280}, use the extinction coefficient and Beer–Lambert equation, because the number of tryptophans impacts the absorbance. Taking the concentrations at 1:10 in water or non-absorbing buffer could provide the most accurate concentrations. Centrifuge the sample at $10{,}000{\times}g$ for 10 min prior to measuring to ensure any aggregated protein is not included in the concentration.

12. If there are no Trp or Tyr in your proteins, then other concentration methods must be used. Amino Acid Analysis is a mass spectrometry method considered to be a gold standard for determining protein concentrations. If AAA is not available at your institution, it may be expensive and impractical. However, it may be useful to determine a stock concentration and then prepare a standard curve for other faster and more practical methods. SDS-PAGE densitometry, BSA, and Bradford methods are all helpful alternatives for determining the concentration using a standard curve. Note that the standard curve should be the same protein because the colorimetric dyes react with specific amino acids, including Trp and Tyr. Also check that your buffer components are compatible with the colorimetric reagent.
13. For proteins that do not contain Tyr or Trp, we measure their concentration using SDS-PAGE densitometry or a Bradford or BCA assay by comparing the band to the standard curve of the same protein with known concentration. We also may determine the absorbance at 205 nm as described by Anthis and Clore, 2013 [35–37]. It is important to note that certain buffers (such as HEPES) or high salt may absorb at the lower wavelengths.
14. If your sample is hydrophobic or tends to stick to plastics, be sure to use low-binding plastics such as microcentrifuge tubes and pipette tips to prepare the protein samples. This will minimize sample loss to ensure accurate protein concentration.

4.2 Setting Up the Experiment

1. Before starting your first experiment, practice your technique using inexpensive chemicals, such as EGTA and $CaCl_2$ in water. Preserve your precious samples until you are confident you can run and analyze an ITC experiment without issues.
2. The heat capacity plays a role in the shape of the binding curve, and it can affect the accuracy of the stoichiometry and affinity measurements. Aim for a C value of 10–100 [28]. This can be done by using the equation:

$$C = n^*[M]_{\text{cell}}/K_{\text{d}}, \tag{4}$$

where C is the heat capacity, $[M]_{\text{cell}}$ is the concentration of molecule in the cell, n is the stoichiometry of binding, and K_{d} is the dissociation constant of the binding complex.

Therefore, estimating the K_{d} beforehand is helpful. If it is not known, we usually start with 5 μM protein in the cell and 50–100 μM protein or ligand in the titration syringe.

3. For unknown affinities, start with a molar ratio of 10:1. This means that the sample concentration in the titration syringe should be at least 10× the concentration in the ITC cell.

4. Decide which protein will be in the cell and which will be in the syringe.
   - (a) If one protein is known to form dimers at high concentration, then that protein may serve better in the ITC cell as the lower concentration.
   - (b) If one protein is limited in its solubility, that may also be best as the lower concentration in the ITC cell.
   - (c) When testing a series of proteins or concentrations to the same ligand, then having that target ligand in the same position each time may help to compare results. For example, we usually place CaM in the cell and titrate peptides with different mutations or modifications to compare results easily.
5. Check the pH of the proteins in the cell and the sample after dilution, and adjust the pH for both buffers to match. A set of pH papers are helpful for this step so only 0.2 μL is needed to check pH.
6. Avoid bubbles in the experiments by de-gassing buffers and proteins. Use technique to ensure there are no bubbles introduced into the cell from the sample loading syringe. We also bring the temperature of our proteins to the desired temperature of the cell prior to starting the experiment. These excess volumes will ensure there is no air introduced to the cell or syringe.
7. Volumes: Prepare more than what the instrument holds, in order to avoid introducing air into the system. For the VP-ITC, the cell holds 1.43 mL, and the titration syringe holds 280 μL, according to the MicroCal VP-ITC Manual. Therefore, we prepare approximately 450–475 μL sample for the titration syringe and 2.0 mL sample for the sample cell. We save and freeze the excess samples to use in later ITC or other experiments. For other ITC instruments, the required volume may be lower, but it is still recommended to prepare samples in excess volume to avoid air bubbles.
8. Make sure the cell and syringe are clean, and the instrument is properly functioning. A good way to test this is to run a water–water or buffer–buffer experiment prior to running the sample. Use fresh buffers, and make sure the reference cell has been changed to fresh, de-gassed pure water within 2–4 weeks. Rinse the cell and titration syringe prior to loading sample. We also pull all remaining droplets of rinsing buffer through the titration syringe prior to loading the sample. This helps to avoid unintended sample dilution.
9. If you have limited samples or they are expensive, it is helpful to save the excess samples. This excess sample can be combined

with a fresh sample in the event there is not quite enough fresh stock to run another experiment. It also allows for additional experiments. For example, if you are interested if the complex becomes precipitated, the product in the cell can be centrifuged to check for pellets. The samples can also be analyzed by SDS-PAGE and Native PAGE or possibly size exclusion chromatography to confirm the stoichiometry, binding, and sample stability.

4.3 Control and Reference Experiments

1. Titrate sample from the titration syringe into buffer, and then titrate buffer into the sample in the cell. This could provide some insight into the background signal. The reference data can be subtracted from the sample data so the curve can be fitted more accurately.
2. Titrate buffer or sample into the saturated complex. This reference experiment may more accurately reflect the background heat signature toward the end of the experiment.
3. Reverse titration (protein A into protein B vs protein B into protein A). Theoretically, you should get the same result. However, if one protein is less stable, dimerizes, or precipitates at higher concentrations, then this may not be a practical option.
4. If your sample amount is very limited, running extra controls may not be practical. In this case, the software tutorial may provide some alternatives for analyzing the data properly. For example, one might define constraints in the curve fitting models or use the "simple math" function in the software to adjust the baseline.
5. Running the experiment at different temperatures or pH could also provide valuable insights to the binding system.

4.4 Troubleshooting

1. In the case where no binding is observed but the samples were expected to form a complex, a few suggestions are offered as a starting place for optimizing the method:
   (a) The temperature at which the experiment is run may also affect the amount of heat absorbed or released. Try running the experiment at 5–10 °C higher or lower.
   (b) The affinity is much higher at a picomolar range, and this experiment may require a displacement or competitive binding experiment, where the protein is incubated with a lower affinity protein and the higher affinity protein is titrated to the low affinity complex.
   (c) The affinity is much lower than the range tested. This lower affinity can be determined at higher concentrations, if enough material is available and the sample is soluble at higher concentrations.

   (d) A protein has precipitated out of solution. Visually inspect the samples for cloudiness at the end of the experiment, and centrifuge to see if a pellet forms. On the one hand, we find that the cloudy sample results after binding for some of our samples. On the other hand, sometimes we find that cloudiness means a sample precipitated before the reaction occurred and inhibited binding. More work should be done to ensure the protein stays in solution.
   (e) Is the correct protein loaded? This might seem a little silly, but simple things like proper labeling or confirming a protein is actually the target protein purified by checking on a Western blot or mass spectrometry can be helpful.
2. The peaks are large but do not appear to reach saturation:
   (a) Buffer mismatch may be causing the problem. Check the pH of the solutions before and after diluting the protein.
   (b) Try subtracting the reference experiment.
3. A binding curve is observed but the peaks are very low:
   (a) Try increasing the concentrations.
   (b) Run the experiment at a different temperature.
   (c) If possible, test the interaction using a complementary method to confirm if the binding is a weak interaction.
4. Oscillations are observed that display both endothermic and exothermic directions of the peaks:
   (a) Increase the reference power.
   (b) Make sure pH and other buffer components are matched.
   (c) Perhaps the experiment does have both endothermic and exothermic components.

References

1. Falkenburger BH, Jensen JB, Dickson EJ, Suh BC, Hille B (2010) Phosphoinositides: lipid regulators of membrane proteins. J Physiol 588(Pt 17):3179–3185. https://doi.org/10.1113/jphysiol.2010.192153
2. Hammond GRV, Burke JE (2020) Novel roles of phosphoinositides in signaling, lipid transport, and disease. Curr Opin Cell Biol 63:57–67. https://doi.org/10.1016/j.ceb.2019.12.007
3. Hilgemann DW, Feng S, Nasuhoglu C (2001) The complex and intriguing lives of PIP2 with ion channels and transporters. Sci STKE 2001(111):re19. https://doi.org/10.1126/stke.2001.111.re19
4. Saimi Y, Kung C (2002) Calmodulin as an ion channel subunit. Annu Rev Physiol 64:289–311. https://doi.org/10.1146/annurev.physiol.64.100301.111649
5. Sheng M, Kim E (1996) Ion channel associated proteins. Curr Opin Neurobiol 6(5):602–608. https://doi.org/10.1016/s0959-4388(96)80091-2
6. Willis BC, Ponce-Balbuena D, Jalife J (2015) Protein assemblies of sodium and inward rectifier potassium channels control cardiac excitability and arrhythmogenesis. Am J Physiol Heart Circ Physiol 308(12):H1463–H1473. https://doi.org/10.1152/ajpheart.00176.2015

7. Wang S, Vafabakhsh R, Borschel WF, Ha T, Nichols CG (2016) Structural dynamics of potassium-channel gating revealed by single-molecule FRET. Nat Struct Mol Biol 23(1): 31–36. https://doi.org/10.1038/nsmb.3138
8. Cooper EC, Abreo T, Tran B (2022) KCNQ channel PIP2 modulation: two loose links, three rings, and a twist. Neuron 110(2): 178–180. https://doi.org/10.1016/j.neuron.2021.12.026
9. Zaydman MA, Cui J (2014) PIP2 regulation of KCNQ channels: biophysical and molecular mechanisms for lipid modulation of voltage-dependent gating. Front Physiol 5:195. https://doi.org/10.3389/fphys.2014.00195
10. Pusch M, Ferrera L, Friedrich T (2001) Two open states and rate-limiting gating steps revealed by intracellular Na+ block of human KCNQ1 and KCNQ1/KCNE1 K+ channels. J Physiol 533(Pt 1):135–143. https://doi.org/10.1111/j.1469-7793.2001.0135b.x
11. Gourgy-Hacohen O, Kornilov P, Pittel I, Peretz A, Attali B, Paas Y (2014) Capturing distinct KCNQ2 channel resting states by metal ion bridges in the voltage-sensor domain. J Gen Physiol 144(6):513–527. https://doi.org/10.1085/jgp.201411221
12. Hou P, Kang PW, Kongmeneck AD, Yang ND, Liu Y, Shi J, Xu X, White KM, Zaydman MA, Kasimova MA, Seebohm G, Zhong L, Zou X, Tarek M, Cui J (2020) Two-stage electro-mechanical coupling of a K(V) channel in voltage-dependent activation. Nat Commun 11(1):676. https://doi.org/10.1038/s41467-020-14406-w
13. Cui J (2016) Voltage-dependent gating: novel insights from KCNQ1 channels. Biophys J 110(1):14–25. https://doi.org/10.1016/j.bpj.2015.11.023
14. Archer CR, Enslow BT, Taylor AB, De la Rosa V, Bhattacharya A, Shapiro MS (2019) A mutually induced conformational fit underlies Ca(2+)-directed interactions between calmodulin and the proximal C terminus of KCNQ4 K(+) channels. J Biol Chem 294(15): 6094–6112. https://doi.org/10.1074/jbc.RA118.006857
15. Baranauskiene L, Kuo TC, Chen WY, Matulis D (2019) Isothermal titration calorimetry for characterization of recombinant proteins. Curr Opin Biotechnol 55:9–15. https://doi.org/10.1016/j.copbio.2018.06.003
16. Draczkowski P, Matosiuk D, Jozwiak K (2014) Isothermal titration calorimetry in membrane protein research. J Pharm Biomed Anal 87: 313–325. https://doi.org/10.1016/j.jpba.2013.09.003
17. Boudker O, Oh S (2015) Isothermal titration calorimetry of ion-coupled membrane transporters. Methods 76:171–182. https://doi.org/10.1016/j.ymeth.2015.01.012
18. Soldovieri MV, Miceli F, Taglialatela M (2011) Driving with no brakes: molecular pathophysiology of Kv7 potassium channels. Physiology (Bethesda) 26(5):365–376. https://doi.org/10.1152/physiol.00009.2011
19. Jepps TA, Barrese V, Miceli F (2021) Editorial: Kv7 channels: structure, physiology, and pharmacology. Front Physiol 12:679317. https://doi.org/10.3389/fphys.2021.679317
20. Zheng Y, Liu H, Chen Y, Dong S, Wang F, Wang S, Li GL, Shu Y, Xu F (2022) Structural insights into the lipid and ligand regulation of a human neuronal KCNQ channel. Neuron 110(2):237–247.e234. https://doi.org/10.1016/j.neuron.2021.10.029
21. Li X, Zhang Q, Guo P, Fu J, Mei L, Lv D, Wang J, Lai D, Ye S, Yang H, Guo J (2021) Molecular basis for ligand activation of the human KCNQ2 channel. Cell Res 31(1): 52–61. https://doi.org/10.1038/s41422-020-00410-8
22. Sun J, MacKinnon R (2020) Structural basis of human KCNQ1 modulation and gating. Cell 180(2):340–347.e349. https://doi.org/10.1016/j.cell.2019.12.003
23. Gamper N, Shapiro MS (2003) Calmodulin mediates Ca2+-dependent modulation of M-type K+ channels. J Gen Physiol 122(1): 17–31. https://doi.org/10.1085/jgp.200208783
24. Xu Q, Chang A, Tolia A, Minor DL Jr (2013) Structure of a Ca(2+)/CaM:Kv7.4 (KCNQ4) B-helix complex provides insight into M current modulation. J Mol Biol 425(2):378–394. https://doi.org/10.1016/j.jmb.2012.11.023
25. Chang A, Abderemane-Ali F, Hura GL, Rossen ND, Gate RE, Minor DL Jr (2018) A Calmodulin C-lobe Ca(2+)-dependent switch governs Kv7 channel function. Neuron 97(4): 836–852.e836. https://doi.org/10.1016/j.neuron.2018.01.035
26. Harvard Medical School CfMI Isothermal Titration Calorimetry. https://cmi.hms.harvard.edu/isothermal-titration-calorimetry
27. Turnbull WB, Daranas AH (2003) On the value of c: can low affinity systems be studied by isothermal titration calorimetry. J Am Chem Soc 125(48):14859–14866. https://doi.org/10.1021/ja036166s
28. Sciences PHIotL Isothermal Titration Calorimetry. https://www.huck.psu.edu/core-facilities/automated-biological-calorimetry-

facility/technical-guides/isothermal-titration-calorimetry

29. Du X, Li Y, Xia YL, Ai SM, Liang J, Sang P, Ji XL, Liu SQ (2016) Insights into protein-ligand interactions: mechanisms, models, and methods. Int J Mol Sci 17(2). https://doi.org/10.3390/ijms17020144
30. Millev S (2015) N-value ITC data. Malvern Panalytical, Malvern
31. Frasca V (2019) What's new in ITC: additions to your (virtual) bookshelf. Malvern Panalytical, Malvern
32. Wen H, Levitan IB (2002) Calmodulin is an auxiliary subunit of KCNQ2/3 potassium channels. J Neurosci 22(18):7991–8001. https://doi.org/10.1523/jneurosci.22-18-07991.2002
33. Yus-Najera E, Santana-Castro I, Villarroel A (2002) The identification and characterization of a noncontinuous calmodulin-binding site in noninactivating voltage-dependent KCNQ potassium channels. J Biol Chem 277(32): 28545–28553. https://doi.org/10.1074/jbc.M204130200
34. Gopalakrishna R, Anderson WB (1982) Ca2+-induced hydrophobic site on calmodulin: application for purification of calmodulin by phenyl-Sepharose affinity chromatography. Biochem Biophys Res Commun 104(2): 830–836. https://doi.org/10.1016/0006-291x(82)90712-4
35. Anthis NJ, Clore GM (2013) Sequence-specific determination of protein and peptide concentrations by absorbance at 205 nm. Protein Sci 22(6):851–858. https://doi.org/10.1002/pro.2253
36. Anthis NJ. Protein parameter calculator. http://nickanthis.com/tools/a205.html
37. Clore GM. Protein calculator. https://www.gmclore.org/clore/Software/A205.html

Index

Simone Furini (ed.), *Potassium Channels: Methods and Protocols*, Methods in Molecular Biology, vol. 2796, https://doi.org/10.1007/978-1-0716-3818-7,

MIX
Papier aus verantwortungsvollen Quellen
Paper from responsible sources
FSC® C105338

If you have any concerns about our products,
you can contact us on
ProductSafety@springernature.com

In case Publisher is established outside the EU,
the EU authorized representative is:
Springer Nature Customer Service Center GmbH
Europaplatz 3, 69115 Heidelberg, Germany

Printed by Libri Plureos GmbH
in Hamburg, Germany